D0712990

ADVANCES IN SPEECH CODING

For Don
from Vladimir
Thank you

Vlod

THE KLUWER INTERNATIONAL SERIES
IN ENGINEERING AND COMPUTER SCIENCE

COMMUNICATIONS AND INFORMATION THEORY

Consulting Editor:
Robert Gallager

Other books in the series:

Digital Communication. Edward A. Lee, David G. Messerschmitt
    ISBN: 0-89838-274-2

An Introduction to Cryptology. Henk C.A. van Tilborg
    ISBN: 0-89838-271-8

Finite Fields for Computer Scientists and Engineers. Robert J. McEliece
    ISBN: 0-89838-191-6

An Introduction to Error Correcting Codes With Applications.
Scott A. Vanstone and Paul C. van Oorschot
    ISBN: 0-7923-9017-2

Source Coding Theory. Robert M. Gray
    ISBN: 0-7923-9048-2

Switching and Traffic Theory for Integrated Broadband Networks.
Joseph Y. Hui
    ISBN: 0-7923-9061-X

ADVANCES IN SPEECH CODING

Editors

Bishnu S. Atal
AT&T Bell Laboratories

Vladimir Cuperman
Simon Fraser University

Allen Gersho
University of California, Santa Barbara

KLUWER ACADEMIC PUBLISHERS
Boston/Dordrecht/London

Distributors for North America:
Kluwer Academic Publishers
101 Philip Drive
Assinippi Park
Norwell, Massachusetts 02061 USA

Distributors for all other countries:
Kluwer Academic Publishers Group
Distribution Centre
Post Office Box 322
3300 AH Dordrecht, THE NETHERLANDS

Library of Congress Cataloging-in-Publication Data

Advances in speech coding / editors, Bishnu S. Atal, Vladimir
  Cuperman, Allen Gersho.
      p. cm. — (The Kluwer international series in engineering and
  computer science ; vol. # 114. Communications and information
  theory)
    Papers based on presentations at the IEEE Workshop on Speech
  Coding for Telecommunications, held in Vancouver, British Columbia,
  Canada, from September 5 to 8, 1989.
    Includes bibliographical references and index.
    ISBN 0-7923-9091-1
    1. Speech processing systems—Congresses. I. Atal, Bishnu S.
  II. Cuperman, Vladimir. III. Gersho, Allen. IV. IEEE Workshop on
  Speech Coding for Telecommunications (1989 : Vancouver, B.C.)
  V. Series: Kluwer international series in engineering and computer
  science ; SECS 114. VI. Series: Kluwer international series in
  engineering and computer science. Communications and information
  theory.
  TK7882.S65A28   1991
  621.382—dc20   90-47146
                                                              CIP

Printed on acid-free paper.

Printed in the United States of America

Contents

ADVANCES IN SPEECH CODING

PART I

INTRODUCTION

Speech coding has been an ongoing area of research for several decades, yet the level of activity and interest in this area has expanded dramatically in the last several years. Important advances in algorithmic techniques for speech coding have recently emerged and excellent progress has been achieved in producing high quality speech at bit rates as low as 4.8 kb/s. Although the complexity of the newer more sophisticated algorithms greatly exceeds that of older methods (such as ADPCM), today's powerful programmable signal processor chips allow rapid technology transfer from research to product development and permit many new cost-effective applications of speech coding. In particular, low bit rate voice technology is converging with the needs of the rapidly evolving digital telecommunication networks.

The IEEE Workshop on Speech Coding for Telecommunications was held in Vancouver, British Columbia, Canada, from September 5 to 8, 1989. The objective of the workshop was to provide a forum for discussion of recent developments and future directions in speech coding. The workshop attracted over 130 researchers from several countries and its technical program included 51 papers.

The workshop was very successful and it was felt that a book representing a cross-section of the topics presented at the workshop will be valuable. The authors were subsequently invited to contribute a chapter based on their presentations in a camera-ready form for publication in this book. This volume contains 35 papers corresponding to Workshop presentations. The papers were prepared and submitted in final form early in 1990 and in many cases they contain more complete and up-to-date material than was originally included in the verbal presentations at the Workshop. Each paper forms a chapter in the book. The chapters are further grouped together under six topics representing central themes in each group.

The chapters in this volume reflect the progress and present the state of the art in low bit rate speech coding primarily at bit rates from 16 kb/s to 2.4 kb/s. Together they represent important contributions from leading researchers in the speech coding community.

The reader will find in this book a good selection of papers covering most of the important research topics in speech coding. The book contains papers describing technologies that were recently adopted or are under consideration as standards for such applications as digital cellular communications (the VSELP 8 kb/s speech codec), secure telephony (the DOD 4.8 kb/s standard), and general network telephony (the emerging 16 kb/s CCITT standard). One of the sections is dedicated to low-delay speech coding, a new research direction which emerged as a result of the CCITT requirement for an universal low-delay 16 kb/s speech coding technology. There are a significant number of papers addressing future research directions. We hope that the reader will find the contributions instructive and useful.

We would like to take this opportunity to thank all the authors for their contributions to this volume and for meeting the very tight deadlines. We wish to thank Renee Leach, at the University of California, Santa Barbara for her valuable help in compiling the material for this volume.

Bishnu S. Atal
Vladimir Cuperman
Allen Gersho

PART II

LOW–DELAY SPEECH CODING

A significant research effort in low delay speech coding was stimulated by the CCITT when it established the requirement that the future 16 kb/s speech coding standard must have very low coding delay, while achieving essentially the same high quality as the 32 kb/s ADPCM standard G.721. Although speech coding algorithms based on Code Excited Linear Prediction (CELP) and other similar configurations are able to provide the required quality at 16 kb/s, these coders introduce a substantial delay, basically due to forward adaptation of the short- and long-term predictors. To meet the low delay constraint, forward adaptation is not feasible, yet backward adaptation at low rates tends to cause degraded quality and severe propagation of bit errors.

This section presents a selection of chapters describing new techniques which have been developed for low delay speech coding. Gibson et al., present a collection of backward adaptive lattice predictors for low-delay speech coding in a tree coding environment. Cuperman et al., present backward adaptive configurations for a low-delay analysis-by-synthesis system called Low-Delay Vector Excitation Coding (LD-VXC). Chen presents a low-delay backward CELP 16 kb/s codec (LD-CELP) using a 50-th order short-term predictor. Be'ery et al., present an approach for achieving variable rate in the LD-CELP environment. Marcellin and Fischer present a low delay codec based on the Trellis Coded Quantization. Finally, Pettigrew and Cuperman present a backward adaptive pitch predictor for the LD-VXC system using recursive coeficient adaptation and pitch tracking.

1

BACKWARD ADAPTIVE PREDICTION ALGORITHMS IN MULTI–TREE SPEECH CODERS

Jerry D. Gibson[†], Yoon Chae Cheong[†],
Hong Chae Woo[†], Wen–Whei Chang[††]

[†]Department of Electrical Engineering
Texas A&M University
College Station, Texas

[††]Department of Communication Engineering
National Chiao–Tung University
Taiwan

INTRODUCTION

To achieve low delay in speech coders, the redundancy removal must be accomplished in a backward adaptive fashion so that both long– and short–term predictors use only the decoder output for parameter adaptation. Numerous backward adaptive algorithms for updating the short–term predictor coefficients have been studied [1–4], and several look promising. In [5] comparative simulation results were presented for a fixed–tap predictor, three gradient–adapted transversal predictors, and two least squares lattice predictors when used in a differential pulse code modulation (DPCM) based code generator for tree coding of speech at 16 kilobits/s (kbits/s).

We extend the work in [5] by comparing the performance of four eighth order, all–pole backward adaptive lattice algorithms for updating the short–term predictors in adaptive predictive coding (APC) based code generators for tree coding of speech at 16 kbits/s. The algorithms studied are the least squares lattice algorithm [2,5], the exponential–window lattice algorithm [4], the signal–driven lattice algorithm [3], and the residual–driven lattice algorithm [3].

The comparisons are based upon frequency weighted signal–to–noise ratio (SNRFW) and informal subjective listening tests.

APC AND TREE CODING

An APC based tree coder transmitter is illustrated in Fig. 1, where $P(z)$ represents the long–term or pitch predictor and is given by

5

$$P(z) = \beta_1 z^{-(M_1-1)} + \beta_2 z^{-M_1} + \beta_3 z^{-(M_1+1)} \tag{1}$$

with M_1 the pitch period length and the weighting coefficients $\{\beta_i, i = 1, 2, 3\}$. We consider only backward updates of the weighting coefficients and M_1. All pole short–term predictor structures are studied exclusively in this paper (some pole–zero results are available in [5]); thus, the short–term or formant predictor has the form

$$A(z) = \sum_{i=1}^{N} a_i z^{-i} \tag{2}$$

where $N = 8$ here and the coefficients $\{a_i, i = 1, 2, ..., N\}$ are calculated using one of the backward adaptive algorithms described in the next section.

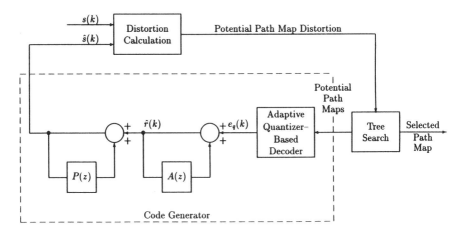

Figure 1. APC Based Tree Coder
(Encoder or Transmitter Only)

The "Adaptive Quantizer–Based Decoder" in Fig. 1 is a device that takes path map inputs and generates an output sequence with values taken from the alphabet of a scalar adaptive quantizer. Thus, for a four–level, integer rate 2 bits/sample tree, the output values for $e_q(k)$ are computed from the output of a four level, MMSE Gaussian assumption quantizer with adaptive step size $\Delta(k)$ computed according to [6],

$$\Delta(k+1) = \Delta^{\gamma_1}(k) F(|I(k)|) \tag{3}$$

where the leakage factor $0 < \gamma_1 < 1$, and $F(\cdot)$ is a time–invariant multiplier function that is 0.8 for inner levels and 1.6 for outer levels.

The code generator output $\hat{s}(k)$ for all possible path map sequences to depth L is compared to the input signal $s(k)$ according to some distortion measure. We employ a weighted squared error criterion,

$$d((s - \hat{s})_w) = d(\varepsilon_w) = \frac{1}{L} \sum_{i=1}^{L} \varepsilon_w^2(i) \tag{4}$$

where $\varepsilon_w(k)$ is generated by passing $\varepsilon(k) = s(k) - \hat{s}(k)$ through a transfer function of the form

$$W(z) = \frac{1 - \sum_{i=1}^{N} a_i z^{-i}}{1 - \sum_{i=1}^{N} \mu^i a_i z^{-i}} \tag{5}$$

and μ is chosen by experiment to be 0.86.

The tree is searched using the (M, L) algorithm investigated by Anderson and Bodie [7] that only retains a fixed number M of paths at any depth. Once the path through the tree to depth L that has the smallest distortion is found, path map symbols describing this path must be sent or released to the receiver. All integer rate trees studied here use single symbol release, while the fractional rate trees release a fixed, small number of path map symbols at any time instant.

Multi–tree codes are described in [8] and allow a varying number of branches to emanate from successive nodes. For example, a rate 3/2 bit/sample multi-tree from [8] interleaves four level and two level trees at alternating sampling instants. When the four level tree occurs, the step size is adapted according to Eq. (3), but when a two level tree occurs, only polarities are available for adaptation, and so the step size is computed according to

$$\Delta(k + 1) = \begin{cases} 1.5\Delta^{\gamma_2}(k), \operatorname{sgn}(e_q(k))\operatorname{sgn}(e_q(k - 1)) = +1, \\ \frac{1}{1.5}\Delta^{\gamma_2}(k), \operatorname{sgn}(e_q(k))\operatorname{sgn}(e_q(k - 1)) = -1, \end{cases} \tag{6}$$

where $\gamma_2 = 127/128$ and $\operatorname{sgn}(\cdot) = +1$ for positive arguments and -1 for negative arguments.

BACKWARD ADAPTIVE PREDICTORS

We adapt the long–term predictor $P(z)$ using the method described in [4]. We focus our studies here on four lattice algorithms, the least squares lattice algorithm [2, 5], denoted later as LATLS8, the exponential window lattice algorithm (LATEW8)[4], the signal–driven lattice algorithm (LATSD8)[3], and the residual–driven lattice algorithm (LATRD8)[3]. For compactness, we simply list the four algorithms, where the notation is with respect to the lattice structure in Fig. 2.

8

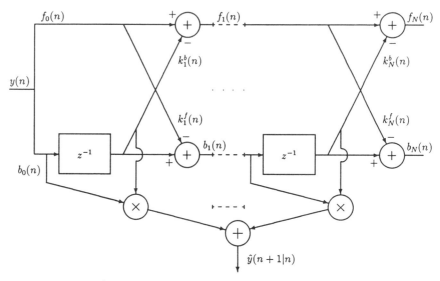

Figure 2. Lattice Form Predictor Structure

Least Squares Lattice [2,5]

Begin with the initial conditions: $\beta_0(n) = 0.0, f_0(n) = b_0(n) = y(n) = \hat{r}(n)$ in Fig. 1, and $R_1^b(n) = R_1^f(n) = 0.99R_1^f(n-1) + f_0^2(n)$. Perform the following recursions, for $\ell = 1, 2, ..., N$:

$$\bar{k}_\ell(n) = 0.99\bar{k}_\ell(n-1) + (f_{\ell-1}(n))b_{\ell-1}(n-1)/(1 - \beta_{\ell-1}(n))$$
$$k_\ell^f(n) = \bar{k}_\ell(n)/R_\ell^f(n)$$
$$k_\ell^b(n) = \bar{k}_\ell(n)/R_\ell^b(n-1)$$
$$R_{\ell+1}^f(n) = (R_\ell^f(n) - k_\ell^b(n)\bar{k}_\ell(n))/0.98^2$$
$$R_{\ell+1}^b(n) = (R_\ell^b(n-1) - k_\ell^f(n)\bar{k}_\ell(n))/0.98^2$$
$$\beta_\ell(n) = \beta_{\ell-1}(n) + b_{\ell-1}^2(n-1)/R_\ell^b(n-1)$$
$$f_\ell(n) = f_{\ell-1}(n) - k_\ell^b(n)b_{\ell-1}(n-1)$$
$$b_\ell(n) = b_{\ell-1}(n-1) - k_\ell^f(n)f_{\ell-1}(n)$$

Exponential Window Lattice [4]

Begin with the initial conditions: $f_0(n) = b_0(n) = y(n) = \hat{r}(n)$ in Fig. 1, and perform the recursions, for $\ell = 1, 2, ..., N$:

$$f_\ell(n) = f_{\ell-1}(n) - k_\ell(n)b_{\ell-1}(n-1)$$
$$b_\ell(n) = b_{\ell-1}(n-1) - k_\ell(n)f_{\ell-1}(n)$$
$$c_\ell(n) = 0.99c_\ell(n-1) + f_{\ell-1}(n)b_{\ell-1}(n-1)$$
$$d_\ell(n) = 0.99d_\ell(n-1) + 0.5f_{\ell-1}^2(n) + 0.5b_{\ell-1}^2(n-1)$$

$k_\ell(n+1) = c_\ell(n)/d_\ell(n)$

Signal–Driven Lattice [3]

Begin with the initial conditions: $f_0(n) = b_0(n) = y(n) = \hat{r}(n)$ in Fig. 1, and perform the recursions, for $\ell = 1, 2, ..., N$:

$$f_\ell(n) = f_{\ell-1}(n) - k_\ell(n)b_{\ell-1}(n-1)$$
$$b_\ell(n) = b_{\ell-1}(n-1) - k_\ell(n)f_{\ell-1}(n)$$
$$c_\ell(n) = 0.96875c_\ell(n-1) + 2f_{\ell-1}(n)b_{\ell-1}(n-1)$$
$$d_\ell(n) = 0.96875d_\ell(n-1) + f_{\ell-1}^2(n) + b_{\ell-1}^2(n-1)$$
$$k_\ell(n+1) = c_\ell(n)/d_\ell(n)$$

Residual–Driven Lattice [3]

Begin with the initial conditions: $f_0(n) = b_0(n) = \bar{f}_0(n) = \bar{b}_0(n) = e_q(n)$ in Fig. 1 and perform the recursions, for $\ell = 1, 2, ..., N$:

$$f_\ell(n) = f_{\ell-1}(n) - k_\ell(n)b_{\ell-1}(n-1)$$
$$b_\ell(n) = b_{\ell-1}(n-1) - k_\ell(n)f_{\ell-1}(n)$$
$$\bar{f}_\ell(n) = \bar{f}_{\ell-1}(n) - k_\ell(n)\bar{b}_{\ell-1}(n-1)$$
$$\bar{b}_\ell(n) = \bar{b}_{\ell-1}(n-1) - k_\ell(n)\bar{f}_{\ell-1}(n-1)$$
$$\bar{c}_\ell(n) = 0.96875\bar{c}_\ell(n-1) + 2\bar{f}_{\ell-1}(n)\bar{b}_{\ell-1}(n-1)$$
$$\bar{d}_\ell(n) = 0.96875\bar{d}_\ell(n-1) + \bar{f}_{\ell-1}^2(n) + \bar{b}_{\ell-1}^2(n-1)$$
$$k_\ell(n+1) = \bar{c}_\ell(n)/\bar{d}_\ell(n)$$

IDEAL CHANNEL PERFORMANCE

We begin by comparing the performance of the four lattice algorithms in two different tree coders for encoding speech at 16 kbits/s. One system uses an integer rate 2 bits/sample tree and a sampling rate of 8000 samples/s, while the second system employs a rate 5/2 bits/sample multi–tree and a sampling rate of 6400 samples/s. Both coders use an APC code generator, the $(M = 8, L = 8)$ tree search algorithm, and the frequency weighted distortion measure. The objective performance results are given in Tables I and II. The speech utterances referred to here are listed in Appendix A, and SNRFW is defined in Appendix B.

Table I
SNRFW(dB) For R=2 Bits/Sample,
APC Code Generator, 8000 Samples/s

Sentence	LATLS8	LATEW8	LATSD8	LATRD8
1	22.19	22.63	22.54	20.31
3	18.54	18.85	18.01	15.12
5	17.71	17.67	17.67	15.79

Table II
SNRFW(dB) For R=5/2 Bits/Sample,
APC Code Generator, 6400 Samples/s

Sentence	LATLS8	LATEW8	LATSD8	LATRD8
1	21.74	21.83	21.75	18.78
3	17.37	17.44	17.89	13.85
5	16.83	17.11	17.27	14.89

From the SNRFW results in Tables I and II, we see that the least squares, exponential window, and signal–driven lattices have approximately the same performance for each coder, while the residual–driven lattice has substantially lower SNRFW values. This latter result is expected since the residual–driven lattice is designed for robustness to errors [3]. Informal subjective listening tests indicate that the least squares, exponential window, and signal–driven lattice algorithms have indistinguishable subjective performance, while the residual–driven lattice has audibly poorer performance.

Comparing the performance of the two 16 kbits/s tree coders, we see that the integer rate tree has slightly higher SNRFW values, but SNRFW comparisons of coders with inputs at different sampling rates are not valid since the reference signal is not the same. Informal subjective listening tests indicate that the performance of both coders is very close, and careful listening seems to reveal a slight subjective preference for the rate 5/2 multi–tree coder.

The objective performance of a rate 3/2 bit/sample multi–tree coder using the least squares lattice, with an APC code generator, $(M = 8, L = 8)$ tree searching, the frequency weighted distortion measure, and a sampling rate of 10,000 samples/s, is given in Table III. This coder has lower SNRFW values than the two 16 kbits/s coders, and informal listening tests indicate audibly poorer performance for this 15 kbits/s system. Perhaps increasing the sampling rate to 10,666 so that the transmitted bit rate is closer to 16 kbits/s would make up the difference, but it does not seem likely here.

Table III
R=3/2 Bit/Sample, APC Code Generator,
10,000 Samples/s, Least Squares Lattice

Sentence	SNRFW
1	17.46
3	15.64
5	14.55

CONCLUSIONS

The least squares lattice, the exponential window lattice, and the signal–driven lattice have essentially identical objective and subjective performance over ideal channels that is clearly superior to the residual–driven lattice. Fractional rate trees allow 16 kbits/s coders to be obtained with various combinations of code rates and sampling rates. The two 16 kbits/s systems and the 15 kbits/s coder have slightly different objective and subjective performance even when they use the same predictor, tree search algorithm, and distortion measure.

Acknowledgement

This research was supported, in part, by BNR, Inc. under their University Interaction Program.

APPENDIX A

The three sentences used in this paper are:
1. "The pipe began to rust while new." (female speaker)
3. "Oak is strong and also gives shade." (male speaker)
5. "Cats and dogs each hate the other." (male speaker)

APPENDIX B

The objective performance measure used in this work is the frequency weighted signal–to–noise ratio (SNRFW) calculated as

$$\text{SNRFW} = 10 \log_{10} \frac{\langle s^2(k) \rangle}{\langle [(s(k) - \hat{s}(k))_w]^2 \rangle}.$$

where $< \cdot >$ denotes time averaging over the entire utterance.

REFERENCES

[1] J. D. Gibson, "Adaptive prediction for speech encoding," ASSP Magazine, Acoustics, Speech, and Signal Processing Society, pp. 12–26, July 1984.

[2] R. C. Reininger and J. D. Gibson, "Backward adaptive lattice and transversal predictors in ADPCM," IEEE Trans. Commun., vol. COM–33, pp. 74–82, Jan. 1985.

[3] P. Yatrou and P. Mermelstein, "Ensuring predictor tracking in ADPCM speech coders under noisy transmission conditions," IEEE J. Selected Areas in Communications, vol. 6, pp. 249–261, Feb. 1988.

[4] V. Iyengar and P. Kabal, "A Low Delay 16 Kbits/sec. Speech Coder," Proceedings, 1988 IEEE Int. Conf. Acoust., Speech, and Signal Proc., New York, NY, Apr. 11–14, 1988, pp. 243–246.

[5] W. W. Chang and J. D. Gibson, "A Comparison of Adaptive Code Generators for Tree Coding of Speech," Proceedings, Thirty–First Midwest Symposium on Circuits and Systems, St. Louis, MO, Aug. 9–12, 1988, pp. 924–927.

[6] N. S. Jayant and P. Noll, <u>Digital Coding of Waveforms: Principles and Applications to Speech and Video</u>, Prentice–Hall, New York, 1984.

[7] J. B. Anderson and J. B. Bodie, "Tree encoding of speech," <u>IEEE Trans. Inform. Theory</u>, vol. IT–21, pp. 379–387, July 1975.

[8] J. D. Gibson and W. W. Chang, "Fractional Rate Multi–Tree Speech Coding," <u>Proceedings</u>, 1989 IEEE Global Commun. Conf., Dallas, TX, Nov. 27–30, pp. 1906–1910.

2

BACKWARD ADAPTIVE CONFIGURATIONS FOR LOW-DELAY VECTOR EXCITATION CODING

Vladimir Cuperman † , Allen Gersho † †, Robert Pettigrew † , John J. Shynk † †, Jey-Hsin Yao † †

† Communications Science Laboratory
School of Engineering Science
Simon Fraser University,
Burnaby, B.C. V5A 1S6

† † Center for Information Processing Research
Department of Electrical and Computer Engineering
University of California
Santa Barbara, CA 93106

INTRODUCTION

Many of the advances in speech coding in the past decade at rates of 4.8 - 16 kbit/s have been based on excitation coding by means of *analysis-by-synthesis*. Excitation coding schemes have a decoder structure consisting of an excitation signal applied to a time-varying synthesis filter to produce the reconstructed, or "synthesized," output speech. In addition to other tasks, the encoder must determine a suitable excitation signal and transmit data that specifies this excitation. In the analysis-by-synthesis technique, the excitation is selected by a *closed-loop* search procedure where a candidate excitation signal segment is applied to the synthesis filter, the synthesized waveform is compared with the original speech segment, the distortion is measured, and the process is repeated for all excitation segments stored in an excitation codebook. The index of the "best" excitation segment is transmitted to the decoder, which retrieves the excitation segment from a codebook identical to that at the encoder. The parameters of the synthesis filter are computed using well-known linear prediction analysis techniques on a frame of buffered input samples and transmitted to the decoder. This coding scheme is often called Vector Excitation Coding (VXC) or Code Excited Linear Prediction (CELP) [1,2].

The general structure of the VXC encoder and decoder are shown in Figs. 1(a) and 1(b). At the encoder side, each candidate vector retrieved from a codebook is first gain-scaled and then applied to the synthesis filter, based on short- and long-term adaptive predictors. A perceptually weighted mean-square error criterion is used in the closed-loop search. The index of the best candidate codevector, the gain, and the synthesis filter parameters are then transmitted to the decoder. At the decoder side, the received index is used to look up the proper codevector in the codebook.

14

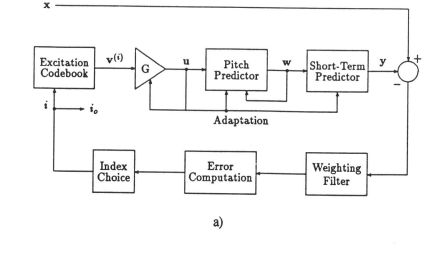

a)

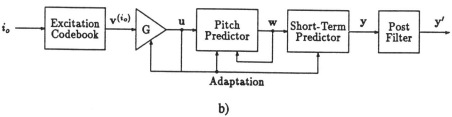

b)

FIGURE 1: General VXC in a) encoder and b) decoder configurations.

This codevector is then gain scaled and applied to the cascade of the long-term (or "pitch") and short-term synthesis filters. The reconstructed speech is postfiltered to reduce the perceived quantization noise.

Ordinarily, the synthesis filter parameters are computed by performing linear prediction analysis on a frame of input speech samples that have been stored in a buffer prior to further processing. In this way, the synthesis filter is adaptively updated every frame using forward adaptation. The use of forward adaptation has two disadvantages: it requires transmission of side information to the receiver to specify the filter parameters and it leads to a large encoding delay of at least one analysis frame due to the buffering of input speech samples. The input buffering and other processing typically result in a one-way codec delay of 50 to 60 ms. In certain applications in the telecommunications network environment, coding delays as low as 2 ms per codec are required. Recently, the CCITT adopted a performance requirement of less than 5 ms delay with a desired objective of less than 2 ms for candidate 16 kbit/s speech coding algorithms to be considered for a new standard. Such a low delay is not feasible with the established coders that are based on forward adaptive prediction coding systems. Although the 32 kbit/s ADPCM algorithm, CCITT Recommendation G.721, satisfies the low delay requirement, it cannot give acceptable quality when the bit rate is reduced to 16 kbit/s.

An alternative solution is based on a recently proposed backward adaptation configuration [3]. In a backward adaptive analysis-by-synthesis configuration, the parameters of the synthesis filter are not derived from the original speech signal, but instead computed by backward adaptation, extracting information only from the sequence of transmitted codebook indices. Since both the encoder and decoder have access to the past reconstructed signal, side information is no longer needed for synthesis filters, and the low-delay requirement can be met with a suitable choice of vector dimension. In order to obtain adequate dynamic range with a modest size excitation codebook, a backward adaptive gain prediction technique is used [3] based on the theory developed in [4]. Backward adaptation is itself not a new idea and has been widely studied in conjunction with an earlier generation of speech coding methods. The classical ADPCM algorithm, although based on scalar rather than vector quantization, also performs backward adaptation and has other similarities to the new configuration, hence the first version of the new approach was called Vector ADPCM [3]. Vector ADPCM, a precursor to LD-VXC, appears to be the first application of backward adaptation to excitation coding. The LD-VXC coder described here was submitted to the CCITT as a candidate for a proposed "universal" standard for 16 kbit/s speech coding.

A comparison of different techniques which may be used for backward adaptation in the analysis-by-synthesis environment will be discussed in this paper. The corresponding analysis leads to a new coding technique for 16 kbit/s called Low Delay Vector Excitation Coding (LD-VXC) which will be presented here.

SYSTEM OVERVIEW

Fig. 2 shows a block diagram of the Low-Delay Vector Excitation Coding (LD-VXC) system, in both encoder and decoder configurations. The main components are the codebook, the gain predictor, the pitch predictor, the short-term predictor, the perceptual weighting filter, and the postfilter.

In the encoder, a codevector is chosen from a codebook using an analysis-by-synthesis technique. Each candidate codevector, $v^{(i)}$, is multiplied by a vector gain calculated using a backward adaptive gain predictor. The resulting gain scaled codevector, u, is input into a synthesis filter, which is a cascade of a pitch predictor and a short-term predictor. The index i of the excitation vector $u^{(i)}$ is omitted for clarity. The components of vector u will be denoted $u(n)$, where n is the time index.

The output of the long term predictor, $w(n)$, is computed by

$$w(n) = u(n) + \sum_{i=-1}^{1} a_i \, w(n - k_p - i) \qquad (1)$$

where $\{a_i\}$ are the coefficients of the pitch predictor and k_p is the pitch period.

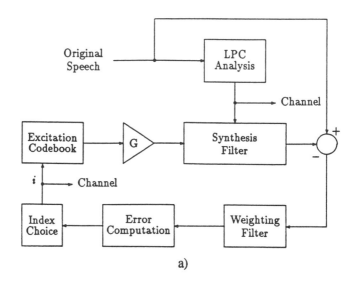

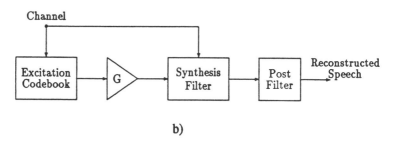

a)

b)

FIGURE 2: LD-VXC in a) encoder and b) decoder configurations.

The output of the short-term predictor, $y(n)$, is computed using the relationship:

$$y(n) = w(n) + \sum_{i=1}^{P} h_i \, y(n-i) + \sum_{i=1}^{Z} g_i \, w(n-i) \tag{2}$$

where $\{g_i\}$ are the coefficients of the all-zero section, and $\{h_i\}$ are the coefficients of the all-pole section.

The output of the short-term predictor is compared to the actual speech signal, and a choice is made of the best candidate codevector using a perceptually weighted minimum squared error criterion. Once the best codevector has been chosen, this codevector is reapplied to the synthesis filter to generate the proper predictor memory. Both the short-term and pitch prediction filters are adapted on a sample-by-sample basis using a backward-adaptive technique. The only information which is then transmitted to the decoder is the index, i_o, corresponding to the chosen codevector.

In the decoder, the received index is used to look up the proper codevector in the codebook. This codevector is then gain scaled using the gain produced by the gain predictor, and input into a synthesis filter identical to that at the encoder. As is done in the encoder, both prediction filters are adapted on a sample-by-sample basis using a backward adaptation algorithm. The output of the synthesis filter is fed into a postfilter, which is designed to reduce the quantization noise perceived by the listener.

BACKWARD ADAPTATION STRATEGIES

Two approaches to backward adaptation will be discussed below, and they can be classified as *block* and *recursive*. In the block algorithms, the reconstructed signal and the corresponding gain-scaled excitation vectors are divided into blocks (frames), and the optimum parameters of the adaptive filter are determined independently within each block. In the recursive algorithms, the parameters are adapted incrementally after each successive excitation and reconstructed vectors are generated.

In a block backward low-delay configuration, the end result is a new set of filter parameters at the end of each frame. These parameters must be used for the duration of the next frame even though the speech statistics are changing from one frame to the next, and the parameters for one frame can often be poorly suited to the next frame. This makes the choice of the frame length a very difficult trade-off: for a long frame, the parameters become obsolete well before the end of the next frame; for a short frame, the estimates of the autocorrelation function used in the Wiener-Hopf equations may become unreliable. A possible alternative is to use a highly overlapped frame structure. This solution, however, leads to a high complexity codec. Recursive adaptation systems are more flexible from this point of view. The adaptation of the parameters can be carried out sample-by-sample, while the update period can be based on complexity considerations [3].

In block adaptive systems, an all-pole short term predictor is traditionally used ($Z = 0$ in (2)). In this case, the coefficients $\{h_i\}$ can be computed by solving the Wiener-Hopf equations

$$\mathbf{R}_{yy}\mathbf{h} = \mathbf{r} \tag{3}$$

where $\mathbf{h} = (h_1, h_2, .., h_P)^T$, $\mathbf{R}_{yy}$ is the autocorrelation matrix of $y(n)$, and $\mathbf{r} = (r_{yy}(1), r_{yy}(2), .., r_{yy}(P))^T$. The element i,j of the matrix $\mathbf{R}_{yy}$ is given by

$$r_{yy}(|i-j|) = \frac{1}{L} \sum_{k=1}^{L-|i-j|-1} y(n_0 + k)y(n_0 + k + |i-j|) \tag{4}$$

where the index n_0 points to the first sample in a block, and L is the block size. This leads to the so-called autocorrelation method and the derived synthesis filter is always stable. Alternatively, the stabilized covariance method can be used. The assumption here is that the coefficients which minimize the MSE of a given block, due to the slowly changing characteristics of speech, will lead to better performance in the next block, where they will actually be used.

Backward block adaptation techniques for the pitch predictor are discussed in [5].

For recursive adaptation, let $[u(n)]^2$ be the objective function in a stochastic gradient approach. Then, using (1) and (2), it can be shown that the corresponding gradient can be approximated by

$$\nabla_{a,g,h} \|u(n)\|^2 \approx$$

$$-2 \, u(n) \left[w(n-k_p-1), w(n-k_p), w(n-k_p+1), w(n-1), \ldots, w(n-Z), y(n-1), \ldots, y(n-P) \right]^T$$

and the corresponding adaptation equations are

$$a_i^{(n+1)} = a_i^{(n)} + \alpha_a u(n) w(n-k_p-i) \qquad (5)$$

$$h_i^{(n+1)} = h_i^{(n)} + \alpha_h u(n) y(n-i) \qquad (6)$$

$$g_i^{(n+1)} = g_i^{(n)} + \alpha_g u(n) w(n-i) . \qquad (7)$$

Equation (5) is used in the adaptation of the pitch predictor [5]. However, an adaptation based on (6) and (7) for the short term predictor is not robust in the presence of transmission errors. Two improvements were found to increase significantly the robustness at the expense of a minor degradation in the performance in the absence of transmission errors. First, the robustness was found to increase by replacing $w(n-i)$ with $u(n-i)$ in (7). This is equivalent to adapting the short- and long-term predictors in parallel rather than in cascade; this approach will be called parallel adaptation. Second, similar to the ADPCM case, it was found that using the all-zero reconstructed signal for adapting $\{h_i\}$, and using leakage factors for all adapted coefficients further improves the robustness. Taking into account the parallel adaptation, the all-zero reconstructed signal, $y'(n)$, is given by

$$y'(n) = u(n) + \sum_{i=1}^{Z} g_i^{(n)} u(n-i) .$$

With these changes, (5-7) become

$$a_i^{(n+1)} = \lambda_a a_i^{(n)} + \alpha_a u(n) w(n-k_p-i) \qquad (5b)$$

$$h_i^{(n+1)} = \lambda_h h_i^{(n)} + \alpha_h u(n) y'(n-i) \qquad (6b)$$

$$g_i^{(n+1)} = \lambda_g g_i^{(n)} + \alpha_g u(n) u(n-i) \qquad (7b)$$

where $\lambda_a, \lambda_h, \lambda_g$ are the corresponding leakage factors.

Equations (6b) and (7b) were used for short-term predictor adaptation. Two versions of the short-term predictor were tried: 3 poles and 3 zeros, and 2 poles and 6 zeros. The difference in performance between these two versions was found to be small. The following experimental results were obtained with 2 poles and 6 zeroes, i.e., $P=2$ and $Z=6$ in (2). For this version, it was found that using the sign algorithm further improves the robustness. Actually, this leads to an adaptation procedure similar to that used in the CCITT 32 kbit/s algorithm. According to this procedure the updates for the all-pole and all-zero coefficients are:

$$g_i^{(n+1)} = \lambda \, g_i^{(n)} + \alpha \, sgn(u(n)) \, sgn(u(n-i)) \tag{8}$$

$$h_1^{(n+1)} = \lambda_1 h_1^{(n)} + \alpha_1 \, sgn(y'(n)) \, sgn(y'(n-1)) \tag{9a}$$

$$h_2^{(n+1)} = \lambda_2 h_2^{(n)} + \alpha_2 \, sgn(y'(n)) \, sgn(y'(n-2)) \tag{9b}$$

$$- \alpha_2 f(h_1^{(n)}) \, sgn(y'(n)) \, sgn(y'(n-1))$$

where

$$f(h_1^{(n)}) = \begin{cases} 4h_1^{(n)} & \text{if } |h_1^{(n)}| \le 0.5 \\ 2 \, sgn(h_1^{(n)}) & \text{if } |h_1^{(n)}| > 0.5 \end{cases}$$

and

$$y'(n) = u(n) + \sum_{i=1}^{Z} g_i^{(n-1)} u(n-i) \,.$$

The step sizes are given by $\alpha = 1/128$, $\alpha_1 = 3/256$, and $\alpha_2 = 1/128$. The same stability conditions for two poles as in standard ADPCM are imposed on the new coefficients. If the stability conditions indicate an unstable filter, then the coefficients will be hard-limited to stay within the stability triangle. In LD-VXC, the coefficients $\{h_i\}$ and $\{g_i\}$ are updated once every 12 vectors (48 samples) using the corresponding $\{h_i^{(n)}\}$ and $\{g_i^{(n)}\}$ values. This "freezing" of the filter coefficients reduces much of the LD-VXC complexity while the degradation in performance due to this update rate is minimal [3].

A comparison between block adaptation and recursive adaptation using the gradient algorithm in the backward analysis-by-synthesis configuration has shown that recursive adaptation leads to a performance that is comparable to that of block adaptation, but with a lower complexity. Furthermore, the recursive adaptation for LD-VXC has a more distributed computational load, and thus it is more amenable to real-time implementations.

PERCEPTUAL WEIGHTING AND POSTFILTERING

A perceptual weighting filter can often improve the subjective quality of coded speech in analysis-by-synthesis coding algorithms. In our case, the short-term predictor provides a simplistic but useful approximation of the spectral shape and may be used to derive the weighting filter.

In this case, the transfer function of the weighting filter is given in terms of the polynomials, $H(z)$ and $G(z)$, of the short-term predictor as:

$$W(z) = \left[\frac{1 - H(z/\gamma_1)}{1 - H(z/\gamma_2)}\right]\left[\frac{1 + G(z/\gamma_2)}{1 + G(z/\gamma_1)}\right] \qquad (10)$$

where $0 < \gamma_2 < \gamma_1 < 1$. We have found that suitable values of γ_1 and γ_2 are 0.98 and 0.6, respectively.

An alternative approach for designing the weighting filter is to compute its coefficients by applying linear predictive analysis directly on the input speech. In this case (10) is still used, but the polynomials $H(z)$ and $G(z)$ are obtained by running a filter identical to the short-term predictor on the input speech. This approach improves the perceived speech quality due to an improved spectral shape approximation at the expense of a small increase in complexity. The use of the original speech for the weighting filter design is acceptable, as the decoder does not need any information about the weighting filter used in the encoder.

At the decoder side, adaptive postfiltering also helps the subjective quality [6]. In LD-VXC, the pole-zero postfilter has a transfer function determined from the short-term predictor and is updated whenever the short-term predictor is updated. A first order high-pass FIR filter is also included to compensate for the muffling effect caused by spectral tilt. The transfer function of the postfilter is given by

$$H_p(z) = \frac{1 + G(z/\beta)}{1 - H(z/\alpha)} \ (1 - \mu z^{-1}) , \qquad (11)$$

where

$$H\left(\frac{z}{\alpha}\right) = \sum_{j=1}^{P} \alpha^j h_j z^{-j}$$

and

$$G\left(\frac{z}{\beta}\right) = \sum_{j=1}^{Z} \beta^j g_j z^{-j} .$$

A suitable set of coefficients (α, β, μ) is $(\,0.3, 0.9, 0.4)$.

The postfilter attenuates the frequency components in the spectral valley regions of the speech spectrum. An automatic gain control (AGC) is incorporated in the postfilter so that the enhanced speech has roughly the same power as the unfiltered speech. The AGC scales the postfiltered speech by the square root of the ratio of the estimated power of the unfiltered and filtered speech. The postfilter is an optional block in the LD-VXC configuration. There are two major arguments against using a postfilter: the distortion accumulates in tandeming configurations, and the performance for voice-band data may degrade. However, there is no reason to avoid a postfilter at the site of the final user of the transmitted information, particularly if such a postfilter can be switched off during voice-band data transmissions.

TRANSMISSION ERRORS

A significant amount of effort was expended to ensure the robustness of LD-VXC in the presence of channel bit errors. Most of this effort was directed at the pitch predictor. The most significant modification involves parallel adaptation of the short-term and pitch predictors. Such an adaptation scheme ensures that errors propagated by the pitch predictor will not affect the adaptation of the short-term predictor. Other modifications include choice of leakage factors to provide the best compromise between system performance and robustness; resetting the pitch predictor after unvoiced regions of speech; and pseudo-gray coding of the codebook [7].

SIMULATION RESULTS

To achieve the low-delay requirement, LD-VXC uses a codebook of dimension 4. With the standard sampling rate of 8 kHz, we are allowed to use a codebook of size 256 at 16 kbit/s. In order to compare different approaches of backward adaptation, the LD-VXC codec which uses recursive adaptation and another version using block adaptation (BA) for the short-term predictor, called LD-VXC-BA, were simulated and tested on a speech file with a duration of approximately 16 s. The codebooks for the two coders were separately trained; speech data used for training the codebooks were not included in the test file. The objective quality was estimated by the Signal-to-Noise Ratio (SNR) and Segmental Signal-to-Noise Ratio (SEGSNR). SEGSNR is calculated by computing the SNR in dB units for each frame of 256 samples (corresponding to 64 successively coded speech vectors), eliminating silence frames, and taking the arithmetic average of these SNR values over the entire speech file. A frame is considered to contain silence if its signal power is 40 dB below the average power level of the entire speech file.

The system uses a complexity reduction technique based on the system zero input response (ZIR) and zero state respone (ZSR), similar to that used in [3]. In Fig. 3 the relation between the ZSR table update rates and the SNR performance of the two coders are illustrated. For LD-VXC-BA, the autocorrelation method with a Hamming window of size 160 was used to obtain 8-pole coefficients at the end of each update period (i.e., $P=8$ and $Z=0$ in (2)). Suitable bandwidth expansion was also incorporated in LD-VXC-BA for robustness to channel errors. However, it is obvious that LD-VXC-BA requires much higher complexity than does LD-VXC. Though LD-VXC-BA shows a slightly higher SNR curve, informal listening tests suggest that the two coders produce similar speech quality. In the figures, it is also shown that the pitch predictor has greatly improved the performance of both coders, and this is also corroborated in our listening tests. The data in these two figures are obtained without perceptual weighting.

The introduction of a perceptual weighting filter into our system results in improved subjective performance. This improvement is at the expense of objective performance. For this reason, it is of interest to give performance data with and

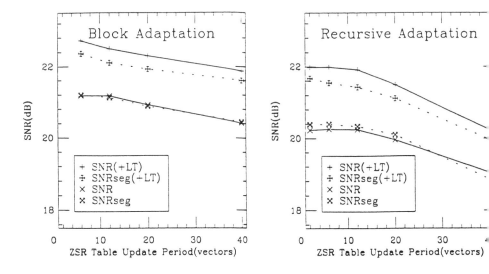

FIGURE 3: LD-VXC and LD-VXC-BA performance as a function of ZSR update period.

without the perceptual weighting filter. At a ZSR update period of 12 vectors, these results are given in Table 1.

Table 2 compares the SEGSNRs of the two coders without perceptual weighting in the presence of transmission errors. The degradation with increased channel errors in the subjective quality of the decoded speech is smoother for LD-VXC. Although the SEGSNR is low for a bit error rate of 10^{-2}, the resulting speech by LD-VXC is intelligible.

Perceptual	LD-VXC		LD-VXC-BA	
Weighting	SNR (dB)	SEGSNR	SNR (dB)	SEGSNR
No	21.92	21.43	22.50	22.11
Yes	19.71	18.55	20.70	19.64

TABLE 1. SNR performance of two coders with and without perceptual weighting.

Bit Error Prob.	LD-VXC	LD-VXC-BA
0	21.43	22.11
10^{-4}	19.70	19.21
10^{-3}	12.68	9.63
10^{-2}	3.87	-0.22

TABLE 2. Comparison of SEGSNR of two coders with transmission errors.

CONCLUSIONS

In this paper, the technique of backward adaptation for achieving low-delay speech coding at low bit rates was described. This technique is applied to both short-term and pitch prediction in an analysis-by-synthesis configuration, leading to the LD-VXC algorithm which produces very good speech quality at 16 kbit/s. Based on the encouraging simulation results and the reasonably low complexity, the authors are now working on a hardware implementation for LD-VXC.

REFERENCES

[1]   G. Davidson, M. Yong, and A. Gersho, "Real-Time Vector Excitation Coding of Speech at 4800 BPS," *Proc. IEEE Int. Conf. Acoustics, Speech, and Signal Processing,* Apr. 1987, pp. 2189-2192.

[2]   B. S. Atal and M. Schroeder, "Code Excited Linear Prediction (CELP): High Quality Speech at Very Low Rates," *Proc. IEEE Int. Conf. Acoustics, Speech, and Signal Processing,* Apr. 1985, pp. 937-940.

[3]   L. Watts and V. Cuperman, "A Vector ADPCM Analysis-By-Synthesis Configuration for 16 Kbit/s Speech Coding," *Proc. IEEE Global Communications Conference,* Nov. 1988, pp. 275-279.

[4]   J. H. Chen and A. Gersho, "Gain Adaptive Vector Quantization with Application to Speech Coding," *IEEE Trans. on Communications,* vol. COM-35, pp. 918-930, Sep. 1987.

[5]   R. Pettigrew and V. Cuperman, "Backward Pitch Prediction for Low-Delay Speech Coding," *Proc. IEEE Global Communications Conference,* pp. 34.3.1-34.3.5, Nov. 1989.

[6]   J. H. Chen and A. Gersho, "Real-Time Vector APC Speech Coding at 4800 bps with Adaptive Postfiltering," *Proc. IEEE Int. Conf. Acoustics, Speech, and Signal Processing,* Apr. 1987, pp. 2185-2188.

[7]   K. Zeger and A. Gersho, "Zero Redundancy Channel Coding in Vector Quantisation," *Electronics Letters,* vol. 23, no. 12, pp. 654-656, June 4, 1987.

3

A ROBUST LOW-DELAY CELP
SPEECH CODER AT 16 KB/S

Juin-Hwey Chen

AT&T Bell Laboratories
600 Mountain Avenue
Murray Hill, NJ 07974

INTRODUCTION

In the past, high-quality speech used to be obtainable only by high-bit-rate coders such as 64 kb/s log PCM or 32 kb/s ADPCM. Currently, several coding techniques can produce high-quality speech at 16 kb/s. These techniques include Code-Excited Linear Prediction (CELP) [1], Multi-Pulse Linear Predictive Coding (MPLPC) [2], Adaptive Predictive Coding (APC) [3], Adaptive Transform Coding (ATC) [4], and Sub-Band Coding (SBC) combined with ADPCM [5], etc. However, all of them require a large coding delay — typically 40 to 60 ms — to achieve high-quality speech. While a large delay is necessary for these coders to buffer enough speech to exploit the redundancy, the delay is undesirable in many applications, especially when echo cancellation is involved. Achieving high-quality speech at 16 kb/s with a coding delay less than 1 or 2 ms has been a major challenge to speech researchers.

Following the G.721 standard of 32 kb/s ADPCM, the CCITT is now planning to standardize a low-delay 16 kb/s speech coder for universal applications by 1991. Because of the variety of applications this standard has to serve, the CCITT has determined a stringent set of performance requirements and objectives for candidate algorithms. The major requirement is that the speech quality should be almost as good as that of G.721 while the one-way encoder/decoder delay should not exceed 5 ms (the objective is ≤ 2 ms) [6].

Because of the low-delay requirement, none of the 16 kb/s coders mentioned above can be used in their current forms. In the past few years, several low-delay 16 kb/s coders have been proposed [7891011121314]. Nearly all of them are predictive coders, and more than half of them rely on postfilters of the type proposed in [7] or [15] to improve the speech quality.

In this chapter, we describe a 16 kb/s Low-Delay CELP (LD-CELP) coder that has an algorithmic delay of 0.625 ms (5 samples) and a one-way coding delay less than 2 ms. This coder has been submitted by AT&T to the CCITT [16] and is now the only candidate algorithm in the CCITT contest. The essence of CELP, which is the analysis-by-synthesis codebook search, is retained in LD-CELP. However, most other parts of CELP have been modified to achieve high quality at a low delay. This LD-CELP coder is quite different from the CELP coder in [14]. The main difference is that the CELP in [14] still uses forward-adaptive predictors, while our LD-CELP uses a purely backward-adaptive predictor. This allows us to have a lower algorithmic delay (0.625 ms rather than 2 ms).

In the conventional forward-adaptive CELP coder [1], shown in Fig. 1, the predictor parameters, the gain, and the excitation are all transmitted to the receiver. In the LD-CELP coder (Fig. 2), on the other hand, only the excitation sequence is transmitted. The predictor coefficients are updated by performing LPC analysis on previously quantized speech. The excitation gain is updated by using the gain information embedded in previously quantized excitation. The excitation to the synthesis filter can thus be considered as being quantized by a backward gain-adaptive vector quantizer [17]. The vector dimension (or block size) of this gain-adaptive VQ is only 5 samples. The one-way coding delay, which is typically 2 to 3 times the block size, is therefore between 1 and 2 ms. This not only surpasses the CCITT delay requirement of 5 ms but actually meets the 2 ms objective.

Formal subjective tests indicated that the speech quality of 16 kb/s LD-CELP was similar to or better than that of G.721 for single encoding. Such a good performance was achieved without the help of postfiltering. While most other low-delay 16 kb/s coders rely on postfiltering to improve speech quality, we avoid it for two reasons. First, the speech distortion introduced by postfiltering will accumulate during tandem coding and may result in severely distorted speech. Second, the postfilter inevitably introduces some phase distortion; this may cause problems when decoding non-voice signals (e.g. DPSK modem signals) which carry information in their phase.

In the following sections, we first describe various parts of the LD-CELP algorithm and then discuss its performance.

HIGH-ORDER LPC PREDICTOR

In LD-CELP, we use an all-pole predictor and update it in a backward-adaptive manner by performing LPC analysis on previously quantized speech. The autocorrelation method is used. Earlier versions of LD-CELP used a 20 ms Hamming window, while newer versions used Barnwell's recursive windowing method [18]. Direct implementation of Barnwell's method gave numerical precision problems in LPC analysis, but the problems were easily avoided by replacing his third-order direct-form filter by three first-order filters in cascade. Compared with the 20 ms Hamming window, the recursive window gives 0.7 to 1.5 dB improvement in overall coder SNR and significant improvement in subjective speech quality. In addition, the recursive window has several advantages in DSP implementations, including smaller memory space requirements, evenly distributed computations, and lower computational complexity for frequent predictor updates.

The conventional CELP coders usually have a forward-adaptive long-term predictor, or pitch predictor, to exploit the pitch redundancy in speech. In LD-CELP, due to the constraint of the small block size, a more natural way to use a pitch predictor is to make it backward-adaptive [12]. However, backward pitch predictor adaptation is very sensitive to channel errors, as it tends to diverge after being hit by bit errors. Artificial resets may somewhat alleviate the problem.

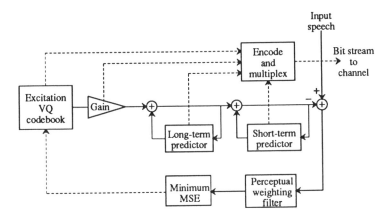

Fig. 1 Conventional Forward-Adaptive CELP Coder

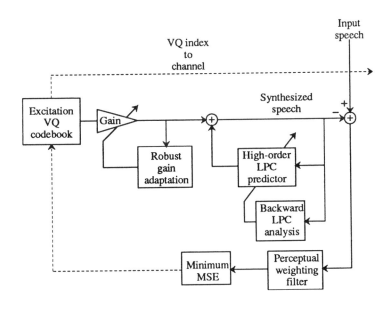

Fig. 2 Low-Delay Backward-Adaptive CELP Coder

However, at a high bit-error rate (BER) such as 10^{-2}, which translates to 160 bit errors per second at 16 kb/s, the reset technique is not likely to work. In contrast, backward LPC analysis appears to be fairly robust to bit errors.

Since the CCITT requires candidate algorithms to be at least as good as G.721 at a BER of 10^{-2}, and since it is not likely that a backward-adaptive pitch predictor can withstand such a high BER, we decided to eliminate the pitch predictor. However, eliminating the pitch predictor significantly affected female speech quality, although the effect on male speech was much less noticeable. To compensate for the loss in speech quality, we increased the LPC predictor order from 10 to 50. The motivation was to use a 50-th order LPC predictor to exploit the pitch redundancy in female speech.

Figure 3 (a) shows the prediction gain versus the LPC predictor order for a male and a female speech utterance. The prediction gain for male saturates at order 20, but for female it does not saturate until order 50. Figure 3 (b) shows the prediction gain for 10 more female speakers. It is clear that most curves indeed saturate around order 40 to 50. The perceptual quality of LD-CELP coded speech seemed to have a similar trend. Thus, we used order 50 in LD-CELP.

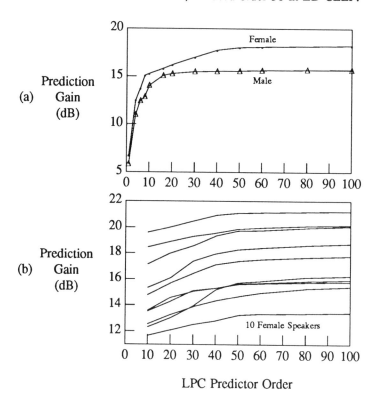

Fig. 3 (a) prediction gain as a function of the backward-adaptive LPC predictor order for a male speaker and a female speaker. (b) prediction gain for orders above 10, for 10 more female speakers.

A common practice in conventional CELP coders is to slightly increase the bandwidth of the LPC spectra so as to smooth very sharp formant peaks. Such bandwidth expansion is also used in LD-CELP. Let $\hat{a}_i$ be the i-th coefficient of the LPC predictor as obtained by LPC analysis. Then, bandwidth expansion is achieved by replacing the predictor coefficients $\hat{a}_i$'s by $a_i = \lambda^i \hat{a}_i$, where the bandwidth expansion factor λ is given by $\lambda = e^{-2\pi B / f_s}$ for speech sampled at f_s Hz and a desired bandwidth expansion of B Hz [1]. In LD-CELP, we chose B to be 15 Hz. This has the effect of scaling all the poles of the LPC synthesis filter radially toward the origin by a factor of $\lambda = 0.9883$. Since the poles are moved away from the unit circle, the formant peaks in LPC spectra are widened. Furthermore, when compared with the original LPC synthesis filter, the bandwidth-expanded filter gives shorter impulse response and shorter propagation of channel error effects in the decoded speech. Therefore, the bandwidth expansion procedure not only avoids occasional chirps in decoded speech due to very sharp formant peaks, but also enhances the coder's robustness to channel errors.

If we update the predictor for each 5-sample speech vector, the computational complexity would be quite high. However, since the spectral envelope of speech does not change rapidly with time, we were able to update the LPC predictor less frequently (e.g. once every 2.5 ms or 5 ms) and have very little degradation in speech quality. Both 2.5 ms and 5 ms predictor update periods gave only barely noticeable degradation in speech quality, although the 2.5 ms update had a slight edge. Initially we had a predictor update period of 5 ms; however, since 2.5 ms updates gave slightly better performance and did not require too much more computations when recursive windowing was used, we later used an update period of 2.5 ms in the LD-CELP coder version submitted to the CCITT.

PERCEPTUAL WEIGHTING FILTER

In conventional CELP, the perceptual weighting filter has the form:

$$W(z) = \frac{1 - Q(z)}{1 - Q(z/\gamma)} , \tag{1}$$

where

$$Q(z) = \sum_{i=1}^{M} q_i z^{-i} , \quad Q(z/\gamma) = \sum_{i=1}^{M} \gamma^i q_i z^{-i} , \quad 0 < \gamma < 1 , \tag{2}$$

and q_i's are quantized LPC coefficients and M is the LPC predictor order. With M as high as 50 in LD-CELP, a 50-th order weighting filter was found to cause artifacts in the synthesized speech, so we used a 10-th order weighting filter.

One natural way to obtain the 10-th order weighting filter is to use the 10-th order predictor that is available in the intermediate stage of the 50-th order backward LPC analysis. However, in this work we derive the weighting filter from a separate 10-th order LPC analysis on the *unquantized* clean speech. The motivation is that backward LPC analysis is based on coded speech which may have inaccurate spectral envelope, whereas a weighting filter derived from unquantized speech always guarantees an accurate coding noise spectrum. (Note that we are allowed to use unquantized speech to obtain the weighting filter, since the LD-CELP decoder does not need to know the weighting filter used in the encoder.)

A further improvement on the weighting filter is to let W(z) have a more general form as proposed in [15].

$$W(z) = \frac{1 - Q(z/\gamma_1)}{1 - Q(z/\gamma_2)} , \quad 0 < \gamma_2 < \gamma_1 \leq 1 . \tag{3}$$

In LD-CELP we let $\gamma_1 = 0.9$ and $\gamma_2 = 0.4$; this combination was found to give a lower perceived noise level than the combination of $\gamma_1 = 1.0$ and $\gamma_2 = 0.8$ typically used in conventional CELP coders.

ROBUST GAIN ADAPTATION

We developed two gain adaptation algorithms which are separately described below. Which one is more suitable depends upon the application.

Jayant Gain Adapter

Our first gain adaptation algorithm is based on the generalization [17] of the robust Jayant quantizer [19, 20] to vector quantization. Let $\sigma(n)$ be the excitation gain at the vector time index n. Then, $\sigma(n)$ is determined as

$$\sigma(n) = M(n-1)\sigma^\beta(n-1) , \tag{4}$$

where β is a leakage factor slightly less than unity, and the gain multiplier $M(n-1)$ is a function of the excitation codevector $y(n-1)$ at time $(n-1)$. If $y(n-1)$ is large, then $\sigma(n-1)$ is probably not large enough, so we let $M(n-1) > 1$ to amplify the gain. Conversely, if $y(n-1)$ is small, we have $M(n-1) < 1$ to reduce the gain. Hence, the adaptation mechanism is similar to the scalar Jayant quantizer.

The leakage factor β gives the algorithm robustness to bit errors. Taking logarithm of Eq. (4) yields

$$\log[\sigma(n)] = \beta \log[\sigma(n-1)] + \log[M(n-1)] . \tag{5}$$

If we consider $\log[M(n-1)]$ as the input and $\log[\sigma(n)]$ as the output, then Eq. (5) represents a first-order all-pole filter with a fixed coefficient β. Since $\beta < 1$, the impulse response of this filter is an exponentially decaying function. In the presence of channel errors, the decoder will have erroneously decoded codevectors which will in turn cause impulsive noise in the $\{\log[M(n-1)]\}$ sequence. The corresponding effects on the output sequence $\{\log[\sigma(n)]\}$ will eventually decay to zero due to the decaying impulse response.

Ideally, each codevector in the excitation codebook should have a dedicated gain multiplier. However, it is difficult to optimize hundreds of gain multipliers for hundreds of codevectors. Hence, we took a simpler approach by forcing the gain multiplier to be a function of the root-mean-square (RMS) value of the selected codevector and then optimizing only a few parameters that control the function.

Let x be the RMS value of the codevector $y(n-1)$, then $M(n-1) = f(x)$ where the function $f(x)$ is chosen to be

$$f(x) = \begin{cases} [(1-x)\sigma_{min}^{1-\beta} + x\sigma_{av}^{1-\beta}]e^{c_2(x-1)} & \text{if } 0 \le x < 1 \\ \sigma_{av}^{1-\beta}e^{c_1(x-1)} & \text{if } 1 \le x < 4 \\ \sigma_{av}^{1-\beta}M_{max} & \text{if } x \ge 4 \end{cases} \quad (6)$$

where suitable parameter values for 16-bit linear PCM input are $\sigma_{min} = 1$, $\sigma_{av} = 100$, $\beta = (31/32)^5 = 0.853$, $c_1 = \ln(M_{max})/3$, $c_2 = -\ln(M_{min})$, $M_{max} = 1.8$, and $M_{min} = 0.8$. Basically, this function was derived so that when $\beta = 1$ (the ideal case with no bit errors), the function is composed of two sections of exponentials $e^{c_2(x-1)}$ and $e^{c_1(x-1)}$ for $0 \le x < 1$ and $1 \le x < 4$, respectively, and the function is clipped at M_{max} for $x \ge 4$. The parameters c_1 and c_2 were chosen such that $f(0) = M_{min}$ and $f(4) = M_{max}$. The term $\sigma_{av}^{1-\beta}$ was then introduced to compensate for the effect of a β less than unity. The term $\sigma_{min}^{1-\beta}$ was introduced to handle coder input signals of very low levels.

Although it is possible to calculate the gain multiplier function in real-time, a simpler approach is to precompute and store the multipler value for each of the codevectors, then it becomes a table look-up procedure during actual coding.

Adaptive Logarithmic Gain Predictor

The second algorithm adapts the excitation gain by using a 10-th order adaptive linear predictor in the logarithmic domain. Let $e(n)$ be the gain-scaled excitation vector at time n, and let $\sigma_e(n)$ and $\sigma_y(n)$ be the RMS values of $e(n)$ and $y(n)$, respectively. Then, we have $e(n) = \sigma(n)y(n)$, or

$$\log[\sigma_e(n)] = \log[\sigma(n)] + \log[\sigma_y(n)]. \quad (7)$$

The goal of the gain predictor is to make $\sigma(n)$ as close to $\sigma_e(n)$ as possible. Therefore, we let $\log[\sigma(n)]$ be a prediction of $\log[\sigma_e(n)]$ based on $\log[\sigma_e(n-1)]$, $\log[\sigma_e(n-2)]$, ..., i.e.,

$$\log[\sigma(n)] = \sum_{i=1}^{10} p_i \log[\sigma_e(n-i)] \quad (8)$$

$$= \sum_{i=1}^{10} p_i \log[\sigma(n-i)] + \sum_{i=1}^{10} p_i \log[\sigma_y(n-i)] . \quad (9)$$

Thus, we have a 10-th order pole-zero filter with $\log[\sigma_y(n-1)]$ as the input and $\log[\sigma(n)]$ as the output.

We update the coefficients p_i's by performing backward LPC analysis on the previous $\{\log[\sigma_e(n)]\}$ sequence (with some gain offset subtracted). The autocorrelation method guarantees the stability of the pole-zero filter defined by Eq. (9). In other words, the impulse response of this pole-zero filter decays to zero eventually. Hence, this gain adapter is also robust to channel errors. To improve the robustness, the coefficients p_i's are modified by the bandwidth expansion procedure with a factor of $\lambda = 0.9$.

In contrast to the fixed-coefficient, first-order filter in the Jayant gain adapter, here we have an adaptive-coefficient, 10-th order filter. Hence, better clear-channel performance is achieved. However, the Jayant gain adapter is more robust to channel errors due to the fixed coefficient and shorter impulse response.

EXCITATION VQ

At a bit-rate of 2 bits/samples and a vector dimension of 5 samples, we have 10 bits to encode each excitation vector. The complexity of searching through a 10-bit codebook (1024 entries) is too high for real-time implementations. To reduce the complexity, we introduced a "gain-shape" structure and decomposed the codebook into the product of a 3-bit gain codebook and a 7-bit shape codebook. The three gain bits consist of a sign bit and two magnitude bits. The complexity can be further reduced by the Zero-State-Response (ZSR) Codebook approach in [15], or by the following method.

Similar to conventional CELP coders, the contribution of previous memory to the filter output in the present vector is subtracted from the speech signal [21] to give a target vector $t(n)$ for the codebook search. In LD-CELP, the backward-adaptive gain $\sigma(n)$ is known before the codebook search at time n starts, so we can compute the normalized target vector $x(n) = t(n)/\sigma(n)$. Let y_j be the j-th codevector in the 7-bit shape codebook, g_i be the i-th levels in the 2-bit magnitude codebook, and let $\eta_k = 1$ or -1 depending on whether the sign bit k is 0 or 1. Then, we seek the best indices i, j, and k that minimize the following distortion

$$D = \| x(n) - \eta_k g_i H y_j \|^2 , \tag{10}$$

where H is a lower triangular matrix [21] with subdiagonals equal to samples of the impulse response of the cascaded LPC filter and weighting filter. This is equivalent to minimizing

$$\hat{D} = -\eta_k b_i p^T(n) y_j + c_i E_j \tag{11}$$

where $b_i = 2g_i$, $c_i = g_i^2$, $p(n) = H^T x(n)$, and $E_j = \| H y_j \|^2$. The b_i and c_i arrays are fixed and thus can be precomputed and stored. The E_j array is time-varying but is fixed over the time period between the updates of predictor and weighting filter coefficients; hence, it can be periodically updated and stored to save computation. The best combination of i, j, and k is obtained by going through $j = 0, 1, ..., 127$ and finding the best (i,k) pair for each j. The amount of computation is greatly reduced due to the precomputation of the E_j array. The advantage of this approach over the ZSR codebook approach is that the E_j array takes only 1/5 of the memory space required by the ZSR codebook.

Rather than populated with random Gaussian numbers, the 7-bit shape codebook is closed-loop optimized by a codebook design algorithm using the same weighted error criterion of the LD-CELP encoder. Thus, the effects of predictor adaptation and gain adaptation are taken into account in the design. Also, to make excitation VQ robust to channel errors, we applied Pseudo Gray Coding [22,23] to the 3-bit gain codebook and the 7-bit shape codebook. With Gray-coded codebook indices, a single bit error will result in a decoded codevector close to the transmitted one. For noisy channels, this technique significantly improved the decoded speech quality when compared with random assignment of codebook indices.

PERFORMANCE

The subjective speech quality of 16 kb/s LD-CELP and G.721 in terms of Mean Opinion Score (MOS)† has been measured and compared in several formal subjective listening tests. With noisy channels, an earlier version of LD-CELP, which used a 20 ms Hamming window, scored slightly higher than G.721 for BER = 10^{-2}, and outperformed G.721 by an MOS margin of 0.5 for BER = 10^{-3}. For BER = 0 (clear channel), a newer LD-CELP version with recursive windowing achieved an MOS of 4.173 for single encoding and 3.474 for 3 asynchronous tandems. In the same test, G.721 obtained 4.165 for single encoding and 3.642 for 4 asynchronous tandems.‡ Although the noisy-channel performance of this newer LD-CELP version has not yet been tested, it is comparable to the earlier version, judging from informal listening.

From these MOS results, the LD-CELP coder appears to have met or exceeded most CCITT requirements for speech quality. The only exception is 3 asynchronous tandems, but even for this condition the MOS is not too far from the target (within 0.168). Furthermore, the coder maintained graceful degradation as the number of tandem stages increased. The difficulty with tandeming is probably a more fundamental problem of coders operating at 16 kb/s or below.

We have also tested LD-CELP under several other conditions. The coder still maintained near-transparency when coding telephone-bandwidth music or multi-speaker voices. With a 1004 Hz test tone, the coder's signal-to-noise ratio as a function of input level exceeded the CCITT requirements for toll-quality coders. The LD-CELP coder successfully passed the DTMF dialing tones. Furthermore, laboratory testing showed that the coder also passed 300, 1200, and 2400 bps voice-band data modem signals with sufficient clarity that the link was error free in the absence of other line impairments. The 2400 bps modems tested included the V.22bis, V.26, and V.29. Decoder convergence was not a problem in this backward-adaptive coder. When the internal states of an LD-CELP decoder were arbitrarily reset in the middle of an utterance, the decoded speech only had a slight disturbance which is not objectionable, then the speech quality quickly returned to normal. This indicated that the decoder states converged to the encoder states within a short period of time.

CONCLUSION

The 16 kb/s Low-Delay CELP coder achieves high-quality speech at a coding delay less than 2 ms. The salient features of this coder includes (1) high-order LPC predictor and its backward adaptation through LPC analysis, (2) backward-adaptive gain and the adaptive log-gain predictor, and (3) closed-loop optimized excitation codebook with Pseudo-Gray code index assignment. The coder is very

---

† The MOS is a 5-point scale subjective measure of speech quality, with 1 and 5 corresponding to "unsatisfactory" and "excellent", respectively.

‡ The CCITT quality requirement is 14 qdu's for 3 tandems of the 16 kb/s candidate; this level of quality is equivalent to 4 tandems of G.721.

robust to channel errors without the help of any forward error protection. It is also robust to several kinds of non-voice signals. Compared with G.721, this coder gives slightly worse performance for asynchronous tandeming but similar or better speech quality for single encoding with clear or noisy channels. In conclusion, the coder achieves performance comparable to G.721 at half the bit-rate.

ACKNOWLEDGEMENT

I am greatly indebted to Richard V. Cox for his invaluable help and contribution throughout the course of this work. I am also grateful to Nikil S. Jayant for making me aware of the CCITT 16 kb/s standardization activity and for many useful discussions. The help from Duane O. Bowker and Horace G. Hagen in conducting MOS tests and modem tests is also greatly appreciated.

REFERENCES

1. M.R. Schroeder and B.S. Atal, "Code-Excited Linear Prediction (CELP): high quality speech at very low bit rates," *Proc. IEEE Int. Conf. Acoust., Speech, Signal Processing*, pp. 937-940 (1985).

2. B.S. Atal and J.R. Remde, "A new model of LPC excitation for producing natural-sounding speech at low bit rates," *Proc. IEEE Int. Conf. Acoust., Speech, Signal Processing*, Paris, France, pp. 614-617 (April 1982).

3. B.S. Atal, "Predictive coding of speech at low bit rates," *IEEE Trans. Communications* **COM-30**(4), pp. 600-614 (April 1982).

4. R. Zelinski and P. Noll, "Adaptive transform coding of speech signals," *IEEE Trans. Acoust., Speech, Signal Processing* **ASSP-25**, pp. 299-309 (1977).

5. F.K. Soong, R.V. Cox, and N.S. Jayant, "A high quality subband speech coder with backward adaptive predictor and optimal time-frequency bit assignment," *Proc. IEEE Int. Conf. Acoust., Speech, Signal Processing*, pp. 2387-2390 (1986).

6. CCITT Study Group XVIII, *Terms of reference of the ad hoc group on 16 kbit/s speech coding (Annex 1 to question U/XV)*, June 1988.

7. N.S. Jayant and V. Ramamoorthy, "Adaptive postfiltering of 16 kb/s-ADPCM speech," *Proc. IEEE Int. Conf. Acoust., Speech, Signal Processing*, pp. 829-832 (1986).

8. M. Berouti, J. Jachner, D. Sloan, and P. Mermelstein, "Reducing signal delay in multi-pulse coding at 16 kb/s," *Proc. IEEE Int. Conf. Acoust., Speech, Signal Processing*, pp. 3043-3046 (1986).

9. T. Taniguchi and et al., "A 16 kbps ADPCM with multi-quantizer (ADPCM-MQ) codec and its implementation by digital signal processor," *Proc. IEEE Int. Conf. Acoust., Speech, Signal Processing*, pp. 1340-1343 (1987).

10. R. V. Cox, S. L. Gay, Y. Shoham, S. R. Quackenbush, N. Seshadri, and N.

S. Jayant, "New directions in subband coding," *IEEE J. Selected Areas Comm.* **6**(2), pp. 391-409 (February 1988).

11. J. D. Gibson and G. B. Haschke, "Backward adaptive tree coding of speech at 16 kbps," *Proc. IEEE Int. Conf. Acoust., Speech, Signal Processing*, New York, pp. 251-254 (April 1988).

12. V. Iyengar and P. Kabal, "A low delay 16 kbits/sec speech coder," *Proc. IEEE Int. Conf. Acoust., Speech, Signal Processing*, New York, pp. 243-246 (April 1988).

13. L. Watts and V. Cuperman, "A Vector ADPCM analysis-by-synthesis configuration for 16 kbit/s speech coding," *Proc. IEEE Global Commun. Conf.*, pp. 275-279 (December 1988).

14. L. Cellario, G. Ferraris, and D. Sereno, "A 2 ms delay CELP coder," *Proc. IEEE Int. Conf. Acoust., Speech, Signal Processing,*, pp. 73-76 (May 1989).

15. J.-H. Chen and A. Gersho, "Real-time vector APC speech coding at 4800 bps with adaptive postfiltering," *Proc. IEEE Int. Conf. Acoust., Speech, Signal Processing*, pp. 2185-2188 (1987).

16. *AT&T contributions to CCITT Study Group XV and T1Y1.2*, October 1988 - November 1989.

17. J.-H. Chen and A. Gersho, "Gain-adaptive vector quantization with application to speech coding," *IEEE Trans. Comm*, pp. 918-930 (September 1987).

18. T. P. Barnwell, III., "Recursive windowing for generating autocorrelation coefficients for LPC analysis," *IEEE Trans. Acoust., Speech, Signal Processing* **ASSP-29**(5), pp. 1062-1066 (October 1981).

19. N. S. Jayant, "Adaptive quantization with a one word memory," *Bell Syst. Tech. J.* **52**, pp. 1119-1144 (September 1973).

20. D.J. Goodman and R.M. Wilkinson, "A robust adaptive quantizer," *IEEE Trans. Comm.*, pp. 1362-1365 (November 1975).

21. I.M. Trancoso and B.S. Atal, "Efficient procedures for finding the optimum innovation in stochastic coders," *Proc. IEEE Int. Conf. Acoust., Speech, Signal Processing*, pp. 2375-2379 (1986).

22. J.R.B. De Marca and N.S. Jayant, "An algorithm for assigning binary indices to the codevectors of a multi-dimensional quantizer," *Proc. IEEE Int. Conf. on Communications*, pp. 1128-1132 (June 1987).

23. K.A. Zeger and A. Gersho, "Zero redundancy channel coding in vector quantization," *Electronics Letters* **23**(12), pp. 654-656 (June 1987).

4

AN EFFICIENT VARIABLE-BIT-RATE LOW-DELAY CELP (VBR-LD-CELP) CODER

**Yair Be'ery †, Zeev Shpiro, Tal Simchony † †,
Leonid Shatz, Joshua Piasetzky † †**

† Department of Electronic Communications,
Control and Computer Systems,
Tel-Aviv University, Tel-Aviv 69978, Israel.

† † ECI Telecom Ltd.,
30 Hasivim St.,
Petach Tikva 49130, Israel.

INTRODUCTION

Two versions [1],[2] of Low-Delay Code-Excited Linear Predictive (LD-CELP) coders have been recently suggested as candidates for the CCITT 16 kbit/s speech coding standard. The goal of this standard is to cover a long list of possible applications like mobile radio, video-phone, Digital Circuit Multiplication Equipment (DCME), etc.. Many of these applications have different requirements so it has been difficult to define the performance requirements and objectives for a unique algorithm which will be suitable for a large variety of these applications. Thus it was suggested [3] and accepted as an objective of the standard that the adopted algorithm will have a *nominal* rate of 16 Kbit/s but can operate at bit rates higher and lower than the nominal rate. This may enable a more optimal implementation for each specific application and provide a better basis for acceptance of the standard by various user groups. For example, Variable-Bit-Rate (VBR) coding can be used to add an error correction/detection information for noisy channel applications like mobile radio. Another example is in DCME systems, where VBR coding can avoid speech clipping during overload traffic periods, and can improve speech quality during underload periods.

Implementing VBR in the proposed LD-CELP algorithms for bit rates lower than the nominal rate can be achieved, with on change in the delay, by using proper subsets of the codebooks or by a re-design of reduced codebooks. For bit rates higher than the nominal rate, a solution based on straightforward codebook expansion is not practical since it significantly increases the computational complexity and memory requirements.

An efficient VBR method for LD-CELP coders will be presented here. This method is mainly based on extending the existing codebooks by means of a lattice code [4], which was appropriately modified to match the statistics of the residual speech signal. A vector from this lattice codebook is added as an offset to the best vector selected from the LD-CELP stochastic codebook. Due to the lattice-offset structure, the search procedure can be performed efficiently and requires a small amount of additional memory.

STRUCTURE OF THE VBR CODER

Block diagrams of the proposed encoder and decoder are shown in Figures 1 and 2, respectively. These are modified versions of the LD-CELP encode-decoder of [1]. The main differences are an extra block *Lattice Codebook* and an additional gain table.

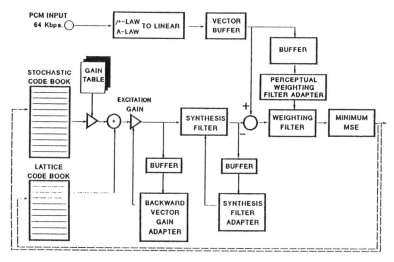

Figure 1. Variable-Bit-Rate LD-CELP Encoder

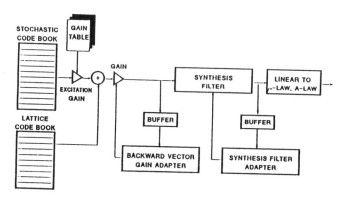

Figure 2. Variable-Bit-Rate LD-CELP Decoder

Two modes of operation, one for higher bit rate (24 kbit/s) and another for lower bit rate (12.8 kbit/s), will be discussed. However, we shall concentrate only on the more interesting case - the higher bit rate mode of operation. The next section details the principles and functional description of a 24 kbit/s extended mode for LD-CELP coder, along with simulation results and complexity figures. The 12.8 kbit/s reduced mode is achieved by taking a proper half of the gain and the shape codebooks. This causes a graceful degradation of the LD-CELP coder speech performance. The parameters of the proposed method has been fine-tuned for the LD-CELP algorithm of [1] with a Jayant gain adapter and the related gain and shape codebooks (see [1], Appendices A,D and E). However, this method can be easily tuned for other LD-CELP algorithms, like the LD-VXC [2], or for improved versions of [1] and [2], e.g. with modified codebooks. The last section details an implementation and simulation results of the proposed algorithm as a continuous VBR-LD-CELP coder over the entire range of rates (12.8 - 24 kbit/s) with 10 msecond VBR frame. The performance of the proposed scheme including channel errors is also addressed.

EXTENDED MODE FOR 24 KBIT/S

To increase the bit rate from 16 kbit/s to 24 kbit/s, with no increase in the delay, 5 bits of index are added to the existing 10 index bits (7 bits for shape plus 3 bits for gain) [1]. In the proposed coder, these 5 extra bits are used as follows: 1 extra bit for the gain codebook and 4 bits for a third new codebook - the *lattice–offset* codebook. *Lattice* means that the selection of vectors for the codebook is based on a lattice structured set of points, which is optimal, under certain assumptions, for quantization [4, Chapter 2]. *Offset* means that a vector from this third codebook is added as an offset to the selected best vector of the shape-gain codebooks, in order to minimize the distortion measure.

There are two main reasons for using a lattice-offset codebook. Firstly, the lattice structure as well as the offset calculation only for the best stochastic shape-gain vector reduce the computational complexity of the extended mode to practical figures.

Secondly, lattice theory indicates which is the optimal lattice quantizer for n-dimensional vectors without going into complex clustering procedures. In our case, an appropriate 5-dimensional lattice quantizer may be composed of the following three sets of vectors, which belong to the first few shells of D_5^*, the dual of the so called *checkerboard lattice* [4]: 1. (0,0,0,0,0); 2. ±(1,0,0,0,0), ±(0,1,0,0,0), ±(0,0,1,0,0), ±(0,0,0,1,0), ±(0,0,0,0,1); 3. ±(1,−1,0,0,0), ±(1,0,−1,0,0), ±(1,0,0,−1,0), ... , ±(0,0,0,1,−1). This means that the offset around the best stochastic shape-gain vector can be either zero (no offset), a single pulse, or a pair of opposite pulses. Thus these vectors can be regarded, in a sense, as multipulse excitation vectors. The optimal subset of 15 vectors and the amplitude of each pulse were selected and tuned empirically over several male and female speech files. The use of these sets of vectors was also motivated by several previous empirical results like those reported in [5]-[8].

40

Additional motivation for the use of such sets of vectors was an observation that the lack of pitch prediction in the LD-CELP coder results in an impulsive shape of the residual error signal. This phenomenon is clearly illustrated in Figure 3, where the error signal for a male voice segment is presented for the nominal 16 kbit/s and for the extended 24 kbit/s modes. The impulses during the pitch periods were significantly reduced by employing the above mentioned vectors. For a female speaker the situation is similar but less severe as shown in Figure 4. This is due to the long synthesis filter (50 taps) which somewhat compensates for a female pitch period but not for a longer period of a male pitch.

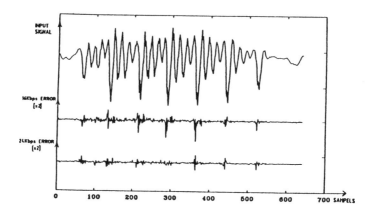

**Figure 3. Pitch Phenomenon for 16 and 24 Kbit/s
LD-CELP with Male Voice.**

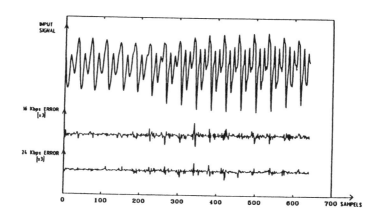

**Figure 4. Pitch Phenomenon for 16 and 24 Kbit/s
LD-CELP with Female Voice.**

Functional Description

Adding one bit to the gain index enlarges the gain codebook by eight extra levels. The new gain codebook contains the original eight levels, six additional intermediate levels and two extreme levels. It is also gray-coded for improved noise immunity. In the encoder, the search procedure for the best gain value is performed with eight different values (excluding the sign), instead of four different values in the 16 kbit/s mode. The computational complexity issues are discussed later in the sequel. In the decoder, the appropriate value is extracted from the extended gain codebook according to the received gain index.

As illustrated in Figure 1, the best offset vector $f(n)$ from the lattice-offset codebook is summed with the best shape-gain vector $y(n)$ to produce $y^*(n)$:

$$y^*(n) = y(n) + f(n) \tag{1}$$

The best shape-gain-offset vector $y^*(n)$ is now fed into the gain and synthesis modules, and influences the related backward adapters.

The search for the best offset vector is performed after the search for the best shape-gain vector has been completed, thus we already have η_{kmin}, g_{imin} and y_{jmin}, which are the best sign, gain and shape vector, respectively, where their indices are given by the subscripts. Let $x(n)$ be the codebook target vector, $\sigma(n)$ be the excitation gain, and $\mathbf{H}$ a 5x5 impulse response matrix of the cascaded synthesis and perceptual weighting filters (for more details see [1]). Then the best offset vector is the one which minimizes the following Mean Square Error (MSE) distortion:

$$D^* = |x(n) - \sigma(n)\mathbf{H}(\eta_{kmin} g_{imin} y_{jmin} + \xi_q f_l)|^2 \tag{2}$$

where f_l is one of the 8 offset vectors listed in Table I (including Gray-code index assignment), and ξ_q is +1 or -1 (the sign of f_l). Expanding the terms of (2) gives us

$$D^* = |x(n)|^2 - 2\sigma(n)x^T\mathbf{H}(\eta_{kmin} g_{imin} y_{jmin} + \xi_q f_l) \tag{3}$$

$$+ \sigma^2(n)|\mathbf{H}(\eta_{kmin} g_{imin} y_{jmin} + \xi_q f_l)|^2$$

Since the terms $|x(n)|^2$, $-2\sigma(n)x^T\mathbf{H}\eta_{kmin} g_{imin} y_{jmin}$, and $\sigma^2(n)|\mathbf{H}(\eta_{kmin} g_{imin} y_{jmin})|^2$ are fixed during the lattice-offset codebook search, minimizing D^* is equivalent to minimizing

$$\bar{D}^* = \left[-2\sigma(n)p^T(n) + 2\sigma^2(n)\eta_{kmin} g_{imin} y_{jmin}^T \mathbf{H}^T\mathbf{H}\right]\xi_q f_l + \sigma^2(n)E_l \tag{4}$$

where

$$p(n) = \mathbf{H}^T x(n) \tag{5}$$

and

$$E_l = |\mathbf{H} f_l|^2 \tag{6}$$

Channel Index	Offset-Vector Components				
0	+0.833	0.00	0.00	0.00	0.00
1	+0.809	-0.715	0.00	0.00	0.00
2	0.00	0.00	+0.964	0.00	0.00
3	0.00	-0.764	+0.752	0.00	0.00
4	0.00	0.00	0.00	0.00	0.00
5	0.00	0.00	0.00	-0.710	+0.971
6	0.00	0.00	+0.753	-0.842	0.00
7	0.00	0.00	0.00	-0.874	0.00

Table I. Lattice-Offset Codebook

Note that E_l and $H^T H$ can be pre-computed and stored for a period of 8 speech vectors (5 msec.), since H is fixed over this period. The structure of the distortion function in (4) and the search for the optimal offset vector f_l, are respectively similar to the structure of the distortion function in eq. (18) of [1] and to the search for the optimal shape vector yj given in [1].

In the decoder, the received 4 bits are split into 3-bits for the lattice-offset codebook index and 1-bit for the sign.

Simulation Results

The proposed algorithm has been tested over three male and three female speech files, each of about 3 seconds length. The segmental SNR ([9] Appendix E) and an experimental perceptual SNR (see Appendix) for 16 and 24 kbit/s modes are summarized in Table II.

Speech File	Seg. SNR (dB)			Percep. SNR (dB)		
	16k	24k	Gain	16k	24k	Gain
Male-1	17.0	21.4	4.4	11.9	15.9	4.0
Male-2	15.7	20.7	5.0	11.4	15.1	3.7
Male-3	15.2	19.5	4.3	10.1	13.8	3.7
Female-1	19.4	23.6	4.2	14.1	17.8	3.7
Female-2	21.2	25.7	4.5	15.7	19.4	3.7
Female-3	20.9	24.6	3.7	15.2	18.5	3.3
Average	18.2	22.6	4.4	13.1	16.8	3.7

Table II. Simulation Results for the Extended Mode

These results imply that an average improvement of more than 4 dB is achieved by the 24 kbit/s extended mode over the nominal 16 kbit/s LD-CELP coder. It was also observed, by looking at the error signal (see Figures 3 and 4), that the proposed lattice-offset codebook partially compensates for the lack of pitch predictor in the LD-CELP coder. Informal listening test indicated substantial subjective improvement.

Complexity

The complexity of the 16 kbit/s LD-CELP encoder is approximately 7.5 MFLOPS. This figure is based on [10] and on an additional complexity reduction of about 1.4 MFLOPS achieved by calculating first the unquantized optimal gain and then quantizing it by a tree structured procedure (see also [2] and [5]). Using this scheme, the increase in complexity for the additional 1-bit of the gain index is only about 0.2 MFLOPS.

The computational complexity of the search for the best offset vector is also about 0.2 MFLOPS (composed of about 0.03 MFLOPS for the pre-computation of (6) and of $H^T H$, and about 0.17 MFLOPS for computing Eq. (4) for all 7 possible offset vectors). Note that ξ_q is determined by the sign of the left-hand-side term in (4), thus it has a negligible influence on the overall complexity.

Altogether, the computational complexity increase is about 0.4 MFLOPS or about 5% of the 16 kbit/s LD-CELP encoder complexity. The total complexity of the 24 kbit/s extended mode is less than 8 MFLOPS. The increase in memory requirements is less than 100 words for all the tables (gain codebook, lattice-offset codebook and pre-computed tables).

CONTINUOUS VBR IMPLEMENTATION

The implementation of continuous VBR over the range of 12.8 - 24 kbit/s was achieved by randomly switching among the three rates (12.8, 16, 24 kbit/s) in proportion to the desired rate, using a 10 msecond VBR frame. This means that the switching from one rate to another may occur every 10 msec. The VBR frame length was selected as the minimal common time frame which is an integral multiple of the LD-CELP time frame (0.625 msec.) and the Intelsat IESS-501 DCME switching frame (2 msec.).

Figure 5 shows the average segmental SNR and perceptual SNR for male and female speakers as a function of average bit rate. Detailed MOS measurements were not performed. However, informal listening tests confirm the monotonic relationship between speech quality and average bit rate. Transients while switching between the three modes could not be perceived. This is also illustrated by an example in Figure 6.

The algorithm performance for various line error rates has been tested, and the results indicates that the VBR-LD-CELP extended and reduced modes degrade under channel errors not more than the nominal 16 kbit/s LD-CELP coder does.

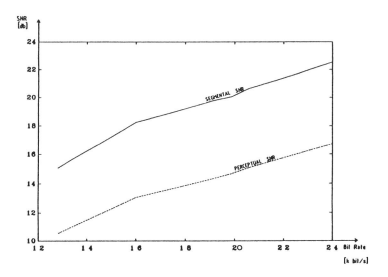

Figure 5. Continuous VBR Performance

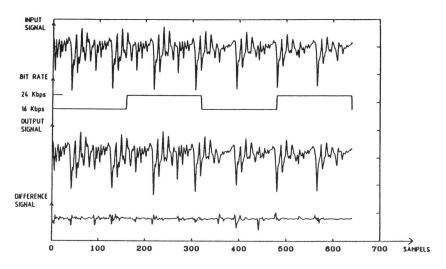

Figure 6. Switching Between 16 and 24 kbit/s
at Rate 0.5 to Produce 20 kbit/s coding

CONCLUSIONS

We have demonstrated that a nominal 16 kbit/s LD-CELP algorithm can be extended to operate at bit rates of 12.8 and 24 kbit/s. The extension of the algorithm to 24 kbit/s provides a significant improvement in speech quality without causing a substantial increase in algorithm complexity. The operation at 12.8 kbit/s shows that an acceptable speech quality can be achieved at this bit rate.

Operating the algorithm at bit rates between 16 and 24 kbit/s, linearly improves the speech quality with bit rate increase, whereas operating at bit rates between 16 and 12.8 kbit/s gracefully degrades the speech quality as bit rate decreases. The intermediate bit rates are achieved by randomly alternating between 12.8, 16 and 24 kbit/s modes according to the desired bit rate. Switching between the three modes does not cause any transient effects.

The proposed VBR method may be regarded as a new hybrid lattice-stochastic CELP coder. This viewpoint can lead to reduced complexity CELP coders even at the nominal bit rate, with a negligible degradation of speech quality, and may have an improved voice-band data (VBD) performance.

REFERENCES

[1]   AT&T, "Description of 16 kbit/s Low-Delay Code-Excited Linear Predictive Coding (LD-CELP) algorithm", *Contribution to CCITT Study Group XV*, No. 2, Geneva, Switzerland, March 1989. (See also in this book.)

[2]   Consortium for Speech Coding, "Description of the Consortium's Low Delay Vector Excitation Coder (LD-VXC)", *Information Document for T1Y1.2*, No. 89-055, April 1989. (See also in this book.)

[3]   Israel, "The Advantage and Feasibility of Variable Bit Rate Speech Coding at a Nominal Rate of 16 kbit/s", *Contribution to CCITT Study Group XVIII*, Altamonte Springs, Florida, December 1988.

[4]   J.H. Conway and N.J.A Sloane, *Sphere Packings, Lattices and Groups*, Springer-Verlag, NY, 1988.

[5]   J.-P. Adoul et. al., "Fast CELP coding based on algebraic codes", *Proc. ICASSP-87*, pp. 1957-1960, April 1987.

[6]   M.R. Schroeder and N.J.A. Sloane, "New permutation codes using Hadamard unscrambling," *IEEE Trans. on Inform. Theory*, vol. IT-33, no. 1, pp. 144-146, January 1987.

[7]   L. Cellario et al., "A 2ms delay coder," *Proc. ICASSP-89*, Glasgow, Scotland, pp.73-76, May 1989.

[8]   A. Bergstrom and P. Hedelin, "Codebook driven Glottal pulse analysis," *Proc. ICASSP-89*, Glasgow, Scotland, pp.53-56, May 1989.

[9]   N.S. Jayant and P. Noll, *Digital Coding of Waveforms*, Prentice-Hall Inc., New Jersey, 1984.

[10] British Telecom, "Study on the complexity of the AT&T LD-CELP coder," *Delayed Contribution to CCITT SG XV*, Turin, Italy, March 1989.

APPENDIX - A PERCEPTUAL SNR MEASURE

The perceptual-segmental SNR is an objective measurement that reflects more accurately (than the conventional segmental SNR) the subjective and objective performance of the LD-CELP algorithm. Figure A.1 describes a block diagram of the proposed measurement. The original and synthesized speech signals pass through adaptive weighting filters. The filters' parameters are derived from the original speech signal. The filters are identical to the one used in the LD-CELP algorithm. The perceptual SNR is a segmental SNR calculated from the outputs of these filters.

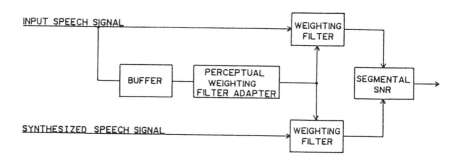

Figure A.1. Block Diagram of the Perceptual SNR Measure

5

A TRELLIS-SEARCHED 16 KBIT/SEC SPEECH CODER WITH LOW-DELAY

Michael W. Marcellin[†] and Thomas R. Fischer[††]

[†]Department of Electrical and Computer Engineering
The University of Arizona
Tucson, AZ 85721

[††]Department of Electrical and Computer Engineering
Washington State University
Pullman, WA 99164

INTRODUCTION

Trellis coded quantization (TCQ) was recently introduced [1] as an efficient method of encoding memoryless sources. Unlike previous trellis coding formulations [2–5], TCQ uses Ungerboeck's notion of set partitioning and trellis branch labeling from trellis coded modulation [6]. When using the Viterbi algorithm [7] for encoding, the computational complexity of the resulting system is fairly modest. Encoding a memoryless source (using TCQ with R bits per sample) requires only 4 multiplications, $2N + 4$ additions, N compares, and 4 rate-$(R - 1)$ scalar quantizations per data sample, where N is the number of states in the encoding trellis. Significantly, this encoding complexity is roughly independent of the encoding rate.

The mean squared error (MSE) performance of TCQ is excellent. For encoding the memoryless uniform source, TCQ achieves a MSE within 0.21 dB of the distortion-rate function at all positive integral rates. This performance is better than that promised by the best lattices known in up to 24 dimensions [8]. In fact, evaluation of the asymptotic quantizer bound [9] indicates that no vector quantizer of dimension less than 69 can exceed the performance of TCQ for encoding the memoryless uniform source.

The performance of TCQ for encoding memoryless Gaussian and Laplacian sources is also good at low encoding rates. In particular, the performance of TCQ is better than that of any scheme we have seen in the literature for encoding the memoryless Gaussian source at encoding rates of 1/2 and 1 bit per sample. Even with a simple 4-state encoding trellis, TCQ outperforms

This work was supported in part by the National Science Foundation under Grant No. NCR-8821764.

47

48

entropy coded quantization (for encoding the memoryless Gaussian source) [10] at encoding rates of 1 and 2 bits per sample.

TCQ has previously been incorporated into a DPCM structure for encoding speech at 2 bits per sample (16 kbps) [11]. The signal-to-noise ratio (SNR) and segmental SNR (SEGSNR) performance of this system is quite good. Specifically, for a variety of speakers and test sentences, values of SNR were obtained between 17.52 dB and 21.66 dB, while the SEGSNR ranged from 19.13 dB to 21.90 dB. The reconstructed speech obtained from this system was judged, in informal listening tests, to be of excellent communications quality.

In the work presented here, we extend the results reported in [11] by incorporating TCQ into a generalized predictive waveform coding structure. This structure is often referred to as a noise feedback coding structure [12]. Similar results for tree coding have been reported by Iyengar and Kabal [13], and Gibson and Chang [14]. Preliminary results of the work described here (with no pitch prediction) were reported in [15,16].

A block diagram of the noise feedback coding structure is shown in Figure 1. The boxes labeled P_L are long term (or "pitch") predictors which are used to model long term redundancies in the sampled speech input-sequence s_i. The boxes labeled P_S are short term (or "formant") predictors which model short term redundancies. The box labeled N is a noise feedback filter which acts to shape the spectrum of the error sequence $(s_i - \hat{s}_i)$ in a perceptually pleasing manner.

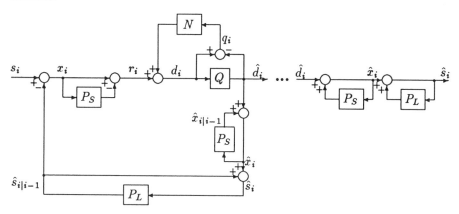

Figure 1. Noise Feedback Coding Structure.

The transfer function of a 3-tap pitch predictor P_L is given by

$$P_L(z) = \sum_{j=-1}^{1} \beta_j z^{-(M+j)}. \qquad (1)$$

For an Lth order formant predictor, the transfer functions of P_S and N are given by

$$P_S(z) = \sum_{j=1}^{L} a_j z^{-j} \quad \text{and} \quad N(z) = \sum_{j=1}^{L} b_j z^{-j}, \tag{2}$$

respectively. It is easily shown that if $b_j = a_j$, $j = 1, 2, \ldots, L$, and the box labeled Q is a standard scalar quantizer, the system of Figure 1 is equivalent to a standard DPCM system with formant predictor P_S and pitch predictor P_L. It is also easily shown that the Z-transform of the error sequence from the system of Figure 1 is given by

$$S(z) - \hat{S}(z) = Q(z)\frac{1 - N(z)}{1 - P_S(z)} \tag{3}$$

where $Q(z)$ is the Z-transform of the quantization error sequence q_i. For a standard DPCM system, $s_i - \hat{s}_i = q_i$ and hence, $S(z) - \hat{S}(z) = Q(z)$. So, with DPCM as a reference, the error transform is shaped by $(1 - N(z))/(1 - P_S(z))$.

$N(z)$ is frequently chosen as a bandwidth expanded version of $P_S(z)$. That is,

$$N(z) = \sum_{j=1}^{L} a_j(\mu z^{-1})^j, \quad \text{for some } \mu \in [0, 1]. \tag{4}$$

The value of μ is usually chosen in the neighborhood of 0.9 for 2 bit per sample (16 kbps) coders [12]. This choice of μ has the effect of suppressing the noise spectrum in frequency bands where the input speech has low energy content, thereby decreasing the audibility of the reconstruction noise. Note that $\mu = 1.0$ corresponds to standard DPCM and that generally, values of μ less than 1.0 lead to lower SNR performance but improved perceptual quality.

NOISE FEEDBACK CODING USING TCQ

Referring to the noise feedback coding structure of Figure 1, several "residuals" can be identified. x_i is the residual obtained after removing the long term prediction $\hat{s}_{i|i-1}$ from s_i, and r_i is the residual remaining after the removal of a short term prediction (computed using unquantized versions of $x_{i-j}, j = 1, 2, \ldots$) of x_i. d_i is the residual that is ultimately quantized and transmitted. It is easily shown that if $N(z) = P_S(z)$, then d_i is the residual obtained by removing $\hat{x}_{i|i-1}$ (the prediction of x_i computed using quantized versions of $x_{i-j}, j = 1, 2, \ldots$) from x_i. In that case, $d_i = x_i - \hat{x}_{i|i-1}$ or, $d_i = s_i - \hat{s}_{i|i-1} - \hat{x}_{i|i-1}$. In any event, the encoded value of the current speech sample is formed (at both transmitter and receiver) by adding both the short and long term predictions to the quantized residual (i.e., $\hat{s}_i = \hat{d}_i + \hat{x}_{i|i-1} + \hat{s}_{i|i-1}$).

Consider now, using TCQ in place of the scalar quantizer of Figure 1. The Viterbi algorithm will be used, and a separate version of each sequence shown in the encoder portion of Figure 1 will be kept for each survivor path [7] in the trellis. Clearly, a version of the residual d_i can be formed for each trellis node using the data sequences associated with the survivor path of that particular node. For each branch emanating from a given node, a scalar quantization is performed to find the element of the corresponding subset which is closest to the residual associated with that node. The distortions incurred by performing these scalar quantizations for each branch are added to the "old" survivor path metrics and the path with the smallest metric at each "next" node is chosen as the "new" survivor.

Given a data sequence to be encoded, say $\mathbf{s} = \{s_1, s_2, \ldots, s_n\}$, the ith step in this encoding process can be described more precisely as follows. Let the survivor path ending at node k (at time $i - 1$) be called *survivor-k* and let the version of each sequence (shown in Figure 1) associated with survivor-k be denoted by appending a superscript k. For example, $\hat{s}_i^k$, $k = 1, 2, \ldots, N$, denotes the encoded value of s_i associated with survivor-k, where N is the number of states in the encoding trellis. Then

$$
\begin{aligned}
x_i^k &= s_i - \hat{s}_{i|i-1}^k \\
&= s_i - \sum_{j=-1}^{1} \beta_j \hat{s}_{i-(M+j)}^k
\end{aligned}
\tag{5}
$$

and

$$
\begin{aligned}
d_i^k &= x_i^k - \sum_{j=1}^{L} a_j x_{i-j}^k + \sum_{j=1}^{L} b_j q_{i-j}^k \\
&= x_i^k - \sum_{j=1}^{L} a_j (x_{i-j}^k - \mu^j q_{i-j}^k).
\end{aligned}
\tag{6}
$$

Finally, let ρ_{i-1}^k be the path metric associated with survivor-k at time $i - 1$.

In all of Ungerboeck's amplitude modulation trellises [6,17], there are two branches entering and leaving each node, each labeled with a subset. We denote the subset associated with the branch leaving node k and entering node l as D_l^k. For each such subset, a scalar quantization is performed to determine the subset element that is closest to d_i^k. This element is denoted by $\overline{D}_l^k$. All the elements of each subset are discarded except the one selected by the scalar quantization operation. After this procedure, there are two branches entering each next node. Clearly, when the next node is node l, these branches are labeled $\overline{D}_l^{k_1}$ and $\overline{D}_l^{k_2}$ where k_1 and k_2 are the nodes from which these two branches emanate. Finally,

$$
\rho_i^l = \min_{k \in \{k_1, k_2\}} \left(\rho_{i-1}^k + (d_i^k - \overline{D}_l^k)^2 \right),
\tag{7}
$$

$$
q_i^l = d_i^{k'} - \overline{D}_l^{k'},
\tag{8}
$$

$$
\hat{x}_i^l = \hat{x}_{i|i-1}^{k'} + \overline{D}_l^{k'},
\tag{9}
$$

and

$$\hat{s}_i^l = \hat{s}_{i|i-1}^{k'} + \hat{x}_i^l \tag{10}$$

where k' is the value of k that achieves the minimum in (7). This recursion is carried out until the end of the data sequence is reached (i.e., $i = n$).

As in the case of memoryless TCQ, 1 bit per data sample is required to specify the sequence of chosen subsets and $R - 1$ bits per sample are used to specify which element from the chosen subset is closest to the corresponding residual. The source decoder uses these bit sequences to produce the sequence of quantized residuals corresponding to survivor-k', where survivor-k' is the survivor with the smallest path metric when the end of the data sequence is reached. This sequence of quantized residuals $\hat{d}_i^{k'}$, $i = 1, 2, \ldots, n$, is passed through the standard DPCM decoder (as shown in Figure 1) to obtain the encoded sequence $\hat{s}_i^{k'}$, $i = 1, 2, \ldots, n$. Note that if backward adaptive prediction is used, then a separate set of predictor parameters (for both formant and pitch) must be maintained for each survivor path. In other words, a_j, β_j, and M will depend on the particular survivor path and hence, should also be superscripted by k.

OBJECTIVE PERFORMANCE

In an attempt to obtain a system with low delay, we have chosen to use backward adaptation of the formant and pitch predictors. The adaptation scheme chosen for the formant predictor was an eighth order least squares lattice algorithm [18,19]. This algorithm was chosen because of its robustness to channel noise [19] and because of its use in previous TCQ-based systems [11,15,16]. This algorithm updates the reflection coefficients of a lattice predictor during each sample time. These reflection coefficients can then be converted to "standard" transversal coefficients for use in equations (2), (4), and (6).

Backward adaptation is also used for the pitch predictor. The value of M^k (the pitch period for survivor-k) is chosen as the value that maximizes the sample correlation coefficient between $\hat{s}_i^k$ and $\hat{s}_{i-M^k}^k$ [20]. That is, M^k is chosen to maximize

$$\rho(M^k) = \frac{\sum_{j=1}^{J} \hat{s}_{i-j}^k \hat{s}_{i-M^k-j}^k}{\sqrt{\sum_{j=1}^{J} (\hat{s}_{i-j}^k)^2 \sum_{j=1}^{J} (\hat{s}_{i-M^k-j}^k)^2}}. \tag{11}$$

The search range for M^k is limited to values between 16 and 160 samples which accommodates most pitch periods found in sampled speech. Thus, an appropriate choice for J is 256.

The values of β_{-1}^k, β_0^k, and β_1^k are then chosen to minimize

$$\sum_{j=1}^{J} (\hat{s}_{i-j} - \hat{s}_{i-j|i-j-1}^k)^2 = \sum_{j=1}^{J} \left(\hat{s}_{i-j} - \sum_{l=-1}^{1} \beta_l^k \hat{s}_{i-j-(M^k+l)}^k \right)^2. \tag{12}$$

This formulation leads to a system of three simultaneous linear equations (involving a covariance matrix) that must be solved for the optimum values of β_{-1}^k, β_0^k, and β_1^k. It was determined in [13] that the pitch predictor parameters obtained by this procedure are too finely tuned to previous data and must be "softened" by adding a small amount to each diagonal term of the covariance matrix mentioned above. For this work, this was accomplished by multiplying each diagonal element by 1.001. Unfortunately, there is no guarantee that the predictor obtained by this procedure will result in a stable synthesis filter and hence, application of the scale factor stabilization method of [21] is necessary. Since the adaptation is backward, the pitch predictor parameters could be updated on a sample-by-sample basis with no penalty in transmission rate. Unfortunately, the computational burden of such a procedure is unmanageable. For this reason, the pitch predictor was only updated every 20 samples.

Adaptive quantization is also incorporated by the use of an exponential-average variance-estimation scheme [22]. For each survivor path, a residual variance estimate is computed as

$$(\hat{\sigma}_i^k)^2 = \alpha(\hat{\sigma}_{i-1}^k)^2 + \gamma(\hat{d}_{i-1}^k)^2. \tag{13}$$

Good values for α and γ for encoding speech at 16 kbps have been found to be $\alpha = 0.70$ and $\beta = 0.36$ [11].

The residual d_i^k is still computed at each node as before, but is normalized by $\hat{\sigma}_i^k$ prior to the scalar quantization operation. The result of the scalar quantization is then multiplied by $\hat{\sigma}_i^k$ to get $\hat{d}_i^k$. Note that the adaptation is done in a backward manner since only quantized values of d_i^k are used in (13).

The system described above was used to encode the set of sentences listed in Table I. These sentences were obtained by sampling speech at 8 kHz and quantizing the resulting samples uniformly at 12 bits per sample. This particular set of data has been used in a number of speech coding studies.

Table I
Sentences Used to Test Source Coders for Sampled Speech

1.	The pipe began to rust while new.	(Female speaker)
2.	Add the sum to the product of these three.	(Female speaker)
3.	Oak is strong and also gives shade.	(Male speaker)
4.	Thieves who rob friends deserve jail.	(Male speaker)
5.	Cats and dogs each hate the other.	(Male speaker)

The signal-to-noise ratio (SNR) and segmental SNR (SEGSNR) were chosen as the objective performance measures for this work. For a speech sequence of length n, the SNR is computed as

$$\text{SNR} = 10 \log_{10} \left(\frac{\sum_{i=1}^{n} s_i^2}{\sum_{i=1}^{n} (s_i - \hat{s}_i)^2} \right) \quad \text{dB}.$$

The SEGSNR is computed by breaking the speech up into $K = n/160$ non-overlapping segments, each containing 160 samples. The SNR's of all segments are then computed and averaged. That is,

$$\text{SEGSNR} = \frac{1}{K} \sum_{i=1}^{K} \text{SNR}_i \ \ \text{dB}$$

where SNR_i is the SNR of the ith speech segment.

The results of encoding each sentence in Table I using the previously described TCQ-based system (for a 4-state trellis with $\mu = 1.0$) are reported in Table II. Each entry is of the form SNR/SEGSNR. Table II lists only the performance for a system with a 4-state trellis because the computational complexity that results when using larger trellises is tremendous. A procedure for bringing the complexity of such systems down to a reasonable level will be discussed in the following section.

Table II
SNR/SEGSNR Performance with 3-Tap Long Term Predictor
(4-State Trellis Only)

Sentence				
1	2	3	4	5
20.76/21.13	23.32/22.11	16.64/16.10	14.42/17.25	16.33/18.15

SYMBOL RELEASE RULE

The results presented in Table II were obtained by partitioning the speech into nonoverlapping blocks of 1024 samples each and then encoding these blocks using the Viterbi algorithm, as discussed previously. Unfortunately, this process results in large encoding delays. In keeping with our desire for a system with low delay, we have devised a variable delay symbol release rule. For $k_r \le k_d$ integers, this symbol release rule is described as follows. At time $i = k_d + mk_r$ ($m \ge 0$, an integer) in the Viterbi encoding process, the best survivor (say survivor-l) is traced back k_d steps in the trellis and k_r encoded residuals are released (i.e., $\hat{d}^l_{i-k_d}$ through $\hat{d}^l_{i-k_d+k_r-1}$). The path metrics of any survivors which, when traced back (from time i), do not merge with the best survivor path by time $i - k_d + k_r$, are set to infinity. Clearly, the buffering delay of a system with this symbol release rule is k_d samples. Intuitively, (for fixed k_r) k_d should be made as large as possible while still incurring an acceptable delay. For fixed k_d, it seems reasonable that $k_r = 1$ should maximize performance. Unfortunately, k_r determines how often we must perform a "trace-back". For $k_r = 1$, we must trace back k_d time steps in the trellis for *every* sample. Clearly, the trace-back burden per encoded sample is proportional to k_d/k_r and may become unmanageable for k_d much greater than k_r.

It was determined in [15,16], that for a system with no pitch predictor, choosing $k_d = 5\log_2 N$ and $k_r = 2.5\log_2 N$ (where N is the number of states in the encoding trellis) is sufficient to achieve virtually all the gain obtained using a block-mode symbol release. It is easily verified that this result also holds when a pitch predictor is present. Note that for these choices of k_d and k_r, we have $k_d/k_r = 2$, which results in a very reasonable trace-back burden and a modest delay. For example, a 256-state encoding trellis yields $k_d = 5\log_2(256)$ and $k_r = 2.5\log_2(256)$ which corresponds to a delay of only 5 msec.

In addition to low delay encoding, this modified symbol release rule provides a solution to the complexity problem encountered in the previous section. Analysis of the TCQ-based noise feedback coding system reveals that (for large encoding trellises) the computational burden is dominated by the algorithm used to compute the pitch predictor parameters. Since, a separate copy of these parameters is required for each survivor path, large trellises result in a prohibitively large computational burden.

The key to the solution of this problem lies in the modified symbol release rule. Recall that every k_r sample times (at time $i = k_d + mk_r$), the survivor paths that have not merged by time $i - k_d + k_r$ are discarded (their path metrics are set to infinity). If the index set of all remaining survivors is denoted by $\mathcal{K}$, then we have that $\hat{s}_{i-k_d+k_r-j}^{k_1} = \hat{s}_{i-k_d+k_r-j}^{k_2}$ for all $k_1, k_2 \in \mathcal{K}$ and all $j \geq 1$. Hence, if the $k_d - k_r$ most recent values of $\hat{s}_i^k$ are not used in the computation of the pitch predictor parameters ($\beta_{-1}^k, \beta_0^k, \beta_1^k$, and M^k), they will be identical for all $k \in \mathcal{K}$ and hence, only need to be computed once. For an N-state trellis, this procedure results in a computational complexity that is roughly $1/N$ that of the system with block mode symbol release discussed previously.

For the sake of simplicity, the following symbol release/pitch predictor adaptation strategy was adopted for all trellis sizes (more efficient schemes are possible). k_d and k_r were chosen to be 40 and 20, respectively. This implies a total buffering delay of 40 samples (5 msec) and that 20 encoded residuals (and hence, speech samples) are released every 20 sample periods. A new set of pitch predictor parameters is then computed using all but the 20 most recent (unreleased) encoded speech samples. Clearly the pitch predictor parameter update period is 5 msec, as before.

The results of encoding each sentence in Table I using the modified symbol release/pitch parameter adaptation strategy are reported in Table III (for various trellis sizes with $\mu = 1.0$). Comparing SEGSNR values from Tables II and III, we see that for a 4-state trellis, the price paid for low delay and reduced complexity ranges from 0.07 dB to 0.32 dB. Note however, that the reduction in complexity allows the use of larger trellises. The improvement in SEGSNR of the 64-state system of Table III over the 4-state system of Table II ranges from 0.18 dB to 1.89 dB, with the largest gain corresponding to sentence 4 (a sentence that is notoriously difficult to encode). It is interesting to note that the 4- and 64-state systems compared here are roughly equivalent in encoding complexity.

Table III
SNR/SEGSNR Performance with 3-Tap Long Term Predictor
(Using Modified Symbol Release Rule)

Sent.	Trellis Size (States)				
	4	8	16	32	64
1	20.44/20.94	21.15/21.44	21.68/21.70	21.92/21.41	20.89/21.31
2	22.54/21.79	23.52/22.31	23.09/22.69	23.91/22.43	23.90/22.53
3	15.78/15.99	17.61/16.99	18.43/17.49	18.73/17.73	19.09/17.66
4	14.22/17.03	15.63/18.04	16.36/18.65	16.47/18.93	16.82/19.14
5	16.24/18.08	17.37/18.96	17.99/19.35	18.02/19.31	19.09/19.86

CONCLUSION

Trellis coded quantization has been incorporated into a generalized predictive (noise feedback) coding structure for encoding sampled speech at 16 kbps. Backward adaptive pitch- and formant-predictors, as well as backward adaptive quantization were used to make low-delay encoding possible. Since delayed decision encoding was used, some amount of buffering delay was unavoidable. This delay was kept to a minimum by the use of a modified Viterbi search and symbol release rule. A pleasant side effect of this modified search procedure was that a significant reduction in the burden of computing pitch predictor parameters was possible. This reduction in complexity resulted in a small penalty in SEGSNR (for a given trellis) but made the use of large encoding trellises practical, which resulted in a net improvement in overall system performance.

A bandwidth expanded version of the formant predictor was used as a noise feedback filter to shape the quantization noise spectrum. This procedure had the effect of decreasing the SNR and SEGSNR of the encoding by roughly 0.5 dB, but provided a noticeable improvement in perceptual quality. Sampled speech encoded at 16 kbps was judged, in informal listening tests, to be of "near toll quality." Often, untrained listeners (students in our lab that have not worked with speech) could not distinguish the encoded speech from the original speech. The encoded speech is natural sounding with very little background noise and is very easy to understand.

REFERENCES

[1] M. W. Marcellin and T. R. Fischer, "Trellis coded quantization of memoryless and Gauss-Markov sources," *IEEE Trans. Commun.*, vol. COM-38, pp. 82–93, Jan. 1990.

[2] R. M. Gray, "Time-invariant trellis encoding of ergodic discrete-time sources with a fidelity criterion," *IEEE Trans. Inform. Theory*, vol. IT-23, pp. 71–83, Jan. 1977.

[3] W. A. Pearlman, "Sliding-block and random source coding with constrained size reproduction alphabets," *IEEE Trans. Commun.*, vol. COM-30, pp. 1859–1867, Aug. 1982.

[4] L. C. Stewart, R. M. Gray and Y. Linde, "The design of trellis waveform coders," *IEEE Trans. Commun.*, vol. COM-30, pp. 702–710, Apr. 1982.

[5] S. G. Wilson and D. W. Lytle, "Trellis encoding of continuous-amplitude memoryless sources," *IEEE Trans. Inform. Theory*, vol. IT-23, pp. 404–409, May 1977.

[6] G. Ungerboeck, "Channel coding with multilevel/phase signals," *IEEE Trans. Inform. Theory*, vol. IT-28, pp. 55–67, Jan. 1982.

[7] G. D. Forney, Jr., "The Viterbi algorithm," *Proc. IEEE* (Invited Paper), vol. 61, pp. 268–278, Mar. 1973.

[8] J. H. Conway and N. J. A. Sloane, "A lower bound on the average error of vector quantizers," *IEEE Trans. Inform. Theory*, vol. IT-31, pp. 106–109, Jan. 1985.

[9] Y. Yamada, S. Tazaki and R. M. Gray, "Asymptotic performance of block quantizers with difference distortion measures," *IEEE Trans. Inform. Theory*, vol. IT-26, pp. 6–14, Jan. 1980.

[10] N. Farvardin and J. W. Modestino, "Optimum quantizer performance for a class of non-Gaussian memoryless sources," *IEEE Trans. Inform. Theory*, vol. IT-30, pp. 485–497, May 1984.

[11] M. W. Marcellin, T. R. Fischer, and J. D. Gibson, "Predictive trellis coded quantization of speech," *IEEE Trans. ASSP*, vol. ASSP-38, pp. 46–55, Jan. 1990.

[12] N. S. Jayant and P. Noll, Digital Coding of Waveforms. Englewood Cliffs, NJ: Prentice-Hall, 1984.

[13] V. Iyengar and P. Kabal, "A low delay 16 kbits/sec. speech coder," *Proc. IEEE* Int. Conf. Acoust., Speech, Signal Processing, pp. 243–246, New York, NY, Apr. 1988.

[14] J. D. Gibson and W. W. Chang, "Fractional rate multi-tree coding," submitted to *IEEE Trans. Commun.*

[15] M. W. Marcellin and T. R. Fischer, "Generalized predictive trellis coded quantization of speech," *Conference Proceedings*, IEEE Int. Phoenix Conf. on Computers and Commun., Phoenix, AZ, Mar. 1989.

[16] M. W. Marcellin and T. R. Fischer, "Generalized predictive TCQ of speech," *Communications of the ACM*, vol. 33, pp. 11–19, Jan. 1990.

[17] G. Ungerboeck, "Trellis-coded modulation with redundant signal sets—Part II: State of the art," *IEEE Commun. Mag.*, vol. 25, pp. 12–21, Feb. 1987.

[18] J. D. Pack and E. H. Satorius, "Least squares, adaptive lattice algorithms," *Technical Report* 423, Naval Ocean Systems Center, San Diego, CA, Apr. 1979.

[19] R. C. Reininger and J. D. Gibson, "Backward adaptive lattice and transversal predictors in ADPCM," *IEEE Trans. Commun.*, vol. COM-33, pp. 74–82, Jan. 1985.

[20] B. S. Atal and M. R. Schroeder, "Adaptive predictive coding of speech signals," *Bell Syst. Tech. J*, pp. 1973–1986, Oct. 1970.

[21] R. P. Ramachandran and P. Kabal, "Stability and performance analysis of pitch filters in speech coders," *IEEE Trans. ASSP*, vol. ASSP-35, pp. 937–946, July 1987.

[22] T. R. Fischer and P. F. Dahm, "Variance estimation and adaptive quantization," *IEEE Trans. Inform. Theory*, vol. IT-31, pp. 428–433, May 1985.

6

HYBRID BACKWARD ADAPTIVE PITCH PREDICTION FOR LOW-DELAY VECTOR EXCITATION CODING*

Robert Pettigrew and Vladimir Cuperman

Communications Science Laboratory, School of Engineering Science
Simon Fraser University, Burnaby, B.C. V5A 1S6

INTRODUCTION

A pitch predictor is used in speech coding systems to remove distant-sample redundancies in speech signals due to the periodicity of voiced segments of speech. This paper describes a backward-adaptive pitch prediction algorithm which operates in conjunction with a backward-adaptive short-term predictor in a low-delay speech coding system operating at 16 kbit/s. Backward prediction in the analysis-by-synthesis environment has been introduced in [1]. The complete low-delay speech codec for which this pitch predictor was designed, Low-Delay Vector Excitation Coding (LD-VXC), will be presented in the companion paper [2]. Backward pitch prediction was been previously mentioned in the literature [3], but in a different environment than what will be discussed in this paper.

The pitch predictor in LD-VXC has three taps. Denoting the pitch predictor input by $u(n)$ and the corresponding output by $w(n)$, the pitch predictor equation is:

$$w(n) = u(n) + \sum_{i=-1}^{1} a_i \, w(n-k_p-i) . \qquad (1)$$

where a_i are the filter coefficients and k_p is the current pitch period.

In a forward-adaptive system, the coefficients are calculated from a buffered frame of input data. These coefficients are transmitted to the decoder as side information, and used to process that frame of data for which they were calculated. Such an algorithm has an inherent delay associated with buffering the input data before it is processed.

In our low-delay environment, such a delay is unacceptable. We are forced, therefore, to use a backward-adaptive algorithm. In a backward-adaptive algorithm, the filter coefficients and the pitch period must be estimated from reconstructed signals which are available in the decoder. This can be done by using either backward block adaptation, backward recursive adaptation, or a combination of the two, called

* This work was supported in part by the National Sciences and Engineering Research Council of Canada and by the Science Council of British Columbia.

58

in this paper hybrid backward adaptation.

BACKWARD BLOCK ADAPTIVE PITCH PREDICTION

Backward block adaptive pitch prediction uses techniques similar to those used in forward block adaptive pitch prediction. However, there are several important differences between the two. First of all, forward block adaptation can use clean speech available in the encoder to compute the filter parameters, while backward block adaptation must only use speech available in the decoder which contains quantization noise. Secondly, forward block adaptation can buffer a frame of speech and compute filter parameters which will then be applied to that same frame of speech. In backward block adaptation, filter parameters are computed from a previous frame of speech, and applied to process a successive frame of speech.

We will discuss separately the computation of the pitch period and filter parameters in a backward block adaptation environment.

Backward Block Pitch Estimation

The pitch period can be estimated from previously reconstructed speech using one of several available algorithms [4]. Our system uses the autocorrelation method, with the output of the pitch prediction filter as the input signal. Alternative methods, such as the average magnitude difference function (AMDF) method [4] may also be used.

To calculate the pitch period, the preceding frame of predictor output, $w(n)$, is buffered and then centre clipped [4], to form the centre-clipped signal $w_{cl}(n)$.

The autocorrelation function of $w_{cl}(n)$, $R_{cl}(k)$, is then calculated at all lags from the minimum, k_{min}, to the maximum, k_{max}, expected pitch periods. Using bounds of $k_{min}=20$ and $k_{max}=125$, corresponding to a frequency range of 64 to 400 Hz, is sufficient for covering the entire range of pitch values.

The pitch period, k_p, is determined by finding the peak in $R_{cl}(k)$. A decision may be made on whether the speech segment contains voiced or unvoiced speech, by applying a threshold to the peak of the normalized autocorrelation function.

Backward Block Coefficient Calculation

Two methods exist for computing the filter coefficients: the autocorrelation method and the covariance method. The covariance method is advantageous in forward adaptation, as it minimizes the prediction residual for a particular frame of data. In backward prediction, however, the coefficients are not applied to the frame of data for which they were optimized, but to the next frame of data. For this reason, in our implementation, the autocorrelation method was used.

The pitch prediction coefficients, a_i, are calculated from the previous frame of predictor output, $w(n)$, using the Wiener-Hopf equations:

$$\begin{bmatrix} (1+\mu)R_{ww}(0) & R_{ww}(1) & R_{ww}(2) \\ R_{ww}(1) & (1+\mu)R_{ww}(0) & R_{ww}(1) \\ R_{ww}(2) & R_{ww}(1) & (1+\mu)R_{ww}(0) \end{bmatrix} \begin{bmatrix} a_{-1} \\ a_0 \\ a_{+1} \end{bmatrix} = \begin{bmatrix} R_{ww}(k_p-1) \\ R_{ww}(k_p) \\ R_{ww}(k_p+1) \end{bmatrix} \qquad (2)$$

where $\mu=0.03$ is a constant softening factor, included because the filter coefficients which are optimized from the previous frame of data are to be applied to a successive frame. $R_{ww}(k)$ is the estimated short-term autocorrelation function of $w(n)$.

BACKWARD RECURSIVE PITCH PREDICTION

Backward recursive adaptation has several advantages over backward block adaptation, including the fact that the computational load is distributed evenly over the duration of a frame, and the fact that changing signal statistics within a frame can be tracked.

In backward recursive adaptation, the filter parameters are adapted on a sample-by-sample basis. Adaptation algorithms, such as the gradient algorithm, can be used for recursively adapting the filter coefficients. A novel pitch tracking algorithm, which is described in this section, has been developed for recursively adapting the pitch period.

Backward Recursive Filter Coefficient Adaptation

The filter coefficients may be recursively adapted using the adaptive step size gradient algorithm based on estimates of the variance of the predictor input and output. The adaptation equation is as follows:

$$a_i^{(n)} = \lambda a_i^{(n-1)} + \frac{\alpha}{\sigma_u \sigma_w} u(n)w(n-kp+i) , \quad i=1,0,-1, \tag{3}$$

where a_i are the filter coefficients, λ is a leakage factor, and α is the constant step size. The leakage factor is incorporated in the algorithm to improve the performance of the system in the presence of transmission errors.

The signal variances are calculated using the following running average algorithm:

$$\sigma_x^2(n) = \delta \sigma_x^2(n-1) + (1-\delta)x^2(n) . \tag{4}$$

A set of stability constraints [5] is applied to the new coefficients. If these constraints are not satisfied, then no update is performed on the coefficients. An alternative would be to perform scaling or bandwidth expansion on the coefficients, as described in [5]. We found that such stabilization does not improve system performance and, therefore, introduces unnecessary complexity.

An alternative adaptation algorithm which has reduced complexity and higher robustness in the presence of transmission errors is the *sign* algorithm. The adaptation equation for this algorithm is as follows:

$$a_i^{(n)} = \lambda a_i^{(n-1)} + \alpha \, sgn(u(n)) \, sgn(w(n-k_p-i)) , \quad i=1,0,-1 , \tag{5}$$

where *sgn* is the sign function which maps each real number onto +1 or -1 according to its sign.

Backward Recursive Pitch Tracking

A pitch tracking algorithm has been developed to track the pitch period in a backward recursive manner. The adaptation algorithm operates in the following manner. A decision is made whether the pitch period has changed since the last adjustment, and whether the current pitch period is greater or less than the current pitch period estimate. If the pitch period has indeed changed, then the pitch period estimate is either incremented or decremented by one. When this occurs, the pitch predictor coefficients are shifted by one in the appropriate direction. The new coefficient, either a_{+1} or a_{-1}, is computed to be a constant fraction of a_0. If this results in an unstable set of coefficients, as determined by the set of stability constraints, then the new coefficient is set to zero.

Two different versions of the pitch tracking algorithm have been developed. The first version, the Autocorrelation Pitch Tracker, uses a running average computation of the autocorrelation function evaluated at three lags to track the pitch period. The second version, the Coefficient Derivative Pitch Tracker, uses the rate of change of the three filter tap gains to track the pitch period.

Autocorrelation Pitch Tracker

The autocorrelation based pitch tracker utilizes the fact that the pitch period should correspond to the lag at which the autocorrelation function of the predictor output is a maximum. An estimate of the autocorrelation function at lags of $k=k_p+1$, k_p, and k_p-1 can therefore be used for pitch tracking.

The estimate of the normalized autocorrelation function, $\rho_{ww}(k)$, can be obtained from the following recursion:

$$\rho_{ww}^{(n)}(k) = \delta \rho_{ww}^{(n-1)}(k) + (1-\delta)\frac{w(n)w(n-k)}{\sigma_w^2(n)} . \tag{6}$$

After each update of the autocorrelation function estimate, a decision is made to increment the pitch period by one if the following are true: $\rho_{ww}(k_p+1) > \rho_{ww}(k_p)$; and $\rho_{ww}(k_p+1) > \rho_{ww}(k_p-1)$; and $\rho_{ww}(k_p+1) > \rho_{min}$. The constant ρ_{min} is a threshold for the autocorrelation term, to avoid tracking in regions of unvoiced speech. We have found that using a value of $\rho_{min}=0.2$ results in good performance.

A decision is made to decrement the pitch period by one in a similar manner if the above conditions are true for $\rho_{ww}(k_p-1)$.

If the pitch period is modified, then the values of the estimate of the autocorrelation function are shifted, in a similar manner as the filter coefficients are shifted. The new autocorrelation value, either $\rho_{ww}(k_p+1)$ or $\rho_{ww}(k_p-1)$, is computed to be a constant fraction of $\rho_{ww}(k_p)$.

Coefficient Derivative Pitch Tracker

An alternative pitch tracking algorithm with lower complexity can be developed by examining the derivative of the pitch predictor coefficients.

It was first thought that the values of the coefficients themselves could be used, in a similar manner as the values of the autocorrelation terms are used in the autocorrelation based tracker. However, because the coefficients are adapted by the gradient algorithm, such a pitch tracker did not have a fast enough response to track a rapidly changing pitch period. If the pitch period increases by one, for example, the rate of change of the coefficient a_{+1} will immediately increase, but it may take some time before a_{+1} is greater than a_0. By this time, the pitch period may have further changed to such an extent that mistracking may occur.

For this reason, the rate of change of the filter coefficients was used as a basis for a pitch tracker. The rate of change of each coefficient is calculated using the simple finite difference:

$$\dot{a}_i(n) = a_i(n) - a_i(n - \Delta n) . \tag{7}$$

We have achieved good results using $\Delta n = 8$.

The rate of change is computed after the coefficients have been adapted. A decision is made to increment the pitch period by one if the following conditions are met: $\dot{a}_{+1}(n) > \dot{a}_0(n)$; and $\dot{a}_{+1}(n) > \dot{a}_{-1}(n)$; and $\dot{a}_{+1}(n) > \dot{a}_{min}$; and $a_{+1}(n) > a_{min}$. The parameter $\dot{a}_{min}$ is a threshold for the coefficient derivative introduced to avoid tracking in steady state conditions. The parameter a_{min} is a threshold for the coefficient itself, to avoid tracking in regions of unvoiced speech. We have achieved good results using $\dot{a}_{min} = 0.01$ and $a_{min} = 0.1$.

HYBRID BACKWARD ADAPTIVE PITCH PREDICTION

The prediction algorithm used in our system is a hybrid combination of backward block adaptation and backward recursive adaptation. This was required by the fact that the pitch tracker needs an accurate starting point in order to track the pitch period. The algorithm operates in the following manner. The block adaptation method is used to compute the pitch period and the filter coefficients at the start of each frame, using data from the previous frame. These values are used to initialize the pitch tracker and the filter coefficients which are then adapted recursively for the duration of the frame.

When the pitch period is computed at the beginning of each frame, a voiced/unvoiced decision is made based on the height of the peak in the autocorrelation function. If the previous frame of data contains unvoiced speech, then the pitch period and filter coefficients are not initialized. This is done in order to prevent the pitch tracker from being initialized with an inaccurate pitch value.

TRACKER PERFORMANCE

Both pitch tracking algorithms give good performance with both the adaptive step size adaptation algorithm and the sign adaptation algorithm. The combination which gives the best performance is the coefficient derivative tracker with the adaptive step size algorithm.

Fig. 1 shows the pitch period of a segment of male speech, corresponding to the $\bar{a}$ vowel sound in the word *crate,* of the sentence *"Open the crate, but don't break the glass."*. The pitch period in this segment roughly doubles in less than 200 ms. The points marked by a "*" correspond to pitch periods which were hand measured. The solid line corresponds to the pitch period computed using the auto-correlation pitch tracker. Finally, the dotted line corresponds to the pitch period computed using backward block adaptation. Since the pitch tracker is updated at the beginning of each frame using the pitch period computed by backward block adaptation, the solid curve and the dashed curve intersect at 32 ms intervals.

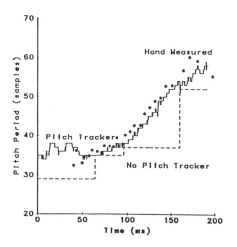

FIGURE 1: Pitch period as a function of time.

It is immediately obvious from this graph that simply using the pitch period computed using backward block adaptation is not sufficient for intervals where the pitch period changes in such a dramatic manner. It is interesting to examine what happened at $t=96$ ms. At this point, the pitch tracker did not get initialized. The reason for this is that the pitch period is changing so rapidly that the voiced/unvoiced decision made by the block adaptive pitch estimator was incorrect. In this circumstance, the pitch initialization curve is grossly incorrect, while the pitch tracker tracks the pitch period quite faithfully.

In the first 32 ms of this graph, the pitch period measured by the pitch tracker is quite different from that measured by the block adaptive pitch estimator. This is because the previous frame of data contained unvoiced speech, so the pitch tracker was not initialized at $t=0$ ms.

While the ability to accurately track the pitch period is important, the entire purpose of utilizing a pitch tracker is to improve the prediction gain of the pitch predictor. Fig. 2 contains three curves showing the gain achieved by a pitch predictor on the same segment of speech used to generate the pitch plots of Fig. 1. The plots were acquired by using the following configurations: (a) backward block adaptation; (b) hybrid backward adaptation without pitch tracking; and c) hybrid backward adaptation with pitch tracking.

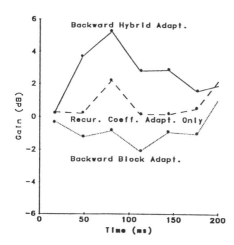

FIGURE 2: Pitch prediction gain as a function of time.

The increase in prediction gain is quite significant when using both pitch tracking and filter coefficient adaptation. The performance with neither of these features is very poor in regions where the pitch period changes rapidly.

TRANSMISSION ERROR CONSIDERATIONS

System robustness in the presence of transmission errors has often been a problem with pitch predictors. This is due to the fact that the long error response of a pitch predictor causes errors to propagate. Several modifications were made to our system to improve performance in the presence of transmission errors.

First, the leakage factor used in the pitch coefficient adaptation λ, was reduced. This had the effect of improving system robustness, at the expense of degrading performance in the absence of transmission errors. A value of $\lambda=63/64$ was chosen as the best compromise between these two requirements.

Secondly, the pitch predictor coefficients are reset to zero and the pitch period to a predefined constant if the block adaptive pitch estimator indicates that the previous frame of data contained unvoiced speech. This resetting operation improves the system performance by limiting the propagation of transmission errors and has little effect on the performance of the system in the absence of transmission errors.

Finally, the short-term predictor and the pitch predictor are adapted in parallel. If the gain of the short-term predictor were to be optimized, it would be adapted using the output of the pitch predictor, $w(n)$. However, the optimal configuration for adapting the cascade of the pitch and short-term predictors is to adapt them in parallel. That is, the short-term predictor is adapted using the excitation signal, $u(n)$, rather than $w(n)$. This approach also dramatically improves system robustness, as errors propagated by the pitch predictor do not affect the adaptation of the short-term predictor. Without parallel adaptation, error propagation through the pitch predictor may lead to negative prediction gains for the short-term predictor. For a bit error rate of 10^{-3}, improvements in frame SNR as large as 18 dB were observed due to the use of parallel adaptation.

SIMULATION RESULTS

The objective system performance was evaluated by running the complete codec on a speech file with a duration of approximately 16 s. No speech material used for training the codebook was included in this file. The objective quality was estimated by the signal-to-noise ratio (SNR) and the segmental signal-to-noise ratio (SEGSNR). The gain of the pitch predictor, Gp, was computed as the ratio of the energy of the pitch predictor output compared to the pitch predictor input. The segmental pitch predictor gain, Gpseg, was computed by averaging the Gp values computed for frames of 256 samples.

For all of the tests involving hybrid backward adaptation, the coefficient derivative pitch tracker along with the adaptive step size gradient algorithm are used. This combination was found to give the best performance, with reasonable robustness.

Table 1 gives the pitch prediction gain and system signal-to-noise ratio for the following three system configurations: (a) no pitch prediction; (b) pitch prediction with backward block adaptation; and (c) pitch prediction with hybrid backward adaptation.

The initialization interval, T, is the interval between updates using backward block adaptation, measured in samples. This should not be confused with the frame size used for computing the pitch period and the filter coefficients. For all of these tests, a frame size of 256 samples was used. Thus, for the higher update rates an overlapping frame structure is used.

With backward block adaptation, using an update period of 32 samples resulted in a 1.1 dB increase in SNR, and a 0.3 dB increase in SEGSNR, over the system without pitch prediction. This comes at a huge complexity increase, as the block computations must be performed every 32 samples. The hybrid backward adaptation algorithm with an update interval of 256 samples achieved a performance improvement of 1.75 dB in SNR and 0.9 dB in SEGSNR. The cost of this improvement is a small complexity increase, as the block computation is only performed after every 256 samples.

Adaptation Algorithm	T (samples)	Gp (dB)	Gpseg (dB)	SNR (dB)	SEGSNR (dB)
-	-	0.00	0.00	17.25	19.68
Block	32	2.67	1.38	18.32	19.98
Block	64	2.50	1.31	18.24	19.93
Block	128	2.32	1.17	18.14	19.86
Block	256	1.95	1.06	17.79	19.70
Hybrid	32	2.42	1.34	18.99	20.69
Hybrid	64	2.41	1.24	19.12	20.69
Hybrid	128	2.32	1.18	19.09	20.66
Hybrid	256	2.09	1.08	19.00	20.59

TABLE 1. System performance for various system configurations, and various initialization intervals.

The increase in SNR resulting from the inclusion of a pitch predictor may not seem to be worth the added complexity. However, the subjective improvement in speech quality is much greater than what may be indicated by the objective improvement.

Fig. 3 shows a typical segment of voiced speech; the quantization error which results from using the LD-VXC codec without pitch prediction; and the quantization error which results from using our codec with pitch prediction. The vertical scale of the quantization noise has been magnified by a factor of five. What becomes immediately apparent are the large spikes in the quantization error at the pitch frequency when the pitch predictor is not used. These spikes are reduced or removed through the use of the pitch predictor. The difference in SNR between these two configurations is only 1.2 dB for this section of speech. It is the removal of these pitch spikes which causes the improvement in subjective quality.

CONCLUSION

A backward-adaptive pitch predictor, based on a hybrid backward-adaptive adaptation algorithm using both backward block adaptation and backward recursive adaptation, improves the prediction gain, and hence the objective performance of the LD-VXC speech codec. Use of recursive adaptation to supplement the block adaptation increases the interval between block updates, thereby reducing the system complexity.

66

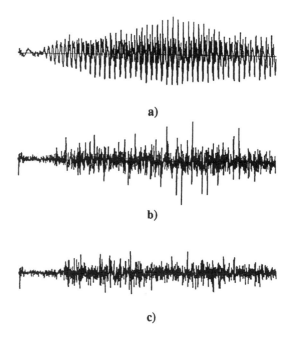

a)

b)

c)

FIGURE 3: a) Typical speech signal and b) quantization error without and C) with pitch prediction.

REFERENCES

[1] L. Watts, V. Cuperman, "A Vector ADPCM Analysis-by- Synthesis Configuration for 16 kbit/s Speech Coding," *Proc. IEEE Global Telecommunications Conference,* Hollywood, Florida, November 1988.

[2] V. Cuperman, A. Gersho, R. Pettigrew, J. Shynk, J. Yao, "Backward Adaptive Configurations for Low-Delay Vector Excitation Coding," this text.

[3] J. D. Gibson, "Adaptive Prediction in Speech Differential Encoding Systems," *Proceedings of the IEEE,* vol. 68, April 1980, pp. 488-525.

[4] L. R. Rabiner, M. J. Cheng, A. E. Rosenberg, C. A. McGonegal, "A Comparative Performance Study of Several Pitch Detection Algorithms," *IEEE Trans. Acoustics, Speech and Signal Processing,* vol. ASSP-24, Oct. 1976, pp. 339-417.

[5] R. P. Ramachandran, P. Kabal, "Stability and Performance Analysis of Pitch Filters in Speech Coders," *IEEE Trans. Acoustics, Speech and Signal Processing,* vol. ASSP-35, July 1987, pp. 937-946.

PART III

SPEECH CODING FOR DIGITAL CELLULAR
AND PORTABLE COMMUNICATIONS

Efficient low bit rate speech coding is important for increasing the capacity of cellular channels. This group of papers focuses on techniques of speech and channel coding for digital cellular and portable radio communication applications at bit rates from 8 to 16 kb/s. Cellular channels encounter fading that produces errors and the speech coders used on these channels are expected to be robust in the presence of channel errors. The papers in this section discuss the design and performance of speech coders in noisy channels.

Block coding of LPC excitation using closed-loop analysis-by-synthesis has provided the major thrust for much of the research in low bit rate speech coding during the past decade and the papers included in this section describe several important speech coders based on this principle. Gerson and Jasiuk present a speech coder (VSELP) that uses structured excitation codebooks to reduce computational complexity and increase robustness to channel errors. Suda, Moriya and Miki describe a transform coder that employs a bit selective error correction scheme for channel error protection. Millar et al., present a multi-pulse speech coder with channel error protection and error detection capability. Varma and Lin discuss the application of two speech coding algorithms, regular-pulse excited LPC and backward-adaptive CELP, at 16 kb/s for portable digital radio communication. Finally, Minde and Wahlberg describe a hybrid coder that combines elements of CELP and multi-pulse coders to reduce complexity, and employs a punctured convolution error correcting code to protect against channel errors.

7

VECTOR SUM EXCITED LINEAR PREDICTION (VSELP)

Ira A. Gerson and Mark A. Jasiuk

Chicago Corporate Research and Development Center
Motorola Inc.
1301 E. Algonquin Road, Schaumburg, IL 60196

INTRODUCTION

Vector Sum Excited Linear Prediction falls into the class of speech coders known as Code Excited Linear Prediction (CELP) (also called Vector Excited or Stochastically Excited) [1,4,5]. The VSELP speech coder was designed to achieve the highest possible speech quality with reasonable computational complexity while providing robustness to channel errors. These goals are essential for wide acceptance of low data rate (4.8 - 8 kbps) speech coding for telecommunications applications.

The VSELP speech coder achieves these goals through efficient utilization of structured excitation codebooks. The structured codebooks reduce computational complexity and increase robustness to channel errors [2,3]. Up to two VSELP excitation codebooks are used to achieve high speech quality while maintaining reasonable complexity. A novel gain quantizer is also employed which achieves high coding efficiency while providing robustness to channel errors. Finally a new adaptive pre/post filter arrangement is used to enhance the reconstructed speech quality. The discussion will be given in the context of two example VSELP coders; one operating at 8 kbps and the other at 4.8 kbps.

BASIC CODER STRUCTURE

Figure 1 is a block diagram of the VSELP speech decoder. The VSELP coder/decoder utilizes up to three excitation sources. The first is from the long term ("pitch") predictor state, or adaptive codebook [4]. The remaining sources are from one or two VSELP excitation codebooks. For the 8 kbps coder described, two VSELP codebooks are used, each containing the equivalent of 128 vectors. The 4.8 kbps coder uses a single VSELP codebook containing the equivalent of 2048 vectors. These two or three excitation sources are multiplied by their corresponding gain terms and summed to give the combined excitation sequence $ex(n)$. After each subframe, $ex(n)$ is used to update the long term filter state (adaptive codebook). The synthesis filter is a direct form 10th order LPC all-pole filter. The LPC coefficients are coded once per 20 msec frame and updated in each 5 msec subframe through interpolation for the 8 kbps system. The excitation parameters are also updated in each subframe. The 4.8 kbps system uses a 30 msec frame and 7.5 msec subframes. The number of samples in a subframe, N, is 40 for the 8 kbps coder and 60 for the 4.8 kbps coder at an 8 kHz sampling rate. The "pitch" prefilter and spectral postfilter are used to enhance the quality of the reconstructed speech.

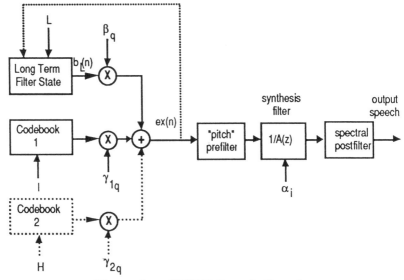

Figure 1 – VSELP Speech Decoder

Tables 1 and 2 show the bit allocations for the 8 kbps and 4.8 kbps VSELP coders. The 10 LPC coefficients are coded using scalar quantization of the reflection coefficients. An energy term, S_q, which represents the average speech energy per frame is also coded once per frame. The three excitation gains are vector quantized to 8 bits (G_S-P_0-P_1 code) per subframe for the 8 kbps coder, while the two excitation gains for the 4.8 kbps coder are vector quantized to 7 bits (G_S-P_0 code).

PARAMETER	BITS/5 MSEC SUBFRAME	BITS/20 MSEC FRAME
LPC coefficients		38
energy - S_q		5
excitation codes - I, H	7 + 7	56
lag - L	7	28
G_S-P_0-P_1 code	8	32
<unused>		1
TOTAL	29	160

Table 1 – Bit Allocations for 8 kbps coder

PARAMETER	BITS/7.5 MSEC SUBFRAME	BITS/30 MSEC FRAME
LPC coefficients		38
energy - S_q		5
excitation code - I	11	44
lag - L	7	28
G_S-P_0 code	7	28
\<unused\>		1
TOTAL	25	144

Table 2 – Bit Allocations for 4.8 kbps coder

VSELP CODEBOOK STRUCTURE

The coders use either one or two VSELP excitation codebooks, each containing 2^M codevectors. Each codebook is constructed from a set of M basis vectors, where M = 7 for the 8 kbps coder and 11 for the 4.8 kbps coder. Defining $v_{k,m}(n)$ as the mth basis vector of the kth codebook and $u_{k,i}(n)$ as the ith codevector in the kth codebook, then:

$$u_{k,i}(n) = \sum_{m=1}^{M} \theta_{im} v_{k,m}(n) \qquad (1)$$

where k = 1 or 2 for the first or second VSELP codebook, $0 \le i \le 2^M$-1, and $0 \le n \le N$-1.

In other words, each codevector in the codebook is constructed as a linear combination of the M basis vectors. The linear combinations are defined by the θ parameters. θ_{im} is defined as:

θ_{im} = +1 if bit m of codeword i = 1
θ_{im} = −1 if bit m of codeword i = 0

Note that if we complement all the bits in codeword i, the corresponding codevector is the negative of codevector i. Therefore, for every codevector, its negative is also a codevector in the codebook. These pairs are called complementary codevectors since the corresponding codewords are complements of each other.

The excitation codewords for the VSELP coder are more robust to bit errors than the excitation codewords for random codebooks. A single bit error in a VSELP codeword changes the sign of only one of the basis vectors. The resulting codevector is still similar to the desired codevector.

SELECTION OF EXCITATION VECTORS

Figure 2 is a block diagram which shows the process used to select the three codebook indices L, I and H for a coder using two VSELP codebooks. For a coder utilizing a single VSELP codebook only two indices are selected, L and I. These excitation parameters are computed every subframe.

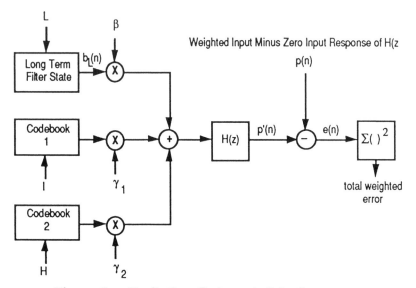

Figure 2 – Excitation Codeword Selection

$H(z)$ is the bandwidth expanded synthesis filter, $H(z) = 1/A(z/\lambda)$, where λ is the noise weighting factor (typical values used for λ are .8 for the 8 kbps coder and .9 for 4.8 kbps coder). Signal $p(n)$ is the perceptually weighted (with noise weighting factor λ) input speech for the subframe with the zero input response of bandwidth expanded synthesis filter ($H(z)$) subtracted out [5].

The three excitation vectors are selected sequentially, one from each of the three excitation codebooks (adaptive codebook and 2 VSELP codebooks). Each codebook search attempts to find the codevector which minimizes the total weighted error.

Although the codevectors are chosen sequentially, the gains of previously chosen excitation vectors are left "floating". First the adaptive codebook is searched assuming gains γ_1 and γ_2 are zero. Then the selection of the codevector from the first VSELP codebook is jointly optimized with both β and γ_1 assuming γ_2 is zero. The selection of the codevector from the second VSELP codebook (if present) is jointly optimized with β, γ_1 and γ_2. These joint optimizations can be achieved by orthogonalizing each weighted (filtered) codevector to each of the previously selected weighted excitation vectors prior to the codebook search. While this task seems impractical in general, for VSELP codebooks it reduces to orthogonalizing only the weighted basis vectors.

The adaptive codebook is searched first for an index L which minimizes:

$$E'_L = \sum_{n=0}^{N-1}\left(p(n) - \beta' b'_L(n)\right)^2 \tag{2}$$

where $b'_L(n)$ is the zero state response of $H(z)$ to $b_L(n)$ and where β' is optimal for each codebook index L.

To perform the VSELP codebook searches, the zero state response of each codevector to H(z) must be computed. From the definition of the VSELP codebook (1), a filtered codevector $f_{k,i}(n)$ can be expressed as:

$$f_{k,i}(n) = \sum_{m=1}^{M} \theta_{im}\, q_{k,m}(n) \tag{3}$$

where $q_{k,m}(n)$ is the zero state response of H(z) to basis vector $v_{k,m}(n)$, $0 \le n \le N-1$ and $1 \le k \le 2$.

The orthogonalized filtered codevectors can now be expressed as:

$$f'_{k,i}(n) = \sum_{m=1}^{M} \theta_{im}\, q'_{k,m}(n) \tag{4}$$

for $0 \le i \le 2^M-1$, $0 \le n \le N-1$, and $1 \le k \le 2$. If k=1 then $q'_{1,m}(n)$ is the component of $q_{1,m}(n)$ which is orthogonal to $b'_L(n)$. If k=2 then $q'_{2,m}(n)$ is the component of $q_{2,m}(n)$ which is orthogonal to both $b'_L(n)$ and $f_{1,I}(n)$.

The codebook search procedure now finds the codeword i which minimizes:

$$E'_{k,i} = \sum_{n=0}^{N-1} \left(p(n) - \gamma'_k\, f'_{k,i}(n) \right)^2 \tag{5}$$

where $k=1$ for the first codebook, $k=2$ for the second codebook and where γ'_k is optimal for each codevector i. In the rest of this section the subscript k indicating the first or second codebook will be dropped. Once we have the filtered and orthogonalized basis vectors, the actual codebook search procedures are identical. Defining:

$$C_i = \sum_{n=0}^{N-1} f'_i(n)\, p(n) \tag{6}$$

and

$$G_i = \sum_{n=0}^{N-1} \left(f'_i(n) \right)^2 \tag{7}$$

then the codevector which maximizes:

$$\frac{(C_i)^2}{G_i} \tag{8}$$

is chosen. The search process evaluates (8) for each codevector. Using properties of the VSELP codebook structure, the computations required for computing C_i and G_i can be greatly simplified. Defining:

$$R_m = 2\sum_{n=0}^{N-1} q'_m(n)\, p(n) \qquad 1 \le m \le M \tag{9}$$

and

$$D_{mj} = 4\sum_{n=0}^{N-1} q'_m(n)\, q'_j(n) \qquad 1 \le m \le j \le M \tag{10}$$

C_i can be expressed as:

$$C_i = \frac{1}{2} \sum_{m=1}^{M} \theta_{im}\, R_m \tag{11}$$

and G_i is given by:

$$G_i = \frac{1}{2} \sum_{j=2}^{M} \sum_{m=1}^{j-1} \theta_{im} \, \theta_{ij} \, D_{mj} + \frac{1}{4} \sum_{j=1}^{M} D_{jj} \tag{12}$$

Assuming that codeword u differs from codeword i in only one bit position, say position v such that $\theta_{uv} = -\theta_{iv}$ and $\theta_{um} = \theta_{im}$ for $m \ne v$ then:

$$C_u = C_i + \theta_{uv} \, R_v \tag{13}$$

and
$$G_u = G_i + \sum_{j=1}^{v-1} \theta_{uj} \, \theta_{uv} \, D_{jv} + \sum_{j=v+1}^{M} \theta_{uj} \, \theta_{uv} \, D_{vj} \tag{14}$$

If the codebook search is structured such that each successive codeword evaluated differs from the previous codeword in only one bit position, then (13) and (14) can be used to update C_i and G_i in a very efficient manner. Sequencing of the codewords in this manner is accomplished using a binary Gray code.

Note that complementary codewords will have equivalent values for (8). Therefore only half of the codevectors need to be evaluated. Once the codevector which maximizes (8) is found, the sign of the corresponding C_i will determine whether the selected codevector or its negative will yield a positive gain. If C_i is positive then i is the selected codeword; if C_i is negative then the one's complement of i is selected as the codeword.

QUANTIZATION OF EXCITATION GAINS

The quantization of the gains is described for the two VSELP codebook case. The quantization of the three excitation gains consists of two stages. The first stage codes the average speech energy once per frame. The quantized value of this energy, S_q, is coded with five bits using 2 dB quantization steps. In the second stage, a G_S-P_0-P_1 code is selected every subframe. This code, when taken in conjunction with S_q and the state of the speech decoder, determines the excitation gains for the subframe. The selection of the G_S-P_0-P_1 code takes place after the three excitation vectors have been determined (L, I and H chosen).

The following definitions are used to determine the G_S-P_0-P_1 code. The combined excitation function, $ex(n)$, is given by:
$$ex(n) = \beta \, c_0(n) + \gamma_1 \, c_1(n) + \gamma_2 \, c_2(n) \qquad 0 \le n \le N-1 \tag{15}$$
where:

$c_0(n)$ is the long term prediction vector, $b_L(n)$

$c_1(n)$ is the codevector selected from codebook 1, $u_{1,I}(n)$

$c_2(n)$ is the codevector selected from codebook 2, $u_{2,H}(n)$

The energy in each excitation vector is given by:
$$R_x(k) = \sum_{n=0}^{N-1} c_k^2(n) \qquad k = 0,1,2 \tag{16}$$

Let R_S be the approximate residual energy at a given subframe. R_S is a function of N, Sq, and the normalized prediction gain of the LPC filter. It is defined by:

$$R_S = N\, S_q \prod_{i=1}^{N_p}(1-r_i^2) \tag{17}$$

where r_i is the ith reflection coefficient corresponding to the set of direct form filter coefficients (α_i's) for the subframe. G_S, the energy offset, is a coded parameter which refines the estimated value of R_S. R, the approximate total subframe excitation energy, is defined as:

$$R = G_S\, R_S \tag{18}$$

P_0, the approximate energy contribution of the long term prediction vector as a fraction of the total excitation energy at a subframe, is defined to be:

$$P_0 = \frac{\beta^2 R_x(0)}{R} \qquad \text{where } 0 \le P_0 \le 1 \tag{19}$$

Similarly, P_1, the approximate energy contribution of the codevector selected from the first codebook as a fraction of the total excitation energy at a subframe, is defined as:

$$P_1 = \frac{\gamma_1^2 R_x(1)}{R} \qquad \text{where } P_0 + P_1 \le 1 \tag{20}$$

Thus β, γ_1, and γ_2 are replaced by three new parameters: P_0, P_1, and G_S. The transformations relating β, γ_1, and γ_2 to G_S, P_0, and P_1 are given by:

$$\beta = \sqrt{\frac{R_S\, G_S\, P_0}{R_x(0)}} \tag{21}$$

$$\gamma_1 = \sqrt{\frac{R_S\, G_S\, P_1}{R_x(1)}} \qquad \gamma_2 = \sqrt{\frac{R_S\, G_S\, (1-P_0-P_1)}{R_x(2)}} \tag{22,23}$$

The three parameters, G_S, P_0 and P_1, are vector quantized. A codebook of 256 vectors is used to quantize these parameters for the 8 kbps coder. The codebook was designed using the LBG algorithm [6] with the normalized weighted error as the distortion criterion. Figures 3 and 4 show the distribution of the gain codebook vectors.

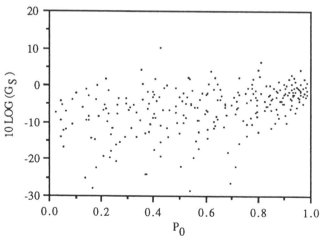

Figure 3 – P_0 vs G_S in dB for gain codebook

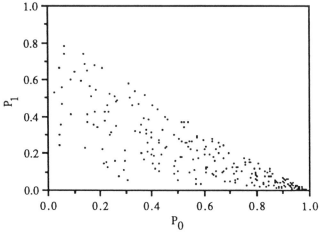

Figure 4 – P_0 vs P_1 for gain codebook

The vector from the gain codebook which minimizes the total weighted error for the subframe is chosen. The codebook search procedure requires only nine multiply-accumulates per vector evaluation.

This technique of quantizing the gains has many advantages. First, the coding is efficient. The coding of the energy once per frame solves the dynamic range issue. The gain quantization will perform equally well at all signal levels within the range of the S_q quantizer. With the average energy factored out, the three gains can be vector quantized efficiently. The vector quantization takes into account the correlations among the three weighted excitation vectors to minimize the weighted error. Second, the values of G_S, P_0 and P_1 are well behaved as can be seen in Figures 3 and 4. Whereas the optimal value for β, the adaptive codebook gain, can occasionally get very large, the sum of P_0+P_1 is always between 0 and 1. Error propagation effects are also greatly reduced with this quantization scheme. Since the energies in the excitation vectors are used to normalize the excitation gains, previous channel errors affecting the energy in the adaptive codebook vector will have very little effect on the decoded speech energy. Channel errors in the LPC coefficients are also automatically compensated for at the decoder in calculating the excitation gains. In fact as long as the code for the average frame energy, S_q, is received correctly, the speech energy at the decoder will not be much greater than the desired energy (see Figure 3 for range of G_S) and no "blasting" will occur.

The single VSELP excitation coder requires only G_S and P_0 to be encoded. Equations 20 and 23 are not required and P_1 in Equation 22 is replaced by $(1-P_0)$. The two parameters, G_S and P_0, are vector quantized using a codebook of 128 vectors for the 4.8 kbps system.

OPTIMIZATION OF THE BASIS VECTORS

The basis vectors for the VSELP codebooks are optimized over a training database. The optimization criterion is the minimization of the total normalized weighted error. For the 8 kbps coder the normalized weighted error for each subframe can be expressed as a function of each of the samples of the 14 basis vectors for the two VSELP excitation codebooks given I, H, $b_L(n)$, $p(n)$, the excitation gains, and the impulse response of $H(z)$ for each subframe of the training data. The optimal basis vectors are computed by solving the 560 (14 basis vectors, 40 samples per vector) simultaneous equations which result from taking the partial derivatives of the total normalized weighted error function with respect to each sample of each basis vector and setting them equal to zero. Since the coder subframes are not independent, this procedure is iterated in a closed loop fashion. Figure 5 shows the improvement in weighted segmental SNR for each iteration.

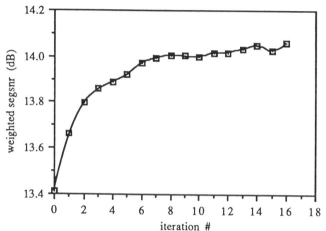

Figure 5 – Basis Vector Optimization

Initially the basis vectors are populated with random Gaussian sequences (iteration 0) which yields a weighted segmental SNR of 13.41 dB. This increases to 14.05 dB after 16 iterations. The subjective quality improvement due to the optimization of the basis vectors is significant. The objective as well as subjective improvements are retained for speech data outside the training data base.

A similar optimization procedure is used for the 4.8 kbps coder. With random basis vectors the weighted segmental SNR is 10.84 dB. This increases to 11.63 dB after 7 iterations. The optimization of the basis vectors for the 4.8 kbps system yields an even greater perceptual improvement than for the 8 kbps system.

ADAPTIVE PRE AND POSTFILTERING

The speech decoder creates the combined excitation signal, ex(n), from the long term filter state and the one/two excitation codebooks. The combined excitation is then processed by an adaptive "pitch" prefilter to enhance the periodicity of the excitation signal (see Figure 1). Following the adaptive pitch prefilter, the prefiltered excitation is applied to the LPC synthesis filter. After reconstructing the speech signal with the synthesis filter, an adaptive spectral postfilter is applied to further enhance the quality of the reconstructed speech. The pitch prefilter transfer function used is given by:

$$H_p(z) = \frac{1}{1 - \xi \, z^{-L}}$$

(24)

where

$$\xi = \varepsilon \, \text{Min}[\, \beta, \sqrt{P_0} \,]$$

(25)

with $\varepsilon = .4$ for the 8 kbps system and $\varepsilon = .5$ for the 4.8 kbps system. Note that the periodicity enhancement is performed on the synthetic residual in contrast to pitch postfiltering which performs the enhancement on the synthesized speech waveform [7]. This significantly reduces artifacts in the reconstructed speech due to waveform discontinuities which pitch postfiltering sometimes introduces. Finally to ensure unity power gain between the input and the output of the pitch prefilter, a gain scale factor is computed and is used to scale the pitch prefiltered excitation prior to applying it the LPC synthesis filter. While the pitch prefilter appears to yield only a minor improvement for the 8 kbps coder, the improvement for the 4.8 kbps coder is significant.

The form of the adaptive spectral postfilter used is:

$$H_S(z) = \frac{1 - \sum_{i=1}^{10} \eta_i \, z^{-i}}{1 - \sum_{i=1}^{10} v^i \, \alpha_i \, z^{-i}} \qquad 0 \le v \le 1$$

(26)

where the α_i's are the coefficients of the synthesis filter. To derive the numerator, the $v^i \alpha_i$ coefficients are converted to the autocorrelation domain (the autocorrelation of the impulse response of the all pole filter corresponding to the denominator of (26) is calculated for lags 0 through 10). A binomial window is then applied to the autocorrelation sequence [8] and the numerator polynomial coefficients are calculated from the modified autocorrelation sequence via the Levinson recursion. This postfilter is similar to that proposed by Gersho and Chen [9]. However, the use of the autocorrelation domain windowing results in a frequency response for the numerator that tracks the general shape and slope of the denominator's frequency response more closely. To increase postfiltered speech "brightness", an additional first order filter is used of the form:

$$H_B(z) = 1 - u \, z^{-1}$$

(27)

The following postfilter parameter values are used for both 8 kbps and 4.8 kbps coders: $v=.8$, $B_{eq}=1200$ Hz, and $u=0.4$. Note that B_{eq} is the bandwidth expansion factor which specifies the degree of spectral smoothing which is performed on the denominator of Equation 26 to generate its numerator.

As in the case of the pitch prefilter, a method of automatic gain control is needed to ensure unity gain through the spectral postfilter. A scale factor is computed for the subframe in the same manner as was done for the pitch prefilter. In the case of the spectral postfilter, this scale factor is not used directly. To avoid discontinuities in the output waveform, the scale factor is passed through a first order low pass filter before being applied to the postfilter output.

CONCLUSIONS

A high quality speech coding algorithm has been described. Examples of the algorithm for both 4.8 and 8 kbps were given. The 4.8 kbps VSELP speech coder is an improved version of the coder which ranked first in an evaluation of 4.8 kbps speech coders conducted by the United States Department of Defense [2]. The 8 kbps VSELP speech coder described was chosen by the Telecommunications Industry Association for the North American digital cellular standard. Both VSELP coders have been implemented in real-time hardware using a single Motorola DSP56001 digital signal processor operating at a clock rate of 27 MHz (13.5 MIPS) requiring only two kilowords of external RAM and eight kilowords of external ROM.

REFERENCES

[1] M. R. Schroeder and B. S. Atal, "Code-Excited Linear Prediction (CELP): High Quality Speech at Very Low Bit Rates", *Proc. IEEE Int. Conf. on Acoustics, Speech and Signal Processing*, pp. 937-940, March 1985.

[2] D. P. Kemp, R. A. Sueda and T. E. Tremain, "An Evaluation of 4800 bps Voice Coders", *Proc. IEEE Int. Conf. on Acoustics, Speech and Signal Processing*, May 1989.

[3] I. Gerson and M. Jasiuk, "Vector Sum Excited Linear Prediction (VSELP)", *IEEE Workshop on Speech Coding for Telecommunications*, pp. 66-68, September 1989.

[4] W. B. Kleijn, D. J. Krasinski and R. H. Ketchun, "Improved Speech Quality and Efficient Vector Quantization in SELP", *Proc. IEEE Int. Conf. on Acoustics, Speech and Signal Processing*, pp. 155-158, April 1988.

[5] G. Davidson and A. Gersho, "Complexity Reduction Methods for Vector Excitation Coding", *Proc. IEEE Int. Conf. on Acoustics, Speech and Signal Processing*, pp. 3055-3058, May 1986.

[6] Y. Linde, A. Buzo, and R. M. Gray, "An Algorithm for Vector Quantizer Design", *IEEE Trans. Comm.*, vol. COM-28, pp. 84-95, Jan. 1980.

[7] P. Kroon and E. F. Deprettere, "A Class of Analysis-by-Synthesis Predictive Coders for High Quality Speech Coding at Rates Between 4.8 and 16 kbits/s", *IEEE J. Select. Areas Commun.*, vol. SAC-6, No. 2, February 1988.

[8] Y. Tohkura, F. Itakura and S. Hashimoto, "Spectral Smoothing Technique in PARCOR Speech Analysis-Synthesis", *IEEE Trans. Acoustics, Speech and Signal Processing*, vol. ASSP-26, pp. 587-596, Dec. 1978.

[9] J. Chen and A. Gersho, "Real-Time Vector APC Speech Coding at 4800 bps with Adaptive Postfiltering", *Proc. IEEE Int. Conf. on Acoustics, Speech and Signal Processing*, pp. 51.3.1-51.3.4, April 1987.

8

AN ERROR PROTECTED TRANSFORM CODER FOR CELLULAR MOBILE RADIO

Hirohito Suda[†],   Takehiro Moriya[‡] ,   Toshio Miki[†]

[†] NTT Radio Communication Systems Labs.
Yokosuka, Kanagawa, 238, JAPAN

[‡] NTT Human Interface Labs.
Musashino, Tokyo, 180, JAPAN

INTRODUCTION

Recently, medium bit rate speech waveform coders have been receiving much attention for use in digital mobile radios. 13 kbit/s RPE-LTP (Regular Pulse Excitation with Long Term Prediction) [1] with 9.8 kbit/s error correction coding has been standardized for the Pan-European digital mobile radio systems to be introduced in 1991. 13 kbit/s speech coding, including 5 kbit/s error correction, has been investigated for the next digital cellular radio systems in the USA.

This chapter shows the design and implementation of an 8 kbit/s transform coder (TC-WVQ: Transform coding with weighted vector quantization) including error correcting codes. Weighted vector quantization of the transform domain linear prediction residue is the main feature of this speech coder, and it employs a bit selective error correction scheme (BS-FEC: Bit Selective Forward Error Correction) for channel error protection. The following section outlines the TC-WVQ coding scheme, as well as its quantization noise and bit error sensitivities. The design methods for an error protected TC-WVQ are then discussed and its performance for cellular radio applications is evaluated. Finally, mean opinion scores of the real time codec is shown.

SPEECH CODING

A block diagram of the TC-WVQ speech coding system is shown in Fig. 1. In this system, input speech is divided into 20 ms coding frames. Line Spectrum Pair (LSP) parameters, pitch gain and pitch period are derived from short term and long term linear prediction analyses. The LSP parameters are vector quantized, and the vector quantization errors for each parameter are scalar quantized. The pitch prediction is performed by a 3-tap filter, and the coefficients are vector quantized. Bit rates for the pitch parameters are lower than those for the usual CELP system. This is helpful in reducing the number of error protection bits since pitch parameters are very sensitive to channel errors.

The prediction residual signal is transformed into the frequency domain by Discrete Cosine Transform (DCT). The DCT coefficients is then reordered based on a block-interleaving method and split into several vectors. Each vector is then quantized by a weighted vector quantizer, whose weighting factors correspond to the speech spectral envelope. The vector quantization codebook is a two-channel-conjugate [2] type in order to improve robustness against channel error and reduce both the complexity and memory requirements. Vector quantization is performed by searching for a codeword pair whose average vector has a minimum distance from the input vector.

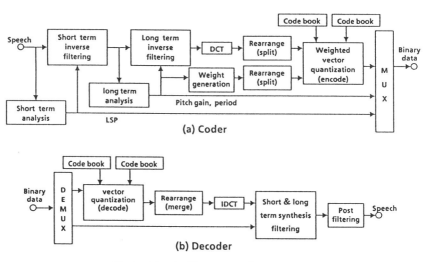

Fig1. Block diagram of TC-WVQ

TC-WVQ is closely related to transform domain CELP [3]. TC-WVQ uses a long-size (128 points) transformation, while transform domain CELP uses a transformation whose size is the same as the dimension of the quantization vector. If computational complexity is limited, the vector dimension must be small in transform domain CELP, causing significantly increased quantization distortion. If a long-size transformation is used, degradation can remain small, even with a small vector dimension [4].

To reduce perceptual quantization noise, an adaptive postfilter is used at the decoder. The structure of this filter is almost the same as one previously proposed [5,6]. The postfilter emphasizes both fine and global spectral peaks due to the pitch and formant. We have confirmed that this filter can improve quality, without degradation in speech intelligibility.

To investigate the objective performance of the TC-WVQ coder, 14 speech samples were used to calculate the SNR, including English and Japanese sentences. In order to change the coding rate, the number of quantization bits for residual and LSP were varied. The results show that the SNR is 12.5 dB for 6 kbit/s TC-WVQ coding, and a 1.2 dB SNR gain can be achieved by a 1kbit/s increase in the coding bit rate.

Error sensitivities of the coded bits are investigated in terms of signal to distortion ratio. The distortion is calculated by a simulation where in each frame an error is introduced in the j-th bit, producing a Bit Error Rate (BER) of 1/160. Roughly speaking, the most sensitive information is power, next is short-term predictor coefficients (LSP), then long term predictor parameters (pitch period and pitch gain), and the least sensitive is residual.

ERROR PROTECTION

A speech transmission method that employs BS-FEC [7] is shown in Fig. 2. The digitized speech is input into the TC-WVQ coder, then the output bits are divided into 4 groups according to their sensitivities. All the groups, except for the least significant one, are protected against channel error by error correcting codes. In Fig. 2, majority logic code (3,1), BCH(15,7) and BCH(31,21) were employed for this protection. Error correction performance and implementation simplicity are the main reasons for employing these error correcting codes. Output bit sequences from the error correction codes and the least significant bit group are multiplexed and bit interleaved, then transmitted over a cellular mobile radio channel. The reverse of this procedure is carried out in the receiver.

With a transmission bit rate constraint of 8 kbit/s, the estimated SNR of the TC-WVQ was calculated from its quantization noise, error sensitivities, and error correction code performance. Three bit allocation designs shown in Table 1 were compared in terms of SNR. We choose the 1.3 kbit/s BS-FEC design (#2) as most suitable for 8kbit/s speech transmission, since its speech performance is nearly equal to that of the 3.9 kbit/s redundancy design (#3) at a BER of 10^{-2}. Moreover, its speech degradation, under an error free condition, is 1.5 dB less than that of the zero redundancy design (#1).

The speech transmission method shown in Fig. 2 was simulated by computer. 6.4 kHz sampled speech is input into the 6.7 kbit/s coder. The output sequence from the coder is merged with 1.3 kbit/s redundancy bits for error protection. To simulate mobile radio channel errors, FSK modulation, a Rayleigh fading channel, and non-coherent demodulation are assumed.

The simulated SNR is shown in Fig.3. Triangles indicate the SNR at the maximum Doppler frequency (f_D) of 160 Hz, rectangles at 40 Hz, and crosses at 10 Hz. As f_D increases, SNR improves. This occurs because the performance of the error correction codes increases in a higher f_D environment.

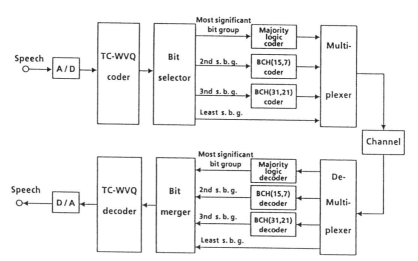

Fig2. Block diagram of speech transmission method
(TC-WVQ with BS-FEC)

Table 1 Three bit allocation designs

Bit allocation design	#1	#2	#3
Speech rate [kbit/s]	8.0	6.7	4.1
FEC rate [kbit/s]	0.0	1.3	3.9
Estimated SNR at BER = 0 [dB]	14.8	13.3	10.4
Estimated SNR at BER = 10^{-2} [dB]	5.2	8.4	8.8

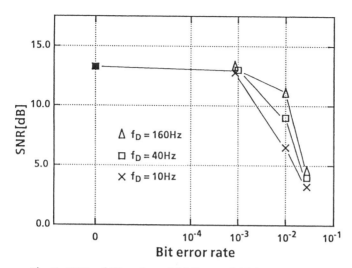

Fig.3 SNR of Simulated (6.7 + 1.3) kbit/s TC-WVQ

HARDWARE IMPLEMENTATION

A real time operating codec was implemented using two DSP chips. One DSP was used for the coder, and the other for the decoder. The typical time required for the coding process is 15 ms and 4 ms for decoding. The sum of the required time for coding and decoding is less than the frame length, which is 20 ms. The coder's required program area is 4 kW, its data ROM area is 5 kW, and its data RAM area is 2 kW. Each DSP processor is a Motorola 56000 with a 100 ns instruction cycle.

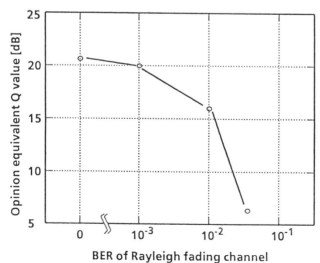

Fig.4 Subjective quality of TC-WVQ

An informal subjective quality test was carried out where the TC-WVQ speech was compared to modulated-noise-reference-unit speech. Two male and two female sources, with a bandwidth of from 300 Hz to 3000 Hz, were used in the test. Fig. 4 shows the opinion equivalent Q value of the error protected 8 kbit/s TC-WVQ. This Q value is 21 dB under an error-free-transmission condition and 16 dB at a BER of 10^{-2} for Rayleigh fading channels with an f_D of 40 Hz.

CONCLUSION

An 8 kbit/s speech coding scheme combining Transform Coding with Weighted Vector Quantization and Bit-Selective Forward Error Correction has been designed. Computer simulations have shown that its SNR is 13 dB under an error-free-transmission condition and 8 dB at an BER of 10^{-2} for Rayleigh fading channels. 8kbit/s speech transmission design was shown in this chapter, the design method is also applicable for other rate speech transmission such as 11~13 kbit/s.

A real-time codec was implemented using two DSPs, and the informal subjective quality test gave an opinion equivalent Q value of 21 dB for an error-free-transmission and 16 dB at a BER of 10^{-2} Rayleigh fading channels.

ACKNOWLEDGMENT

The authors wish to thank Tetsuo Ohi, Dr. Sadaoki Furui, and Dr. Masaaki Honda of NTT Labs for their helpful guidance and discussion throughout this work.

REFERENCE

[1] J.E.Natvig, "Evaluation of Six Medium Bit-Rate Codecs for Pan-European Digital Mobile Radio System," IEEE J. Selected Areas in Commun., vol.6, pp.346-352, 1988.
[2] T.Moriya and H.Suda, "An 8kbit/s Transform Coder for Noisy Channels," Proc. ICASSP'89, pp.196-199
[3] I.M.Trancoso and B.S.Atal, "Efficient Procedures for Finding the Optimum Innovation in Stochastic Coders," Proc. ICASSP'86, pp.2375-2378.
[4] T.Moriya and M.Honda, "Transform Coding of Speech Using a Weighted Vector Quantization," IEEE J. Selected Areas in Commun., vol.6, pp.425-431, 1988.
[5] P.Kroon and B.S.Atal, "Quantization Procedures for the Excitation in CELP Coders," Proc. ICASSP'87, pp.1649-1652.
[6] J.H.Chen and A.Gersho, "Vector Adaptive Predictive Coding of Speech at 9.6kb/s," Proc. ICASSP'86, pp.1693-1696.
[7] H.Suda and T.Miki, "An Error Protected 16kbit/s Voice Transmission for Land Mobile Radio," IEEE J. Selected Areas in Commun., vol.6, pp.346-352, 1988.

9

A MULTIPULSE SPEECH CODEC
FOR DIGITAL CELLULAR MOBILE USE

D. Millar, R. Rabipour, P. Yatrou and P. Mermelstein

Bell-Northern Research Ltd.
3 Place du Commerce
Verdun, Quebec H3E 1H6

INTRODUCTION

There is much current interest in the development of a speech coding algorithm suitable for use in a digital cellular mobile radio telephone communications system. There are a number of potential advantages in converting the analogue FM cellular system in North America to a digital system:

a) The capacity of the cellular system depends on spectrum utilization and the frequency reuse pattern. Conversion to digital transmission will initially increase the spectrum utilization by a factor of three. Digital speech compression allows three virtual channels to be transmitted over a single 30 KHz bandwidth radio channel using a TDMA (time-division multiple access) format. Improved voice coding techniques are expected to increase the capacity by an additional factor of two in the next few years.

b) The digital transmission technique will support many new customer services such as data and FAX transmission, encryption and anti-fraud features. Additional services may be based on the ability to determine vehicle locations by triangularization from the base stations.

The design goal for the speech codec is good speech quality at a gross transmission bit rate of 13 Kbit/s. The quality must remain acceptable for noisy channels with up to 4% average bit error rate in a Rayleigh fading environment. The channel fading statistics correspond to vehicle speeds from zero up to 60 mph and a carrier frequency of 900 MHz. This gives a maximum doppler shift of 170 Hz. Coding delay of about 100 ms is acceptable since the radio link is at the termination of the telephone network. That is, the radio link is not considered network transmission equipment and can appear at most twice in an end-to-end connection: this occurs if both ends are mobile telephones. The introduction of this much delay will, however, require echo cancellation on every circuit. Another objective is to minimize complexity. Complexity is difficult to measure because it is a function of the particular implementation and involves memory, processing power, gate counts, cycle times, etc. It is desirable to implement the codec on a single DSP chip using current technology. Further cost reductions can be achieved if there is processing power left over for encryption, echo cancellation and other functions.

The codec described here represents BNR's proposal for application in the new generation of digital cellular mobile communication systems planned for North America. The algorithm is called LPC-multipulse [1,2] and uses LPC analysis, pitch prediction, and multipulse residual modeling. Reed-Solomon block coding is used to provide channel error correction and error detection capability.

SPEECH CODEC

Figure 1 shows a high-level block diagram of the speech and channel encoders. The speech coder uses the multipulse algorithm [1,2] tailored to this particular application. The analysis frame is 24 ms long. LPC analysis consists of finding an LMS optimal 10-pole analysis filter. The speech frame is Hamming windowed, the autocorrelation coefficients are calculated and the optimal reflection coefficients are calculated using Levinson's recursion. The reflection coefficients are uniformly quantized in the log-area-ratio domain. The outer output levels were heuristically determined through a combination of statistics collection and listening tests to maximize the speech quality. The values used are given in Table 1.

Coefficient	Minimum value	Maximum value
k_1	-.9829	.416
k_2	-.6833	.9815
k_3	-.9152	.6194
k_4	-.439	.917
k_5	-.6736	.5815
k_6	-.4971	.7377
k_7	-.734	.4951
k_8	-.5196	.6795
k_9	-.6802	.4962
k_{10}	-.3533	.5609

Table 1   Coefficient quantizer outer levels

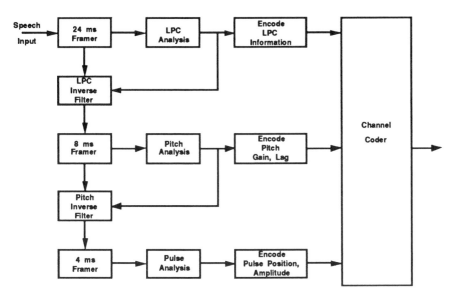

Figure 1   Block diagram of encoder

Pitch analysis is performed on the residual waveform from the LPC analysis. The 24 ms frame is divided into three 8 ms subframes for pitch analysis and for each subframe an optimal one-tap pitch predictor filter is determined using analysis by synthesis. The output of the pitch inverse filter is equal to the LPC residual minus the pitch excitation. The synthesized excitation signal is stored in a sample buffer called the pitch memory. The pitch excitation is given by a scaled, delayed sequence from the pitch memory. The scale factor is called the pitch gain and the delay is called the pitch lag. To avoid the need for recursive analysis, the pitch lag is doubled for any excitation samples which would otherwise be taken from within the current 8 ms analysis subframe. The pitch lag and gain are determined to minimize the mean square error. This error is defined as the energy in the waveform obtained by passing the pitch analysis filter output through a bandwidth-expanded synthesis filter with an expansion factor of .85 That is, all 10 poles are moved radially towards the origin of the z-plane with magnitudes multiplied by .85. The pitch lag is restricted to the range from 32 through 95 samples, allowing a 6-bit representation. The optimal pitch coefficient value is quantized with a 3-bit non-uniform quantizer.

The multipulse model for the residual waveform after pitch prediction is applied to 4 ms subframes within the original 24 ms analysis frame. Four pulses are placed, one at a time, within each 32-sample subframe. The position and amplitude of the first pulse are determined optimally using the same bandwidth-expanded error criterion as in the pitch predictor. In practice, this requires finding the lag value which produces maximum cross-correlation between the impulse response of the bandwidth-expanded synthesis filter and the "desired" signal (the response of the same filter to the residual waveform being modelled). After placing the first pulse, its amplitude is quantized using a 3-bit uniform quantizer The bandwidth-expanded synthesis filter response to the quantized first pulse is now subtracted from the original desired signal and the process is repeated for each of the three remaining pulses. Since no two pulses are allowed to be placed at the same position, there are 32-choose-4 possible pulse positionings. The pulse positions are jointly coded with 15 bits (3192 of the possible position patterns are forbidden). There is also a 5-bit, logarithmically quantized frame gain. The frame gain is used to normalize each of the 24 pulses in the analysis frame before it is quantized. Its value is determined from the speech energy in the current analysis frame, as well as from the ratio of the unquantized pulse energy to the speech energy in the previous analysis frame. The overall coded bit rate from the speech encoder, including LPC, pitch, pulse, and frame gain information, is 10 Kbit/s.

The speech decoder is shown in Figure 2. It consists primarily of three blocks:

a) The excitation waveform generator re-creates the multipulse train from the received pulse position and amplitude codes.

b) The pitch synthesis filter adds pitch excitation according to the received pitch lag and gain coefficient. The pitch excitation comes from a memory containing past samples of the LPC synthesis filter input signal.

c) The received reflection coefficient values are used to implement the 10-pole LPC synthesis filter. The output of the pitch synthesis filter is then passed through the LPC synthesis filter to produce the reconstructed speech waveform. The replicator block is discussed later under channel error protection.

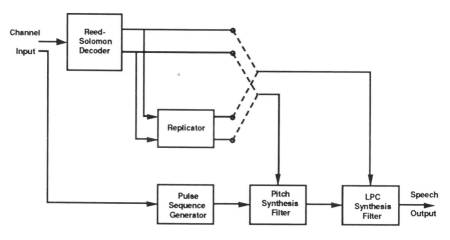

Figure 2   Block diagram of decoder

The multipulse waveform model can be compared conceptually to the CELP [3] or vector quantization technique. The set of possible pulse positions and amplitudes can be considered to form a particular CELP codebook. This would correspond in fact to a sparse codebook since only four out of 32 samples are non-zero. The fundamental difference between the two approaches is that in multipulse the pulses are determined one at a time in a sequentially optimal manner, while in CELP the codebook is exhaustively searched to find the joint optimal pulse positions and amplitudes. The advantage of multipulse is considerably reduced computational complexity at the cost of slightly inferior speech quality.

CHANNEL CODEC

The techniques used to protect the speech quality against the effects of transmission channel bit errors are Reed-Solomon (R-S) block error-correcting codes [4], interleaving, and speech replication. The sensitivity of speech quality to bit errors is not uniform over all the coded bits. Subjective listening tests reveal that the speech quality is least sensitive to the pulse position and amplitude codewords, and also to some of the least significant reflection coefficient codeword bits. The best protection strategy was found to be: protect the sensitive bits with a 100% overhead R-S block code and transmit the insensitive bits without protection. The coding block contains the sensitive bits from one 24 ms analysis frame. A (24,12) reduced R-S code consists of 72 bits of data with 72 bits of parity. Table 2 summarizes the bit allocations. For each 24 ms analysis frame there are 72 protected speech data bits, 72 parity bits, and 168 unprotected speech data bits for a total of 312 bits. This gives a transmitted bit rate of 13 Kbit/s.

Field	Protected by R-S	Unprotected
LPC	40	6
Frame gain	5	
Pitch lag	18 (3 x 6)	
Pitch coefficient	9	
Parity	72	
Pulse position		90 (6 x 15)
Pulse amplitude		72 (6 x 4 x 3)
Totals = 312	144	168

Table 2   Bit allocations

Interleaving is done at two levels: first with the other virtual channels, then within the code bits for a 24 ms analysis frame. The TDMA transmission time frame is 24 ms long, and is divided into six transmission bursts or slots. Two slots are assigned to each virtual speech channel (slots 1&4 or 2&5 or 3&6). This has the advantage of providing time diversity: if one slot falls into a Rayleigh fade, the other one may not. This format also allows future transition to a half-rate (6 times capacity increase) by assigning one slot per virtual channel. Further time diversity for the R-S symbols is achieved by interleaving the unprotected bits between the R-S symbols. This maximally separates the R-S symbols from each other thereby maximizing the length of a Rayleigh fade required to defeat the block error correcting code. The actual instantaneous transmission rate is 24000 baud or 48000 bits/s in the 30 KHz radio channel. This allows for overhead bits in each slot, allocated to inter-slot guard time, power amplifier ramp-up, equalizer training and synchronization sequences.

Speech replication is used to minimize the audio discontinuity created when the R-S decoder fails to correct the channel errors. The R-S decoder is designed in such a way as to provide the capability of detecting failure of the correction process. When failure of the error correction is detected, the speech decoder is put into replication mode (see Figure 2). Once the protected bits are known to be wrong, values from a previous correctly received frame are used. During voiced speech, the pitch synthesis filter effectively copies the excitation waveform from one pitch period in the past to fill in the missing segment. The process is similar to packet speech replication techniques [5], except that it is performed on the residual waveform instead of the speech waveform. The LPC filter smoothes the transitions at the block edges and eliminates the need for an explicit waveform smoothing algorithm.

Erasure detection may be used to increase the correcting power of the R-S block code. The R-S algorithm normally requires two parity symbols to correct each symbol error: one to locate the error and one to correct it. If an incorrect symbol is marked as erased, its location is known and only one parity symbol is needed to correct it. Accurate erasure information can therefore potentially double the correcting power of the code. The channel coding scheme used here is DQPSK with two data bits per channel symbol. As each channel symbol is decoded, a measure of decoding confidence is obtained, based on the inter-symbol phase shift. If the confidence falls below a threshhold the symbol is marked as erased. If any of the three channel symbols comprising an R-S symbol is erased, the entire R-S symbol is marked as erased. Erasure detection is not included in the results given below. The effect of erasure was found to be greatest at 14 dB C/I and 30 mph, where the block error rate is reduced by one half. The effect is less at higher or lower speeds.

RESULTS

Figure 3 shows the simulation results for the interleaved R-S channel codec (without erasure). The channel model consists of Rayleigh fading and added Gaussian white noise. Dispersion due to multi-path reception is not modelled. The carrier frequency is 900 MHz. The transmission format is DQPSK and TDMA, as described above. The horizontal axis is the average carrier to interference ratio (channel SNR) in dB. The block error rate is the fraction of blocks (protected bit field only) which contain one or more errors after channel decoding. The four curves give the rates for the above protection scheme at vehicle speeds of 5, 30, and 60 mph, and also for the case of no error correction. These curves demonstrate the overall improvement due to error protection. The error rate climbs rapidly below 14 dB C/I, indicating rapid quality degradation under very heavy interference conditions. The error correction is less effective at low speed because the error bursts become long relative to the R-S symbol separation.

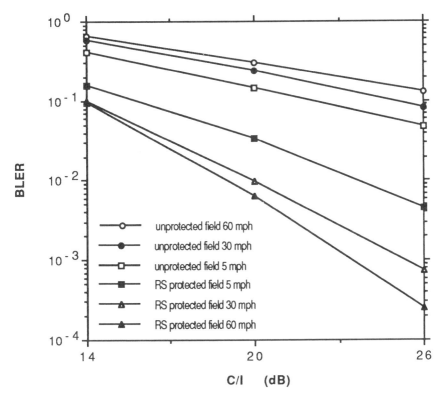

Figure 3   Performance of R-S channel coder/decoder

The effect of block errors on speech quality depends not only on the block error rate but also on the average number of consecutive blocks in error. This is because the speech replicator is able to replicate short periods of missing data with little perceived quality degradation. Block error bursts exceeding several consecutive blocks, however, cannot be replicated accurately and are more objectionable. Figure 4 is a histogram showing the burst duration statistics at 14 dB C/I and three different vehicle speeds. The histogram shows that not only is the average block error rate higher at low speed, but the burst durations are also longer. Degradation of speech quality becomes more noticeable at low vehicle speed as a greater number of consecutive speech frames have to be replicated.

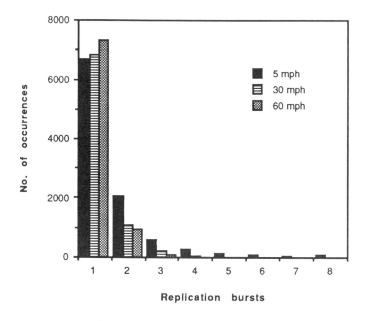

Figure 4   **Burst duration statistics**

Figure 5 shows the sinewave SNR of the speech codec without channel errors. This is reasonably good performance for this transmission bit rate and is of interest mostly for transmission of signalling tones. SNR falls off at high frequency because the LPC quantizer was designed for speech signals.

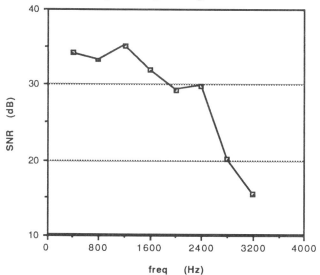

Figure 5   **Sinewave SNR performance**

Figure 6 gives the results of a mean opinion score (MOS) subjective test evaluation of the speech quality under a number of different simulated channel conditions. There were 8 test sentences: 4 male and 4 female. There were 16 subjects, each of whom listened to 2 male and 2 female speakers under all conditions. The short vertical lines give the results for each C/I and vehicle speed condition. The upper and lower ends of the line represent the 95% confidence values for the error in the mean, averaged over all source files and subjects. The line joining the squares represents the MOS scores obtained for log-PCM coded versions of the same source files which were mixed randomly amongst the test files. By locating the codec results on the log-PCM line, the PCM equivalent quality can be determined for each test condition. For example, the average result for 26 dB, 30 mph is 3.8 MOS score and 5.8-bit PCM equivalent.

The processing complexity for a Motorola DSP56001 implementation of the speech and channel codec is 8.7 MIPs for coder and decoder, equivalent to 87% of the chip with a 10MHz instruction rate. The off-chip memory requirements are 7900 words of ROM and 4600 words of RAM. The end-to-end delay (including coding and transmission) is about 60 ms.

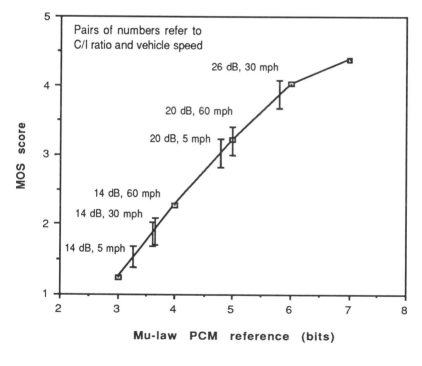

Figure 6   **MOS test results**

CONCLUSION

The multipulse / Reed-Solomon codec provides good speech quality over the range of channel conditions encountered in the cellular mobile network in North America. Formal testing indicates that the subjective speech quality of this algorithm is slightly less than that of analogue FM under error-free conditions, but better than FM for noisy channels. It is of moderate complexity and can be implemented in a single DSP chip. Signal delay is minimized but echo cancellation is necessary.While the speech quality of the multipulse codec is slightly inferior to a CELP codec with the same transmission rate, its complexity is significantly lower. It can therefore serve as a cost-effective alternative for digital cellular use.

REFERENCES

[1]   B. S. Atal and J. R. Remde, "A new model of LPC excitation for producing natural-sounding speech at low bit rate", Proc. Int. Conf. on Acoustics, Speech and Signal Processing, Paris, France, 1982, pp. 614-617

[2]   M. Berouti, H. Garten, P. Kabal, and P. Mermelstein, "Efficient computation and encoding of the multipulse excitation for LPC," *Proc. Int. Conf. on Acoustics, Speech and Signal Processing*, San Diego, California, 1984, pp. 10.1.1-10.1.4.

[3]   M. R. Schroeder and B. S. Atal, "Code -excited linear prediction (CELP): high quality speech at very low bit rates," *Proc. Int. Conf. on Acoustics, Speech and Signal Proc.*, March 1985, pp. 937-940.

[4]   S. Lin and D. J. Costello, *Error Control Coding: Fundamentals and Applications*, Prentice-Hall Inc., 1983.

[5]   O. J. Wasem, D. J. Goodman, C. A. Dvorak, and H. G. Page, "The effect of waveform substitution on the quality of PCM packet communications," *IEEE Trans. on Acoustics, Speech and Signal Proc.*, vol. 36, no. 3, March 1988.

10
TRANSMITTING DIGITAL SPEECH IN A PORTABLE RADIO ENVIRONMENT

Vijay K. Varma and David W. Lin

Bellcore
331 Newman Springs Road
Red Bank, NJ 07701

INTRODUCTION

There is an increasing demand for providing voice communications to a person away from his/her wireline telephone. The popularity of cordless telephones, mobile radio telephones, radio paging, and other emerging portable communications technologies clearly demonstrates this demand. By extending the capabilities of the existing telephone network, portable communications can be provided to roving users.

Many of the existing approaches to portable telephony use analog transmission. However, digital transmission is employed in newer designs. The advantages of digital transmission over analog are well known. In addition to these known advantages, digital transmission in portable communications environment has the potential for better spectrum utilization. Properly designed digital codes are more resistant to impairments typical of the portable communications environment, thereby improving the spectral efficiency of such systems.

Digital speech coding forms an important and integral part of any digital portable communication system. The design of a speech codec for such an application needs to consider many of its unique characteristics and requirements. In this chapter, we discuss some of the considerations in transmitting digital speech in a portable radio environment and illustrate the concepts using two 16 kbps coders as examples.

OBJECTIVES OF A SPEECH CODER FOR PORTABLE RADIO

The important parameters of a speech coder are voice quality, bit rate, complexity, robustness, and delay. A brief summary of the objectives of an ideal speech coder for portable communications includes:

- Speech quality comparable to that provided by 64 kbps μ-law PCM
- Bit rates 16 kbps or below
- Low complexity and power consumption
- Robustness to transmission errors
- Low to medium coding delay

A voice quality comparable to that of 64-kbps μ-law PCM assures a quality equal to or exceeding today's wireline telephones. The bit rate has an important bearing on the amount of radio spectrum necessary to transmit a voice channel,

98

which in turn decides the spectral efficiency of the system. A bit rate of 16 kbps or below is required to justify efficient use of scarce radio spectrum. Since personal portable sets must be lightweight and lightweight implies small batteries, low power consumption is a very important consideration. The power consumption is largely decided by the complexity of the algorithm and hardware implementation. The coder should employ some form of error control and error recovery techniques to improve its robustness to errors caused by the portable radio environment. Speech coding delay is a part of an overall delay budget. Since other system parameters of the portable communication system introduce delay which necessitates use of echo cancelers, a low delay (<5 ms) is not a requirement. However, delay should be minimized and a reasonable goal is that one-way coding-decoding delay should not exceed about 30 ms.

PORTABLE RADIO ARCHITECTURE AND ENVIRONMENT

Universal Digital Portable Communications (UDPC) is one proposal for providing ubiquitous personal portable communications [1]. It is an advanced form of low-power widespread communication system which improves the state-of-the-art of low-power cordless telephones. Figure 1 illustrates the use of radio links for accessing the telephone network in residential and business environments. Fixed radio equipment (Port) would function as an entry/exit point to the network via radio. A regular arrangement of ports (spaced every 1000 to 2000 feet in residential areas or every 100 to 200 feet in large buildings) integrated to the network allows a user to carry a pocket-sized telephone unit so that he/she may originate and receive telephone calls wherever the service has been implemented. Other designers have proposed approaches with the same general goal in mind.

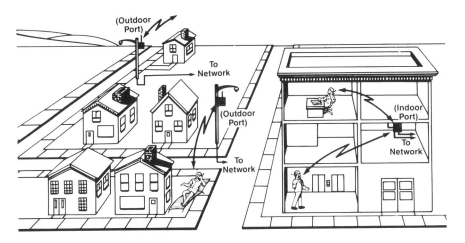

Figure 1. Digital Portable Radio Communications

Time division multiple access (TDMA) technology has received increasing attention in digital portable communications. The TDMA architecture simplifies the radio hardware and provides flexibility in accommodating variable user data rates.

In this approach, several fixed-rate bit streams would be time-division multiplexed (TDM) onto a radio carrier for transmission from a port to several active portable sets. In the reverse direction, transmission from portable sets to ports would be time sequenced and synchronized on a common frequency for time division multiple access. Such a plan is illustrated in Fig. 2, where H indicates bits comprising a frame header, and the numbered time-slots indicate block of bits from fixed-rate bit streams. A choice of speech coding rates could be accommodated by using multiple time-slots.

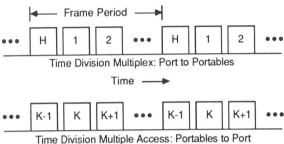

Figure 2. Time-Division Multiple Access Radio Link Architecture

A portable radio environment exhibits very complex and variable behavior. In such an environment, the direct path between any pair of transmitter and receiver is usually blocked by walls, ceilings and other objects. Different propagation paths are produced by reflections from various objects. Each path may have a different time delay and a different attenuation. Therefore, the portable radio channel experiences signal fluctuations caused by multipath signal additions and delay spread caused by the time smearing of different propagation paths. Errors are produced in bursts when the received signal envelope fades below some noise related threshold. The delay spread results in intersymbol interference which affects the system performance and limits the transmission rates. Antenna diversity, where two antennas separated by at least half of the average null spacing or having different polarizations, is a key technique in improving performance in this environment [2].

In a TDMA system, the active transmission duration of each block of speech bits is only a fraction of the TDMA frame. The higher transmission rate associated with time division multiplexing makes the transmission quality sensitive to frequency selective fading caused by delay spread.

The combined effect of slow fading, delay spread, and a TDMA architecture for the system is that the portable radio system exhibits error characteristics very different from that exhibited by a Gaussian channel. As a portable user moves at slow speeds in a system employing a carrier frequency of about 1 GHz and a transmission rate of several hundred kbits/s, there are of the order of at least 10^4 bits transmitted between significant changes in impulse response [3]. Therefore, a typical portable communication channel can be characterized as a quasi-static channel since the fading duration and interfade duration are much longer than the time-slot duration.

ERROR CONTROL IN SLOW FADING TDMA CHANNELS

For speech and data transmission in a portable radio environment, error control of some kind needs to be employed to make the system more robust to channel impairments. When errors are random or could be randomized by bit interleaving, forward error correction (FEC) is a very useful technique. In a portable radio environment, the fading produces a channel that is either very good (hence FEC is not needed) or very bad (hence FEC is not very effective). Because of the slow fading nature of the channel, the substantial amount of bit interleaving required to randomize errors results in unacceptable delay. Error detection schemes can be incorporated with fewer redundancy bits than error correction schemes, thus requiring less bandwidth. In communications systems using TDMA, synchronization of the individual time-slot is an important issue. A combined synchronization and error detection scheme can be implemented to achieve greater efficiency than by using separate processes [4]. Therefore, for slow fading TDMA channels, error detection is the more appropriate technique.

Error detection has another advantage when we consider the use of portable radio for data communications. When used with an automatic repeat request (ARQ) protocol for data, we may need only one type of channel coding to serve both speech and data traffic. This has the advantage of simplicity and hence lower cost. For speech, error detection combined with speech estimation techniques can be used to mitigate undesirable subjective effects caused by transmission errors.

SPEECH CODING FOR PORTABLE RADIO

To design a speech codec consistent with the objectives outlined earlier for the proposed portable radio system, we need to know certain system parameters. The important parameters are the frame duration and the lowest multiplexing rate used in the time division multiple access scheme. For the system under consideration, the TDMA frame period is 16 ms and the lowest multiplexing rate is 8 kbps. A TDMA frame period of 16 ms means that the system has an inherent minimum delay of 16 ms. Therefore, a submultiple of the TDMA frame period for the speech coder frame is desirable to minimize the overall delay. Since the lowest multiplexing rate of the TDMA scheme is 8 kbps, a multiple of 8 kbps for the speech coder bit rate assures full use of multiple time-slots.

Two 16 kbps coding algorithms are examined, one in the RPE-LPC (regular-pulse excited linear predictive coding) [5] family and the other a backward-adapted CELP (code-excited linear predictive) coder. The latter yields somewhat higher speech quality, but needs approximately 5 times as many multiplications per second as the former. Description of the CELP coder can be found elsewhere in this book [6] and in [7]. Considered in this work is its earlier version given in [8]. The RPE-LPC algorithm is briefly described below. More details can be found in [9].

The RPE-LPC Algorithm

RPE-LPC has been known to yield good-quality speech at medium rates with moderate delay and complexity. A variant of it has been chosen as a European mobile radio system standard [10]. Our coder yields a bit rate of exactly 16 kbps. It has an inherent delay of 20 ms and requires approximately 1.8 million multiplications and 1.8 million additions and subtractions per second. And it is designed such that the excitation pulses are as dense as possible, so as to result in minimum distortion in coding high-pitched voice.

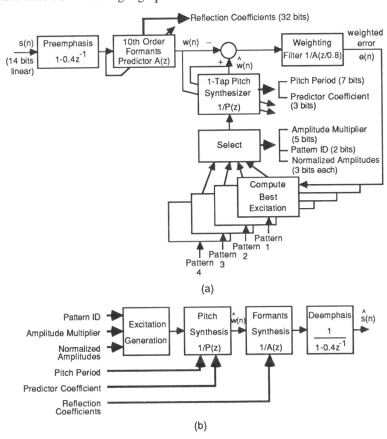

Figure 3. RPE-LP Coder and Decoder. (a) Coder. (b) Decoder.

The coder-decoder pair has the structure shown in Fig. 3. Since the decoder structure is fairly straightforward, we concentrate the following discussion on the coder. The formants predictor is computed once per TDMA frame length, i.e., 16 ms, using the autocorrelation method [11] of linear prediction on 24 ms of Hamming-windowed speech which encompasses the 16-ms block in its middle. Due to the 24-ms window, the inherent coding delay is 20 ms. The pitch predictor (or synthesizer) is updated once every 8 ms. The candidate pitch period ranges from 16 to 143 samples. Each excitation vector is 16 samples in duration and contains 7 pulses. It is limited to have one of the four patterns obtained by confining either the odd or the even samples in a 16-sample sequence to zero and, further, confining either the fourth or the eighth of the remaining samples also to zero. Quantization of the pulse amplitudes is done once per two excitation vectors. First, the maximum amplitude (henceforth called the "amplitude multiplier") in the 14 pulses is clamped to between 1 and 4096 and logarithmically quantized. Then each pulse is normalized to this quantized amplitude multiplier and quantized using a Max quantizer [12] for Gaussian random variables with a standard deviation 1/2.2. (Although the normalized amplitudes do not follow a Gaussian distribution, we have found Gaussian quantizers to yield good results.)

Table I. SSNR (in dB) of Coded Speech

		Speech #1	Speech #2	Speech #3
RPE-LPC	Single Coding	17.0	14.5	13.2
	Tandem Coding	14.1	11.9	10.6
CELP	Single Coding	18.3	17.0	15.2
	Tandem Coding	15.2	14.1	12.2

Speech #1: Two phonetically balanced sentences each spoken by a male
and a female, 6.528 s in total duration.
Speech #2: Conversational speech, female, 19.2 s.
Speech #3: Conversational speech, high-pitched male, 19.2 s.

Informal listening reveals that the above algorithm yields near toll quality speech. The distortion due to coding is noticeable with trained ears (especially in the case of high-pitched voices), but slight. Table I gives the average SSNR (segmental signal-to-quantization noise ratio) values from coding several pieces of speech, for both the RPE-LPC and the CELP coders. Also shown are the results from one case of asynchronous tandeming of two identical coders.

RECONSTRUCTION OF SPEECH DURING DEEP FADES

The effect of random bit errors on the quality of speech has received a great deal of attention. However, earlier work on the effects of burst errors on digitized speech has been limited to packet communications. Jayant and Christenson report zero substitution and odd-even sample interpolation on PCM and DPCM speech [13]. Goodman *et al.* suggest using correlation properties of speech in substituting lost packets in PCM coded speech [14].

Frame substitution techniques can be applied to compressed speech to make the coder more robust to frames dropped during deep fades. The advantage of such an approach is that all the processing takes place in the receiver, thus avoiding any overhead in the transmission rate. Some form of error detection is necessary to detect errors in the received frame of speech. In a simple frame repeat, when a frame loss is detected, speech samples from the previous frame are repeated. This results in better perceptual quality than either letting the errors in or simply blanking the frame. By making use of the correlation properties of speech, the annoying effects can be further reduced. In this case, when a frame loss is detected, speech samples from the previous frames with a proper delay (a pitch period or a multiple of it for voiced segments, or a random delay for unvoiced segments) are substituted for the lost frame. This technique can be performed either in base band, where speech samples are estimated from the decoded speech, or in parameter domain, where speech parameters are estimated from their values from previous frames. Since speech exhibits only short-term periodicity, care should be taken to allow waveforms to decay to zero during long fades. Informal listening tests have shown that, using frame substitution with proper pitch phasing, up to 10% frame losses can be tolerated for a 16 kbps sub-band coder [15]. The speech codec in the European cellular mobile system also uses frame repeat techniques to make the coder more robust to the fading environment [16].

Error Recovery for RPE-LP Coded Speech

In linear predictive coders transmitting filter coefficients explicitly, it is easier to perform frame repeat in the parameter domain. Before examining ways of performing this after an error, let us first look at the effects of errors on speech quality. Objectively, we first examine the SSNR sensitivities to different bit errors in the coded speech, using the experimental method described in [17]. In this method, the same bit in every block is set in error and the resulting SSNR of the decoded speech is measured. Figure 4 shows the result of such an analysis for the RPE coder for the three pieces of speech described in Table I. Besides the SSNR for the reflection coefficients bits, all other SSNR figures are averaged values for the marked type of quantities over a block of coded speech. For example, there are eight pattern ID's per block. So the value shown in Fig. 4 for the first bit in the pattern ID is the average of the eight SSNR values corresponding to the eight first bits in the eight ID's. The reason why we have taken averages is because there does not seem to be any significant systematic difference between the SSNR sensitivities to errors in these bits.

Subjectively, we noticed that errors in the more significant bits of the amplitude multipliers and the reflection coefficients can cause loud and annoying pops and pings in the decoded speech. Errors in pitch periods and pitch predictor coefficients, though degrading the speech quality and yielding great SSNR loss, do not result in such annoying pops and pings or affect significantly the speech intelligibility. Our focus in error control is thus on mitigating the undesirable subjective effects from errors in the more significant bits of the amplitude multipliers and the reflection coefficients.

Our basic approach is the following: if, for a certain block, the received reflection coefficients or amplitude multipliers are found to be in error, then they are discarded and replaced, in the case of reflection coefficients, by a scaled version of the previous block's reflection coefficients and, in the case of amplitude multipliers, by a scaled version of the last amplitude multiplier in the previous block. If this is the second or a subsequent block in a row that the same group of quantities (the group of reflection coefficients or the group of amplitude multipliers) is in error, then the above-described substitution of values is performed using a decreased scaling

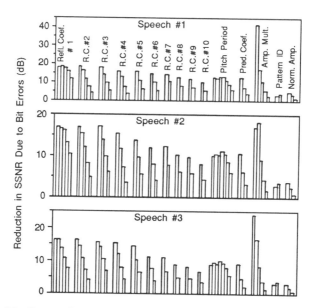

Figure 4. Bit Error Sensitivities. The SSNR values for each parameter are arranged from left to right in the order of descending bit significance in natural binary representation.

factor to obtain a faster decay toward zero. The reasons for this accelerated decay toward zero are that the successive appearance of errors may signify that the received signal is in a deep fade and that an accurate estimation of the true reflection coefficients or the true amplitude multipliers may be more difficult as time elapses. Errors in other quantities are left unattended.

After experimentation with some simple binary numbers, we chose 1 to be the scaling factor for substitution of the first errored set of reflection coefficients and 0.875 that for the first errored set of amplitude multipliers. For the second and subsequent errored sets, these factors were scaled down by 0.9375. We assumed flat fading and that the fading envelope was constant over the two TDMA time slots constituting a coded block of speech. Figure 5 shows a typical fading envelope for a walking speed of one mile per hour with two-branch antenna diversity, which we assumed in our simulation study. We also assumed binary antipodal signaling and that the noise was additive white Gaussian. A reasonable point of operation is at 18.1 dB or higher average SNR at the receiver input [18]. However, to increase the number of faded time slots for easier examination of the performance of the error recovery method, we assumed a 7 dB average receiver input SNR. It was not our intention to be concerned with specific error control codes. Thus we assumed that errors were detected perfectly. This assumption is not unreasonable since the probability of an undetected codeword error using a well-behaved error-correcting code for error detection is at most 2^{-r}, where r is the number of parity bits [19].

Now that each TDMA time slot contains 128 bits, a natural arrangement is thus to do error detection on a whole time slot as a unit. Since the reflection coefficients

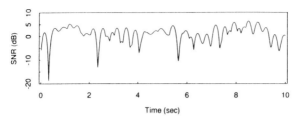

Figure 5. A Typical Fading Envelope. At 1 GHz for a speed of one mile per hour, with two-branch antenna diversity and an average receiver input SNR of 0 dB.

and the amplitude multipliers total 52 bits, we first considered placing them into one time slot. Any bit error in this time slot would trigger a substitution of the critical parameters (the reflection coefficients and the amplitude multipliers) as described earlier. Errors in the other time slot were ignored. For the 76 bits to be packed with the 52 critical ones in the same time slot, we selected those associated with higher SSNR sensitivities in Fig. 4. Results from an arbitrary selection of 76 other bits showed a slightly lower average SSNR.

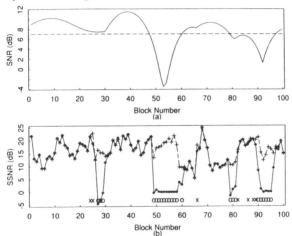

Figure 6. Objective Quality of A 100-Block Section of Speech After Error Recovery. (a) Fading envelope. Dashed line shows the nominal average of 7 dB. (b) Solid line shows the SSNR after error recovery. Dashed line shows the error-free SSNR. Near the bottom, "o" indicates where parameter substitution was performed and "x" indicates where error occurred only in the time slot not containing the critical parameters (and hence parameter substitution was not performed).

With the above arrangement, the loud and annoying pops and pings which would otherwise show up at 7 dB SNR were eliminated. Instead, silence or low level buzzes were heard in the faded segments. Figure 6 compares the SSNR values of a section of speech transmitted through a fading channel with that transmitted free of errors, where each SSNR value corresponds to one 16-ms block of speech. Experience shows that the SSNR of the noise-disturbed speech quickly recovers to that of the error-free within a few blocks after an error condition.

Placing the reflection coefficients and the amplitude multipliers, respectively, into two time slots in a certain way (and performing error detection on both slots) yielded slightly lower average SSNR than placing them into one time slot. The resulting speech also seemed a little more degraded. However, the difference in speech quality was too slight to facilitate a clear-cut conclusion on which way of packing is superior. Other error-detection arrangements are possible depending on the availability of bits for error-control coding. Some discussion can be found in [9].

CELP Coded Speech Over Slow Fading Channels

The CELP coder was found to be fairly robust to channel errors. Simulated transmission of coded speech through the same fading channel as above resulted in some audible clicks in the decoded speech. But the clicks are not as loud as that of the RPE coder with no error control. The situation is even better at higher SNRs. Parameter substitution techniques, following error detection, can probably be devised to further reduce the subjective effects of errors.

CONCLUSIONS

The high quality and low complexity requirements for the speech coder make 16 kbps a desirable bit rate for application to portable radio communications. Both RPE-LP and CELP coders give good quality speech at 16 kbps, though the CELP coder is much more complex than the RPE-LP coder. The quasi-stationary nature of the channel and other system considerations make error detection more desirable than forward error correction. Error detection combined with speech estimation techniques can be used to mitigate undesirable subjective effects caused by transmission errors. For the RPE-LPC, substitution of critical parameters in the lost frame is found to replace annoying pops and pings with low level buzzes, which were more preferable subjectively. Speech estimation techniques make speech coders more robust to lost frames in a fading environment.

REFERENCES

[1]   D.C. Cox, "Universal digital portable radio communications," *Proc. IEEE,* vol. 75, pp. 436-477, April 1987.

[2]   D.C. Cox, "Antenna diversity performance in mitigating the effects of portable radiotelephone orientation and multipath propagation," *IEEE Trans. Commun.,* vol. COM-31, pp. 620-628, May 1983.

[3]   J.C-I. Chuang, "The effect of time delay spread on portable radio communication channels with digital modulation," *IEEE J. Sel. Areas Commun.,* vol. SAC-5, pp. 879-889, June 1987.

[4]   L.F. Chang and N.R. Sollenberger, "The performance of a TDMA portable radio system using block code for burst synchronization and error detection," *Conference Record, IEEE Global Telecom. Conf.,* Globecom'89, pp. 1371-1376, Dallas, TX.

[5]   P. Kroon, E. F. Deprettere, and R. J. Sluyter, "Regular-pulse excitation – a novel approach to effective and efficient multipulse coding of speech," *IEEE*

Trans. Acoust., Speech, Signal Processing, vol. ASSP-34, no. 5, pp. 1054-1063, October 1986.

[6]   J.-H. Chen, "A robust low-delay CELP speech coder at 16 kbit/s," this book.

[7]   J.-H. Chen, "A robust low-delay CELP speech coder at 16 kbit/s," *Conference Record, IEEE Global Telecom. Conf.*, Globecom'89, pp. 1237-1241, Dallas, TX.

[8]   AT&T, "Description of 16 kbit/s low-delay code-excited linear predictive coding (LD-CELP) algorithm," CCITT Study Group XV, Document 1, (Question 21/XV), March 1989.

[9]   D. W. Lin, V. K. Varma, and J. L. Dixon, "Design of a medium rate linear predictive speech coder for digital portable radio communications," *IEEE Vehicular Technol. Conf.*, VTC'90, Orlando, FL.

[10] P. Vary *et al.*, "Speech codec for the European mobile radio system," *Proc. 1988 IEEE Int. Conf. Acoust., Speech, Signal Processing*, vol. 1, pp. 227-230.

[11] N. S. Jayant and P. Noll, *Digital Coding of Waveforms*. Englewood Cliffs, New Jersey: Prentice-Hall, 1984.

[12] J. Max, "Quantizing for minimum distortion," *IRE Trans. Info. Theory*, vol. IT-6, pp. 7-12, March 1960; reprinted in D. Slepian, ed., *Key Papers in the Development of Information Theory*. New York: IEEE Press, 1974, pp. 267-272.

[13] N.S. Jayant and S. Christensen, "Effect of packet Losses in waveform coded speech and improvements due to an odd-even sample interpolation procedure," *IEEE Trans. Commun.*, vol. COM-29, pp. 101-109, February 1981.

[14] D.J. Goodman *et al.*, "Waveform substitution techniques for recovering missing speech segments in packet voice communications," *IEEE Trans. Acous., Speech, Signal Processing*, vol. ASSP-34, pp. 1440-1448, December 1986.

[15] V.K. Varma *et al.*, "Performance of sub-band and RPE coders in the portable communications environment," *Proc. Int. Conf. Land Mobile Radio*, pp. 221-227, Coventry, U.K., December 1987.

[16] C.B. Southcott *et al.*, "Voice control of the Pan-European digital mobile radio system," *Conference Record, IEEE Global Telecom. Conf.*, Globecom'89, pp. 1070-1074, Dallas, TX.

[17] N. Rydbeck and C.-E. Sundberg, "Analysis of digital errors in nonlinear PCM systems," *IEEE Trans. Commun.*, vol. COM-24, pp. 59-65, January 1976.

[18] V. Varma and L. F. Chang, "Use of error control coding and antenna diversity to improve performance of sub-band coding," *IEEE Vehicular Technol. Conf.*, VTC'88, pp. 153-157, Philadelphia, PA.

[19] G. C. Clark, Jr., and J. B. Cain, *Error-Correction Coding for Digital Communications*. New York: Plenum Press, 1981.

11

A STUDY ON
SPEECH AND CHANNEL CODING
FOR DIGITAL CELLULAR

Tor Björn Minde, Ulf Wahlberg

Research & Development
Ericsson Radio Systems AB
S–164 80 Stockholm
Sweden

INTRODUCTION

In the present mobile telephone systems, the fast growing number of subscribers has lead to a lack of capacity. This has forced the cellular telecommunication operators and equipment suppliers to find a solution for a second generation mobile telephone system. For the North American market, the specification work has been done during 1989 and this system, as well as the European GSM–system, will be put into operation in the early 1990's.

For this purpose, Ericsson Radio Systems proposed a 30 kHz TDMA system, with 3 users per carrier, for the North American market [1]. As part of the work for this system proposal, a study has been conducted on speech and channel coding using a total bit rate of 13 kb/s.

This study is described in this chapter. After an introductory discussion on speech and channel coding for digital cellular, a speech coding algorithm at 8–9 kb/s is described. The radio channel and the study on channel error robustness, for the combination of speech and channel coding, are also discussed.

SPEECH AND CHANNEL CODING FOR DIGITAL CELLULAR

Efficient low bit rate speech coding is needed to increase channel capacity in the near future mobile radio systems. Narrower bandwidth and more robust error correction are needed. Narrower bandwidth is needed to increase the number of users per carrier and more robust error control to enhance the spectrum efficiency, by the possibility to operate at lower Carrier–to–Interference ratio (C/I).

Demands On Speech And Channel Coding

The main selection criteria, for the choice of speech and channel coding, is the speech quality versus the carrier–to–interference ratio for a given total bit rate. It is important to achieve good speech quality at cell boundries. This puts constraints on possible bit rates for the speech coding, to accomodate the needed amount of error correction. The design has to be a close interaction between the source and channel coding, to achieve high robustness against bit errors.

The demands on the speech coding are high, in the mobile telephone environment. The coding scheme should provide high quality for the user in clear channel and robustness against background noise. Quality is important since the mobile phone could become the user's personal telephone. The complexity of the speech coding system needs to be low. The speech coding chip will share available space and power with other signal processing and RF units, in a 160 g and 160 cm^3 handheld pocketphone [2].

Low complexity should however be evaluated relative to available bit rate and needed speech quality. Lower bit rates require higher complexity to achieve the needed speech quality. The demand on reasonably short delays, including the time of processing, will put constraints on possible interleaving depth of data bits and thereby lower the possibility of increased bit error robustness.

Choice Of Speech And Channel Coding

To reach the required channel robustness for the proposed TDMA system, about 1/3 of the total bit rate of 13 kb/s is needed for the channel coding [1]. The level of distortion on the channel is the same as for the present analog system. Therefore the speech coding bit rate is 8–9 kb/s.

To achieve the required speech quality at this bit rate, relatively complex algorithms, compared with the GSM codec [3], are needed. Closed–loop Analysis–by–Synthesis LPC coders such as MPE [4] and CELP [5] are promising approaches for high quality and low bit rate speech coding. An important aspect is the use of a closed–loop pitch predictor implemented as an adaptive codebook [6]. This is essential for high speech quality. The complexity can be kept low if the quantization of the residual, as described here, is done with a multipulse type of analysis.

To increase the robustness, without increasing the delay significantly, diagonal interleaving over 2 frames is used. Since the data bits from the speech coder have different sensitivity for bit errors, a punctured convolutional code [7] is used for error correction. For the detection of error bursts from deep fades, a Cyclic Redundancy Check (CRC) code is used for the most sensitive bits.

SPEECH CODING

The speech coder consists of three parts: short–term correlation analysis (LPC - Linear Predictive Coding), long–term correlation analysis (LTP - Long–Term Prediction) and residual quantization analysis (Fig. 1).

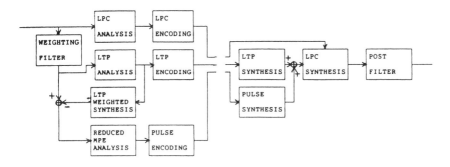

Figure 1: Speech Coder And Decoder.

LPC Analysis

The LPC analysis is performed open–loop with the autocorrelation method [3]. A symmetrical 25 millisecond (ms) Hamming window is applied on the data for every 20 ms frame and the Schur recursion is used to determine the reflection coefficients, from the estimated autocorrelation function.

Log–area–ratios are used for scalar quantization. Using non–symmetrical uniform quantizers, at least 38 bits/ 10 coefficients (Table 1) are needed. Even better performance is realized with a nonuniform quantizer with 36 bits/ 10 coefficients [8]. The reconstructed coefficients are used in the rest of the analysis. In the first subframe the coefficients are linearly interpolated.

LTP Analysis

The LTP analysis is performed closed–loop, single tap for every 5 ms subframe, using a recursive end–point correction algorithm [9]. The past excitation signal is searched for the segment which gives the best match, using the bandwidth expanded mean–squared–error criteria.

Extending the search lag below the subframe length provides a much better performance for high pitched voices and using the past excitation as an adaptive codebook gives the same level of complexity as for search lags above the subframe length [6]. For multipulse excitation with three pulses per subframe the high pitched voices are enhanced 0.3–0.6 dB in segmental SNR.

The long–term correlation coefficient is quantized with a non–symmetrical quantizer using 3 bits and for the lag 7 bits is used (Table 1).

A delta searched lag can be incorporated for every other subframe using a range [–32,31] around the lag from the previous subframe. This marginally affects the performance.

Residual Quantization Analysis

The residual quantization analysis which is a reduced search multipulse analysis, is performed closed–loop with the autocorrelation method [10]. The positions and the amplitudes of 3 pulses per subframe are determined sequentially to match the residual signal.

The reduced pulse position search is done by excluding a regular grid of positions for the second and third pulse, depending on the previous pulse positions in the subframe [11]. The pulse positions are described by a phase position for the grid and a sub–block number for the position in the grid.

By doing, this the pulse position encoding is split into different codewords, instead of using enumerative coding for all positions. Different codewords and different error sensitivity for bits in the codewords, favor the channel coding described below.

The three pulse positions in one subframe can now be coded with 13 bits using the sub–block numbers and the phase positions. The maximum pulse amplitude is logarithmically quantized using 5 bits and each normalized pulse amplitude is uniformly quantized with 2 bits (Table 1). With 3 pulses and 2 bits per amplitude, a sequential search with inloop quantization performs quite well.

Parameters	bits/frame	kb/s
LPC	38	1.9
LTP gain	4 * 3	0.6
LTP lag	4 * 7	1.4
Blockmax	4 * 5	1.0
Pulseamp	4 * 6	1.2
Pulsepos	4 * 13	2.6
TOTAL	174	8.7

Table 1: Speech Coder Parameters

Decoder

The speech decoder consists of the corresponding three parts: reconstruction of the excitation, long–term synthesis and short–term synthesis (Fig. 1). Post–filtering is also done on synthesized speech. The post–filter consists of a bandwidth expanded all–pole filter and a preemphasis filter.

Glottis Pulse Excitation

One way to improve the quality of a CELP coder and at the same time reduce the complexity is to use a two stage codebook, the first one consisting of a shifted whitened glottis pulse and the second of random sequences [12].

To further reduce the complexity, the random codebook can be exchanged with multipulses. If the glottis codebook is divided into a multipulse "codebook" and a truncated glottis function impulse response the same technique as for multipulses can be used to search for the position of the whitened glottis pulse. This hybrid coder of a CELP and a MPE has been called DCELP, when just incorporating deterministic functions.

The glottis pulse and the multipulses can be searched sequentially. The best performance is achieved if no restriction is made in which order the glottis pulse and the multipulses are determined. The pulse amplitude for either a glottis pulse or a multipulse which gives the minimum mean–squared–error are given by the relations below. The pulse positions are p and p_k respectively and the number of the multipulse to be determined is k. The amplitudes of the multipulses are a_k and the amplitude of the glottis pulse is g. If no glottis pulse has been placed g is set to zero in the relation for the multipulses. M is the number of multipulses.

$$g = \frac{C_g(p) - \sum_{i=1}^{k-1} a_i \cdot C_t(p - p_i)}{R_g(0)} \qquad 1 \leq p, p_k \leq N$$

$$a_k = \frac{C(p_k) - \sum_{i=1}^{k-1} a_i \cdot R(p_k - p_i) - g \cdot C_t(p - p_k)}{R(0)} \qquad 1 \leq k \leq M$$

$C_g(p)$ and $C(p_k)$ are the crosscorrelation functions between the weighted speech and the weighted impulse responses. For the multipulses, the impulse response is the weighted synthesis filter and for the glottis pulse, the convolution of the glottis function and the weighted synthesis filter. $C_t(p)$ is the crosscorrelation between the two impulse responses. $R_g(p)$ and $R(p_k)$ are the autocorrelation functions of the impulse responses.

Using a glottis pulse instead of one of the multipulses in the previous described coder, the performance is enhanced 0.3–0.5 dB in segmental SNR. The onset and pitch of the speech is improved. The level of complexity is the same as using only multipulses. One extra bit is needed in the phase position codeword to describe the position of the glottis pulse.

Implementation

The described speech coder has been implemented on both a floating–point DSP TMS20C30 and a fix–point DSP TMS20C25. Table 2 contains the complexity of the algorithm and of the two different implementations. The numbers relate to the version of the coder described in Table 1. The code is not fully optimized, but gives a good indication of the coder complexity.

The main complexity is in the LTP analysis, which is caused by the recursive update of the synthesized speech from past excitation for each lag. Both the LPC and the residual analysis are of low complexity.

For fix–point implementation, both the LPC analysis and the residual quantization analysis perform well with fix–point precision. Double precision is needed in just a few places, for the computation of the correlation functions before normalization.

Two approaches could be taken to achieve performance equal to floating–point in the closed–loop fix–point pitch predictor. In both approaches block floating–point

		TMSC30	TMSC25
	Mops	Mips	Mips
CODER			
LPC Analysis	0.3	0.7	1.0
LTP Analysis	3.2	4.8	6.6
Residual Analysis	0.4	0.8	1.2
I/O		0.7	0.3
DECODER			
Synthesis	0.25	0.6	0.9
I/O		0.8	0.3
TOTAL	4.15	8.4	10.3
RAM		1k	1.5k
ROM		3k	3.6k

Table 2: Complexity Of The Speech Coder

normalization is used for the vectors. One way is to use double precision computation when evaluating the error criteria, the other is to use single precision floating–point values. The former gives full precision for the full dynamics of the error function with the cost of huge complexity. The other approach gives a trade–off between performance and complexity, depending on the range of the exponent. Even with the full range, 4 bits for a 32 bits accumulator, the complexity is much lower than for double precision. The same performance as with floating–point is reached at about 3 bits, see Fig. 2.

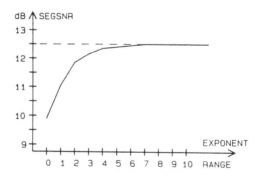

Figure 2: Performance Of Fix–point LTP Analysis

CHANNEL CODING

The Mobile Radio Channel

The mobile radio channel is a difficult channel for radio transmission. The received signal is affected by both long–term fading and short–term fading [13]. The long–term fading is mainly caused by the terrain configuration between the base station and the mobile unit. The short–term fading is caused by multi–path reflections. If the time delays between the received rays are small, compared to the bit time, the reflections will cause flat amplitude fading with Rayleigh distribution. Longer time delays will cause time dispersion, also called frequency selective fading.

The fade duration and interval are dependent on the carrier frequency, the channel bandwidth, the vehicle speed and the geography. For the speech and channel coding scheme which segments the coded speech into frames, these channel characteristics can be described by the frame error rate (FER), the bit error rate (BER) and the internal frame bit error rate (IBER). Figure 3 describes the characteristics of the mobile radio channel. It is neither a Gaussian random nor a typical on/off channel. For high fading rates the error distribution is more random and for low fading rates it is more burst–like.

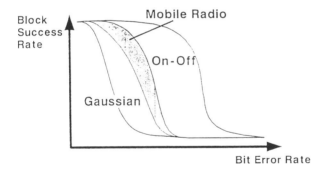

Figure 3: The Mobile Radio Channel

To combat fading, different techniques can be employed. Error correcting codes can efficiently take care of random distributed bit errors. By addding interleaving in the time domain, the bit errors become more randomly distributed and the error correcting code becomes more efficient. Due to system delay aspects, the interleaving depth, in the proposed system, can be no longer than about 40 ms, which corresponds to two speech coder frames.

At slow fading rates, other actions must be taken such as frequency hopping and/or antenna diversity. By using adaptive channel equalizers, which analyze the channel characteristics for every data burst, the effect of time dispersion can be used to enhance the quality of the detected bits instead of reducing it.

116

Study on Error Robustness

The channel error robustness can be studied by separating the random and the burst error sensitivity of the speech decoder. First the random error sensitivity for each speech coder bit is studied, then procedures to cope with burst errors are studied in the speech decoder. Finally the trade off between error correction, burst error detection and the number of protection classes needed in the coding scheme are studied for a number of different channel situations. Knowing the random and burst error sensitivities of the speech decoder, the requirements on correction and detection can be derived respectively.

For random bit errors the speech coder robustness itself is important. This robustness is increased by smoothing of parameters both prior to transmission and in the receiver during intervals of low quality recieved data. The encoding of the parameters can also be done to lower the sensitivity to errors. When combining the speech and channel coding, separable sensitivity to errors in the codeword bits is desirable, to make use of the channel coding ability to apply different protection on the data bits.

For burst errors, the ability to do speech extrapolation in the speech decoder is important. This can be done by smoothing and repeating parameters from the previous frames. Selective repetition of the para meters require more error detection than repetition of the whole speech coder frame. The detection can be done within block codes or outside the channel coding.

A mix of error correction and detection is needed in the mobile radio environment. During deep fading dips, the bit error rate might be so high that there is no use correcting the data but only detecting the burst errors. Elsewhere, the cost of increased bandwith for error correction will pay off generously with increased robustness.

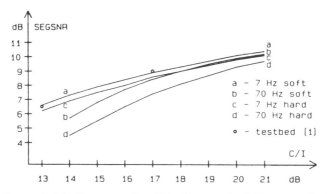

Figure 4: Performance For The Source And Channel Coding

For the feasibility of VLSI implementation, the ability to take full benefit of the channel state information and the ability of puncturing for different protection classes, a convolutional coder is chosen for error correction [1]. Coding rates of 1/2 and 2/3 are applied for correction. The most sensitive bits are also protected by a five bit CRC code for error detection.

Figure 4 shows the performance in segmental SNR on a simulated Rayleigh fading channel for different C/I. The performance with and without channel state information is compared for both low and high fading rates. A greater benefit from channel state information is achieved at high fading rates due to the coding schemes ability to correct more random-like distributed errors.

Lower fading rates gives a higher segmental SNR compared with higher fading rates when the bit error rate is increased. This is due to the bit errors gather in the already detected frames. The bit errors for higher fading rates are spread out over more frames which affects the SNR more.

The performance of a two-split TDMA testbed [1], with the GSM RPE-LTP speech codec at 13 kb/s [3] and a 6.5 kb/s error correcting code giving a total bit rate of 19.5 kb/s, is also shown in Fig. 4.

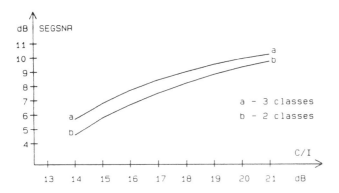

Figure 5: Comparison Of 2 And 3 Protection Classes

Figure 5 shows a comparison between two and three protection classes. For high fading rates the error correction code performs better which in this case leads to better performance for three protection classes than two, since the differentiated protection is more adapted to the bit error sensitivity.

CONCLUSIONS

The combination of the source and the channel coding scheme is important to survive in the severe error conditions in a mobile radio environment.

A novel choice to meet the requirements for source and channel coding in a mobile cellular system is a combination of a multipulse–like coder with a punctured convolutional error correcting code. The described 8.7 kb/s speech coder is quite feasible to implement in one DSP. The combination, of the speech and channel coding at 13 kb/s, performs well in a three-split TDMA system.

We feel that this choice of speech and channel coding is a good trade-off between quality and complexity.

118

REFERENCES

[1]   K. Raith, B. Hedberg, G. Larsson, R. Kåhre, "Performance of a digital cellular experimental testbed", VTC–89, San Fransisco, 1989.

[2]   N. Rydbeck, B. Hedberg, J. Uddenfeldt, "Implementation of hand–portables for a narrowband TDMA system", Int. conf. on Digital Land Mobile Radio, Venice, 1987.

[3]   P. Vary, K. Hellwig, R. Hoffman, R. Sluyter, C. Galand, M. Rosso, "Speech codec for the European mobile radio system", ICASSP–88, New York, 1988.

[4]   B. Atal, J. Remde, "A new model of LPC excitation for producing natural sounding speech at low bit rates", ICASSP–82, Paris, 1982.

[5]   B. Atal, M. Schroeder, "Stochastic coding of speech signals at very low bit rates", ICC–84, Amsterdam, 1984.

[6]   W. Kleijn, D. Krasinski, R. Ketchum, "Improved speech quality and efficient vector quantization in SELP", ICASSP–88, New York, 1988.

[7]   J. Hagenauer, N. Seshadri, C–E Sundberg, "The performance of rate–compatible convolutional codes for future digital mobile radio", 38th IEEE VTC, Philadelphia, June, 1988.

[8]   H. Hermansson, "Speech coding for the GSM half–rate codec?",2nd Eurasip workshop on medium to low rate speech coding,Hersbruck,1989.

[9]   T. Tremain, J. Campbell, V. Welch, "A 4.8 kbps code excited linear predictive coder", Proc. Mobile Sattelite Conf., May, 1988.

[10]   K. Ozawa, S. Ono, T. Araseki, A study on pulse search algorithms for multipulse excited speech coder realization", IEEE J. SAC–4, No. 1, Jan., 1986.

[11]   T.B. Minde, U. Wahlberg, "An implementation of a 8.7 kb/s speech coder and a study on the channel error robustness for the combination of the speech and channel coding", IEEE Workshop on speech coding, Vancouver, 1989.

[12]   A. Bergström, P. Hedelin, "Codebook driven glottal pulse analysis", ICASSP–89, Glasgow, 1989.

[13]   W. Lee, Mobile communications design fundamentals, Howard W. Sams & co, Indianapolis, 1986.

ACKNOWLEDGEMENT

We want to thank the director of ERA R&D J. Uddenfeldt for the permission to publish this study and we also want to thank prof. P. Hedelin for valuable discussions.

PART IV

SPEECH CODING BELOW 8 KB/S USING CODE EXCITED LINEAR PREDICTION

High quality speech coders at rates of 4.8 and 6 kb/s are becoming very important for supporting increased demand for cellular telephones and for providing secure voice communication over the public switched telephone network. This Section present papers describing speech coders below 8 kb/s based on code excited linear prediction (CELP). Campbell, Tremain and Welch present the 4.8 kb/s speech coding algorithm that forms the basis for the Federal Standard 1016 and discuss its performance. Tzeng describes analysis-by-synthesis linear predictive coding algorithms at rates of 2.4 and 4.8 kb/s. Salami present a new Transformed Binary Pulse Excitation approach for representing the stochastic excitation in CELP coders that eliminates the need for the exhaustive search of an excitation codebook. Taniguchi, Tanaka, and Gray discuss dynamic bit allocation techniques for improving the speech quality in CELP coders. Galand et al., describe a new CELP coder, called Adaptive CELP, that reduces the complexity by employing global perceptual weighting of the codebook. Finally, Delprat, Lever, and Gruet describe a low complexity CELP coder at 6 kb/s that uses binary regular pulse codebook.

12

THE DOD 4.8 KBPS STANDARD (PROPOSED FEDERAL STANDARD 1016)

Joseph P. Campbell, Jr.
Thomas E. Tremain
Vanoy C. Welch

U.S. Government
Department of Defense
Fort Meade, Maryland 20755-6000
USA

INTRODUCTION

In 1984, the U.S. DoD launched a program to develop a third-generation secure telephone unit (STU-III) capable of providing secure voice communications to all segments of the Federal Government and its contractors. In 1988, the DoD conducted a survey of 4800 bit per second (bps) voice coders to select a standard for use in an upgrade of the STU-III to supplement its 2400 bps LPC-10e vocoder. A code excited linear predictive (CELP) coder, jointly developed by the DoD and AT&T Bell Laboratories, was selected in this survey [1]. Listening tests and Dynastat's diagnostic rhyme test (DRT) and diagnostic acceptability measure (DAM) show this revolutionary coder outperforming all U.S. Government standards operating at rates below 16,000 bps; it's even comparable to 32,000 bps continuously variable slope delta modulation (CVSD) and is robust in acoustic noise, channel errors, and tandem coding conditions.

The Proposed Federal Standard (PFS) 1016 [2] is based on an enhanced version of the coder selected in the survey and described in our ICASSP '89 paper [3]. These enhancements maintain or improve the coder's speech intelligibility and quality, channel robustness, and implementation and do not restrict Government usage.

PFS-1016 has been endorsed for use in the STU-III. It is embedded in the proposed Land Mobile Radio standard (PFS-1024) that includes signaling and forward error correction to form an 8000 bps system [4]. A real-time implementation of a 6400 bps system with an embedded PFS-1016 coder is under consideration for the International Maritime Satellite Organization's (INMARSAT) Standard-M system. PFS-1016 is expected to be finalized in 1990.

Fed-Std 1016 could have many far-reaching applications. It will be proposed for a U.S. Military Standard, a NATO Standardization Agreement (STANAG), a public safety standard (APCO Project 25), and a microcellular personal communications network (PCN) standard. The Jet Propulsion Laboratory plans to use Fed-Std 1016 in their Mobile Satellite program. We expect Fed-Std-1016-based systems to replace many existing systems now based on 16,000 bps CVSD.

CELP ALGORITHM DESCRIPTION

CELP coding is a frame-oriented technique that breaks a sampled input signal into blocks of samples (i.e., vectors) that are processed as one unit. CELP coding is based on analysis-by-synthesis search procedures, perceptually weighted vector quantization (VQ), and linear prediction (LP). A 10th order LP filter is used to model the speech signal's short-term formant structure. Long-term signal periodicity is modeled by an adaptive code book VQ (also called pitch VQ because it often follows the speaker's pitch in voiced speech). The error from the short-term LP and pitch VQ is vector quantized using a fixed stochastic code book. The optimal scaled excitation vectors from the adaptive and stochastic code books are selected by minimizing a time varying, perceptually weighted distortion measure that improves subjective speech quality by exploiting masking properties of human hearing.

The major enhancements to our ICASSP '89 coder [3] are changes to the code books. The stochastic code book is now ternary valued (-1, 0, +1), as suggested by Dan Lin [5] and Jean-Pierre Adoul, and half as large (512 codewords). The adaptive code book is now twice as large (256 codewords) and contains 128 noninteger delays, as suggested by Peter Kroon [6], in addition to the original 128 integer delays.

CELP's computational requirements are dominated by the two code book searches. The computational complexity and speech quality of the coder depend upon the search sizes of the code books. Any subset of either code book can be searched to fit processor constraints, at the expense of speech quality.

We use an 8 kHz sample rate and a 30 ms frame size with four 7.5 ms subframes [7]. CELP analysis consists of three basic functions: 1) short-term linear prediction, 2) long-term adaptive code book search, and 3) innovation stochastic code book search. CELP synthesis consists of the corresponding three synthesis functions performed in reverse order with the optional addition of a fourth function, called a postfilter, to enhance the output speech. The transmitted CELP parameters are the stochastic code book index and gain, the adaptive code book index and gain, and 10 line spectral parameters (LSP). Our entire PFS-1016 transmitter and receiver are shown in reference [8]. The following description of our CELP coder represents only one of many possible implementations that would comply with PFS-1016.

Synthesis

The CELP synthesizer, shown in Figure 1, is used in the receiver and transmitter to generate speech by a parallel gain-shape code excitation of a linear prediction filter. The excitation is formed using a fixed stochastic code book and an adaptive code book in parallel. The stochastic code book contains sparse, overlapping, ternary valued, pseudorandomly generated codewords. Both code books are overlapped and can be represented as linear arrays, where each 60 sample codeword is extracted as a contiguous block of samples. In the stochastic code book, the codewords overlap by a shift of -2 (each codeword contains all but two samples of the previous codeword and two new samples). The adaptive code book has a shift of one sample or less between its codewords. The codewords with shifts of less than one sample are interpolated and correspond to noninteger pitch delays. The linear prediction filter's excitation is formed by adding a stochastic code book vector, given by index i_s and scaled by g_s, to an adaptive code book vector, given by index i_a and scaled by g_a. The adaptive code book is then updated by this excitation for use in the following subframe. Thus, the adaptive code book contains a history of past excitation signals, and the delay indexes the codeword containing the best block of excitation from the past for use in the present. The number of samples back in time in which this vector is located is called the pitch delay or adaptive code book index. For delays less than the vector length, a full vector does not exist and the short vector is replicated to the full vector length to form a codeword. Finally, an adaptive postfilter may be added to enhance the synthetic output speech.

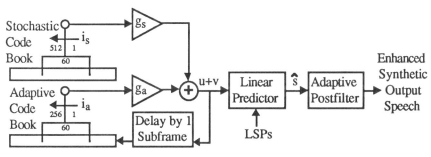

Figure 1. CELP Synthesizer

Analysis

The transmitter's CELP analyzer, shown in Figure 2, contains a replica of the receiver's synthesizer (minus the postfilter) that, in the absence of channel errors, generates speech identical to the receiver's. This approximation, $\hat{s}$, is subtracted from the input speech and the difference is perceptually weighted. This perceptually weighted error is then used to drive an analysis-by-synthesis (closed-loop) error minimization gain-shape VQ search procedure. The search procedure finds the adaptive and stochastic code book indices and gains that minimize the perceptually weighted error. The linear prediction filter can be determined by conventional open-loop short-term LP analysis techniques on the input speech.

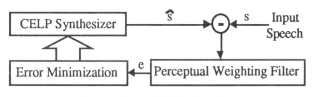

Figure 2. CELP Analyzer

Linear Prediction Analysis. The short-term LP analysis is performed once per frame by open-loop, 10th order autocorrelation analysis using a 30 ms Hamming window, no preemphasis, and 15 Hz bandwidth expansion. The bandwidth expansion operation replaces the LP analysis predictor coefficients, a_i, with $a_i \gamma^i$, which shifts the poles radially toward the z-plane origin by the weighting factor γ, for $0 < \gamma < 1$. These expanded coefficients form the LP filter $1/A(z)$, using $\gamma = 0.994$ for 15 Hz expansion. Besides improving speech quality, the 15 Hz bandwidth expansion is beneficial to LSP quantization [3] and also to Lionel Wolovitz's fast direct conversion of predictor coefficients to quantized LSPs [9]. (We form the perceptual weighting filter, $A(z)/A(z/\gamma')$, by a bandwidth expansion of the denominator filter polynomial using a weighting factor of $\gamma' = 0.8$.) Our coder's algorithmic delay, as opposed to its total voice processor delay, is determined by the LP analysis. The algorithmic delay of the coder is 15 ms because, as suggested by Bastiaan Kleijn, the analysis window is centered at the end of the last subframe. The linear predictor is coded using 34 bit, independent, nonuniform scalar quantization of line spectral pairs, as suggested by George Kang. Because the LSPs are transmitted only once per frame, but are needed for each subframe, they are linearly interpolated to form an intermediate set for each of the four subframes.

Adaptive Code Book Search. The adaptive code book search [10] is performed by closed-loop analysis using modified minimum squared prediction error (MSPE) criteria of the perceptually weighted error signal. We reserve 8 bits to allow up to a 256 codeword adaptive code book search. However, to reduce computational complexity, interoperable analyzers may search any subset of this code book. For every odd subframe, the coding consists of 128 integer and 128 noninteger delays ranging from 20 to 147. For every even subframe, delays are delta searched and coded with a 6 bit offset relative to the previous subframe. This greatly reduces computational complexity and data rate while causing no perceivable loss in speech quality. This adaptive code book index and gain are searched and coded four times per frame (every 7.5 ms). The gain is coded between -1 and +2 using absolute, nonuniform, scalar, 5 bit quantization, as specified in PFS-1016.

The MSPE search criteria is modified to check the match score at submultiples of the delay to determine if it is within 1/2 dB of the MSPE. The shortest submultiple delay is selected if its match score satisfies our modified criteria. While maintaining high quality speech, this results in a smooth "pitch" delay contour that is crucial to delta coding and the receiver's smoother in the presence of bit errors.

Use of noninteger delays in the analyzer is optional; however, the complete adaptive (and stochastic) code book is required for interoperable receivers. Noninteger values of delay can be obtained without increasing the 8 kHz sample rate by resampling or polyphase filtering the integer delay codewords to generate interpolated noninteger delay codewords [6]. We recommend using a set of five 8-point Hamming windowed sinc interpolating functions corresponding to the five allowable fractional parts of the noninteger delays specified in PFS-1016. Noninteger delays provide the following benefits by reducing: reverberant distortion; the roughness of some high pitched speakers; noise, because improved pitch prediction reduces the noisy stochastic excitation component; and pitch doubling and tripling, which improve delta coding. This high resolution delay is also beneficial to long-term "pitch" prefiltering or postfiltering. The delay coding specified in PFS-1016 is nonuniform, with 1/4, 1/3, 1/2, or 1 sample resolution, depending on the region of the delay. This coding was designed to gain the greatest improvement in speech quality by providing the highest resolution for typical female speakers and lower resolution for typical male and child speakers.

Stochastic Code Book Search. The stochastic code book search is performed by closed-loop analysis using conventional MSPE criteria of the perceptually weighted error signal. We reserve 9 bits to allow up to a 512 codeword stochastic code book search. However, to reduce computational complexity, interoperable analyzers may search any subset of this code book. The code book index and gain are searched and coded four times per frame. The gain (positive and negative) is coded using 5 bit, absolute, nonuniform scalar quantization, as specified in PFS-1016.

A special form of stochastic code book containing sparse, overlapped shift by -2, and ternary valued samples (-1, 0, +1) is used to allow fast convolution and energy computations by exploiting recursive end-point correction algorithms [7, 10, 11]. This code book, specified in PFS-1016, contains samples of a zero-mean, unit-variance, white Gaussian sequence center clipped at 1.2 and ternary level quantized, resulting in approximately 77 percent sparsity (zero values). This form of a code book is unambiguous, regardless of arithmetic; compact; has potential for fast search procedures; causes no degradation in speech quality relative to other types of code books; and significantly reduces search computation.

Postfilter. The postfilter is a short-term pole-zero type with adaptive spectral tilt compensation, as suggested by J.-H. Chen [12]. Cautious application of postfiltering at the receiver's output is recommended. The ear's masking properties are exploited to trade off speech distortion vs quantizing noise. Usually, the postfilter enhances the synthesized speech by the variances of the DRT and DAM tests. However, in some noisy environments where the LP analysis models the noise, the noise is enhanced because the postfilter is controlled by the LP analysis. In addition, if not taken into consideration, postfiltering can be detrimental to tandem coding. If a postfilter is used in every stage, the signal becomes degraded through multiple tandems of CELP. Optimum performance is obtained when the postfilter is used in only the first stage; however, this is usually impractical. A practical solution is to remove all postfilters, except for the final stage's postfilter.

A standard is required to provide 4800 bps voice coding today. To prevent this standard from becoming obsolete, we provide a bit per frame for future expansion, as suggested by Bishnu Atal. As long as each expansion provides another expansion, the standard may continue to branch many times. For example, this bit could allow: adaptive bit allocation [13, 14], adaptive postfiltering, new LP coding, or new code book designs. Our CELP coder's characteristics are summarized in Table 1.

Table 1. CELP Characteristics

	Linear Predictor	Adaptive CB	Stochastic CB
Update	30 ms	30/4 = 7.5 ms	30/4 = 7.5 ms
Parameters	10 LSPs (independent)	1 gain, 1 delay 256 codewords	1 gain, 1 index 512 codewords
Analysis	open loop 10th order autocorrelation 30 ms Ham window no preemphasis 15 Hz expansion interpolated by 4	closed loop 60 dimensional mod MSPE VQ weighting = 0.8 delta search range: 20 to 147 noninteger delays	closed loop 60 dimensional MSPE VQ weighting = 0.8 shift by -2 77% sparsity ternary samples
Bits Per Frame	34 (3,4,4,4,4,3,3,3,3,3)	index: 8+6+8+6 ±gain: 5x4	index: 9x4 ±gain: 5x4
Rate	1133.33 bps	1600 bps	1866.67 bps
Miscellaneous	The remaining 200 bps are used as follows: 1 bit per frame for synchronization, 4 bits per frame for forward error correction, and 1 bit per frame for future expansion.		

ERROR PROTECTION

Parameter coding and continuity are the basis for our error protection strategy. We developed an integrated adaptive error protection system combining: forward error correction (FEC), smoothers, parameter coding, and intraframe interleaving. We employ adaptive smoothers based on estimates of the channel error rate, which are largely responsible for our coder's robust performance [3]. The channel error rate is estimated by time averaging the syndrome detection from an FEC code. This allows the smoothers to be disabled in error-free conditions. Forward error correction is accomplished with a Hamming (15,11) single error detecting and correcting code that protects 10 bits of the adaptive code book index (pitch delay) and gain. Robust pitch delay protection is provided by jointly optimizing the FEC and the channel symbol assignment of delays. The absolute pitch delays each have 3 bits protected by FEC. The channel symbols are assigned to minimize the perceptual distortion due to single bit transmission errors in the unprotected bits using simulated annealing, similar to the technique in reference [15]. We chose to place all the FEC power on the absolute delay, since correct absolute delays are crucial for correct delta delay decoding. The adaptive code book gain has many nonsmooth regions where smoothing is ineffective; therefore, we protect its most perceptually sensitive bit. For future expansion, we reserved 1 bit per frame. This is the 11th bit protected by the Hamming code.

CODE BOOK SEARCH METHODS

The search procedures for the stochastic and adaptive code books are virtually identical, differing only in their code books and target vectors. To reduce computation, a sequential two-stage search of the code books is performed. The target for the first stage adaptive code book search is the weighted linear prediction residual plus encoding errors introduced in previous frames that affect the present frame. The second stage stochastic code book search target is the first stage target minus the filtered adaptive code book VQ excitation.

Let the $L = 60$ dimensional row vectors $\mathbf{s}$, $\hat{\mathbf{s}}$, and $\mathbf{e}$ represent the original speech signal, the synthetic speech signal, and the weighted error signal, respectively. Let $\mathbf{v}$ represent the excitation vector being searched for in the present stage and let $\mathbf{u}$ be the excitation vector of the previous stage. The excitation sequence for an N size code book within a subframe of size L is characterized by a code book index i, $1 \leq i \leq N$, and a corresponding optimized gain parameter g_i. The excitation vector $\mathbf{v}^{(i)}$ can be written as:

$$\mathbf{v}^{(i)} = g_i \, \mathbf{x}^{(i)}, \tag{1}$$

where the superscript denotes the code book index of the code book vector $\mathbf{x}^{(i)}$.

Let $\mathbf{H}$ and $\mathbf{W}$ be $L \times L$ matrices whose j-th rows contain the truncated impulse response caused by a unit impulse $\delta(t-j)$ of the LP filter and error weighting filter, respectively. As shown in Figure 1, the synthetic speech can be expressed as the LP filter's zero input response, $\hat{\mathbf{s}}^{(0)}$, plus the convolution of the LP filter's excitation and impulse response:

$$\hat{\mathbf{s}}^{(i)} = \hat{\mathbf{s}}^{(0)} + (\mathbf{u} + \mathbf{v}^{(i)})\mathbf{H}, \qquad 1 \leq i \leq N \tag{2}$$

where $\mathbf{u}$ is a zero vector in the first stage search or the scaled adaptive excitation vector in the second stage search. As shown in Figure 2, the weighted error signal is:

$$\mathbf{e}^{(i)} = (\mathbf{s} - \hat{\mathbf{s}}^{(i)})\mathbf{W} \tag{3a}$$

$$= \mathbf{e}^{(0)} - \mathbf{v}^{(i)} \, \mathbf{HW}, \tag{3b}$$

where the target is:

$$\mathbf{e}^{(0)} = (\mathbf{s} - \hat{\mathbf{s}}^{(0)})\mathbf{W} - \mathbf{uHW}. \tag{4}$$

Thus, the weighted error, $\mathbf{e}^{(i)}$, is the target minus the scaled filtered codeword:

$$\mathbf{e}^{(i)} = \mathbf{e}^{(0)} - g_i\mathbf{y}^{(i)}, \tag{5}$$

where $\mathbf{y}^{(i)}$ represents the filtered codeword:

$$\mathbf{y}^{(i)} = \mathbf{x}^{(i)}\mathbf{HW}. \tag{6}$$

Let E_i represent the norm or total squared error for codeword i:

$$E_i = \|\mathbf{e}^{(i)}\| = \mathbf{e}^{(i)} \, \mathbf{e}^{(i)T} \tag{7a}$$

$$= \mathbf{e}^{(0)} \, \mathbf{e}^{(0)T} - 2g_i\mathbf{e}^{(0)}\mathbf{y}^{(i)T} + g_i^2\mathbf{y}^{(i)}\mathbf{y}^{(i)T}, \tag{7b}$$

where T denotes transpose. E_i is a function of both the gain factor g_i and the index i. For a given value of i, the optimal gain can be computed by setting the derivative of E_i with respect to the unknown gain value to zero:

$$\frac{\partial E_i}{\partial g_i} = -2e^{(0)}y^{(i)T} + 2g_iy^{(i)}y^{(i)T} = 0. \tag{8}$$

Therefore, the optimum gain is the ratio of the crosscorrelation of the target and filtered codeword to the energy of the filtered codeword:

$$g_i = \frac{e^{(0)}y^{(i)T}}{y^{(i)}y^{(i)T}}. \tag{9}$$

The gain can now be quantized to jointly optimize the search for gain and index:

$$\hat{g}_i = Q[g_i]. \tag{10}$$

Our objective is to minimize the squared error at the receiver given by Eq. (7b) with $\hat{g}_i$ substituted for g_i. Minimizing E_i with respect to i is equivalent to maximizing the negative of the last two terms in Eq. (7b); the first term, $e^{(0)}e^{(0)T}$, is independent of the codeword i. This corresponds to maximizing the match score:

$$match_i = \hat{g}_i(2e^{(0)}y^{(i)T} - \hat{g}_iy^{(i)}y^{(i)T}). \tag{11}$$

If the gain quantization is ignored, Eq. (9) can be substituted in Eq. (11), and the match score approximation is the familiar normalized squared crosscorrelation:

$$match_i \approx \frac{(e^{(0)}y^{(i)T})^2}{y^{(i)}y^{(i)T}}. \tag{12}$$

Thus, as shown in Eq. (11) and Figure 3, the code book search procedure for the MSPE excitation sequence for analysis-by-synthesis weighted gain-shape vector quantization is to find the codeword, i, and gain, $\hat{g}_i$, that maximize the match score, $match_i$. To reduce computational complexity, the searches are performed sequentially, first using the adaptive code book and its target, followed by the stochastic code book and its target. The optimal excitation vectors are entirely characterized by each of their indices and corresponding gain factors.

We found joint optimization of index and gain to be subjectively similar to searching twice as large a stochastic code book without joint optimization (ignoring gain quantization). Adaptively attenuating the stochastic code book gain during sustained voiced segments [16] reduces roughness and quantizing noise and further improves our CELP coder's speech quality.

In Figure 3, there are four main operations: convolution (represented by *), a crosscorrelation inner-product, an energy auto inner-product, and a quantized divide. A discussion of their computation follows.

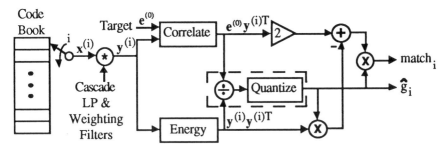

Figure 3. Code book search with joint index and gain optimization

IMPLEMENTATION AND COMPUTATIONAL ESTIMATES

CELP's computations, excluding the code book searches and including the receiver, require approximately 2 million instructions (multiply, add, multiply-accumulate, or compare) per second (MIPS) [9]. CELP's major computational requirements are dominated by the transmitter's code book searches. For a specific implementation to achieve its highest speech quality, tradeoffs must be made within each code book search and between the two code book searches. The complexity of the code book searches can vary tremendously depending on the procedures used.

The adaptive code book search is especially ripe for nonexhaustive search procedures. The match score function of the adaptive code book search in voiced speech is well behaved and shows prominent harmonic pitch structure. This well-behaved match score function lends itself to many nonexhaustive search procedures, including: subsampling, interpolation, reduced search based on prior information, open-loop search, and hierarchical searches. Any combination of these methods may be used. We recommend a two-stage hierarchical search of integer delays (with preference to submultiples), followed by searching neighboring noninteger delays.

The following estimates assume an adaptive code book search without joint optimization of index and gain (Eq. 12), using cross-multiply match score calculation (gain is not calculated in the search), delta searching even numbered subframes, recursive end-point correction computation of the convolution with truncated overlap and add, and circular buffering. The combination of these techniques speeds the pitch search by a factor of 18 relative to brute force methods. For integer delays, the first codeword requires 0.0576 MIPS, plus each additional codeword requires an average of 0.00651 MIPS per subframe [9]. Therefore, to search 128 integer pitch delays twice per frame and 32 integer pitch delays twice per frame requires 2.3 MIPS. The estimates for searching noninteger delays are not given because they are optional and vary depending on the chosen search procedure (e.g., interpolation vs exhaustive).

Our stochastic code book search computations are reduced by a factor of 20 relative to brute force techniques. The overlapped ternary valued code book allows computation of the convolution end-point correction recursions with a simple add or subtract of the truncated impulse response for each codeword's two new samples. The following estimates assume a stochastic code book search using a code book representation that requires no instructions to test its samples, recursive end-point correction computation of the convolution and energy, circular buffers, and joint optimization of index and gain (Eq. 11). As suggested by Bastiaan Kleijn, the quantized gain can be determined very rapidly by a simultaneous division and quantization using a binary-tree search with cross-multiplication. For each subframe, the first codeword requires 0.0270 MIPS, plus each additional codeword requires 0.00399 MIPS [9]. Therefore, to search the whole 512 size stochastic code book four times per frame requires 8.3 MIPS.

As shown in Table 2, the total upper bound CELP transmitter and receiver computation estimate is 12.6 MIPS, neglecting noninteger delays and using our current search procedures. This is an upper bound because we are not claiming to know the fastest code book search procedure. Other search procedures (e.g., energy recursion) [10], alternate search domains (e.g., autocorrelation) [17, 18], and code book transformations [19] may yield faster methods.

Our MIPS values shouldn't be confused with DSP chip peak MIPS ratings. Today's state-of-the-art DSP chips require approximately twice our estimated MIPS values, as shown in the last column of Table 2, because of overhead and breaks in the pipeline. Although only power-of-2 size code book searches are shown, to achieve the highest quality speech, we expect designers to optimize their implementations to search as many codewords as possible without exceeding their real-time limit.

Table 2.   CELP Computational Complexity

Stochastic CB Size	Stochastic Search	Total (full-duplex)	DSP Chip Rating
64	1.1 MIPS	5.5 MIPS	11 MIPS
128	2.1 MIPS	6.5 MIPS	13 MIPS
256	4.2 MIPS	8.5 MIPS	17 MIPS
512	8.3 MIPS	12.6 MIPS	25 MIPS

The proof of these estimates lies in real-time implementation. Many firms are implementing PFS-1016 coders, including: AT&T, DSP Software Engineering, GE/RCA, Motorola, and Spectron Micro Systems. For example, DSP Software Engineering's real-time full-duplex implementation uses Texas Instruments' TMS320C30 16.6 MIPS DSP chip and TLC32044 A/D and D/A with filters chip [20]. They search half the stochastic code book and hierarchically search the adaptive code book's integer delays followed by neighboring noninteger delays. This confirms our 17 MIPS DSP chip estimate and demonstrates the feasibility of a high quality and simple hardware PFS-1016 implementation.

PERFORMANCE

CELP coders do not exhibit the usual vocoder problems in background noise because they use a more sophisticated excitation model than the classical vocoder's pitch and voicing (e.g., LPC-10). Background noise, including multiple speakers, is faithfully reproduced. Informal listening tests indicate that our 4800 bps CELP coder's speech intelligibility and quality are comparable to 32,000 bps CVSD!

We formally measure speech intelligibility and quality subjectively using Dynastat's diagnostic rhyme test (DRT) and diagnostic acceptability measure (DAM). Typical test variances are 1 point for DRT and 2 points for DAM scores. The PFS-1016 coder outperformed our ICASSP '89 coder [3] by 3 DAM points and produced identical DRT scores. DRT performance is reported in references [3] and [21].

Figures 4 and 5 show subjective quality scores evaluated in different environments for: input speech (whose quiet DAM is off the scale at 84), long distance plain old telephone service (POTS), and common narrowband U.S. Government standard coders. Coder performance is shown in Figure 4 for quiet, office, E-3A/E-4B Airborne Command Post compartment environments, and a 1 percent uniform random bit error rate condition. In Figure 5, tandem coding of each coder into 2400 bps LPC-10e is represented by ->LPC and 16,000 bps CVSD into each coder is shown by CVSD->. As shown in the figures, CELP's performance is outstanding among other coders. CELP, at 4800 bps, breaks the performance barrier of most Government standards, providing Consortium ratings of "very good" intelligibility and "excellent" quality, comparable to 32,000 bps CVSD. CELP will usher in a new era of narrowband speech coders capable of receiving wide user acceptance by providing very high quality speech.

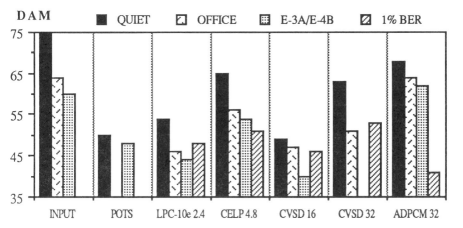

Figure 4. **Government standard speech coder quality comparison**

132

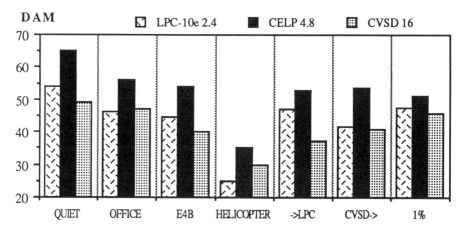

Figure 5. Speech coder quality for LPC-10e, CELP and CVSD 16

CONCLUSIONS

Our implementation of PFS-1016 is robust in real-world conditions (noisy environments, nonspeech input, tandem coding, and transmission errors). Most importantly, PFS-1016 is practical to implement and sets an expandable 4800 bps coding standard that will soon provide high quality speech in many applications, including secure voice, Mobile Satellite, and Land Mobile Radio.

ACKNOWLEDGMENTS

We are grateful for all the cited suggestions and to our families. We wish to give special recognition to Bishnu Atal, George Kang, W. Bastiaan Kleijn, Peter Kroon, and Dan Lin for their outstanding contributions.

REFERENCES

[1] Kemp, D., R. Sueda and T. Tremain, "An Evaluation of 4800 bps Voice Coders," *Proceedings of the IEEE International Conference on Acoustics, Speech, and Signal Processing (ICASSP)*, 1989, p. 200-203.

[2] Fenichel, R., *Proposed Federal Standard 1016 (Second Draft)*, National Communications System, Office of Technology and Standards, Washington, DC 20305-2010, 13 November 1989.

[3] Campbell, J., V. Welch and T. Tremain, "An Expandable Error-Protected 4800 bps CELP Coder," *Proceedings of ICASSP*, 1989, p. 735-738 (and *Proceedings of Speech Tech*, 1989, p. 338-341).

[4] Rahikka, D., T. Tremain, V. Welch and J. Campbell, "CELP Coding for Land Mobile Radio Applications," submitted to *Proceedings of ICASSP*, 1990.

[5] Lin, D., "New Approaches to Stochastic Coding of Speech Sources at Very Low Bit Rates," *Signal Processing III: Theories and Applications (Proceedings of EUSIPCO-86)*, 1986, p. 445-448.

[6] Kroon, P. and B. Atal, "On Improving the Performance of Pitch Predictors in Speech Coding Systems," *Abstracts of the IEEE Workshop on Speech Coding for Telecommunications*, 1989, p. 49-50.

[7] Tremain, T., J. Campbell and V. Welch, "A 4.8 kbps Code Excited Linear Predictive Coder," *Proceedings of the Mobile Satellite Conference*, 1988, p. 491-6.

[8] Campbell, J., T. Tremain and V. Welch, "The Proposed Federal Standard 1016 4800 bps Voice Coder: CELP," submitted to *Speech Technology Magazine*, April/May 1990.

[9] Campbell, J., V. Welch and T. Tremain, "The New 4800 bps Voice Coding Standard," *Proceedings of Military and Government Speech Tech*, 1989, p. 64-70.

[10] Kleijn, W., D. Krasinski and R. Ketchum, "An Efficient Stochastically Excited Linear Predictive Coding Algorithm for High Quality Low Bit Rate Transmission of Speech," *Speech Communication*, October 1988, p. 305-316.

[11] Lin, D., "Speech Coding Using Efficient Pseudo-Stochastic Block Codes," *Proceedings of ICASSP*, 1987, p. 1354-1357.

[12] Chen, J.-H. and A. Gersho, "Real-Time Vector APC Speech Coding at 4800 bps with Adaptive Postfiltering," *Proceedings of ICASSP*, 1987, p. 2185-2188.

[13] Jayant, N. and J.-H. Chen, "Speech Coding with Time-Varying Bit Allocations to Excitation and LPC Parameters," *Proceedings of ICASSP*, 1989, p. 65-68.

[14] Taniguchi, T., S. Unagami, Y. Tanaka, F. Amano and Y. Ohta, "Distortion Measure for Mode Decision of Multimode CELP Coder," *Abstracts of the IEEE Workshop on Speech Coding for Telecommunications*, 1989, p. 45-48.

[15] Cox, R., W. Kleijn and P. Kroon, "Robust CELP Coders for Noisy Background and Noisy Channels," *Proceedings of ICASSP*, 1989, p. 739-742.

[16] Shoham, Y., "Constrained-Excitation Coding of Speech," *Abstracts of the IEEE Workshop on Speech Coding for Telecommunications*, 1989, p. 65.

[17] Trancoso, I. and B. Atal, "Efficient Procedures for Finding the Optimum Innovation in Stochastic Coders," *Proceedings of ICASSP*, 1986, p. 2375-2378.

[18] Xydeas, C., M. Ireton and D. Baghbadrani, "Theory and Real Time Implementation of a CELP Coder at 4.8 and 6.0 kbits/second Using Ternary Code Excitation," *Proceedings of the Fifth International Conference on Digital Processing of Signals in Communications*, 1988, p. 167-174.

[19] Salami, R. and D. Appleby, "A New Approach to Low Bit Rate Speech Coding with Low Complexity using Binary Pulse Excitation (BPE)," *Abstracts of the IEEE Workshop on Speech Coding for Telecommunications*, 1989, p. 63-65.

[20] Macres, J., "The First Real-Time Implementation of U.S. Federal Standard 4800 bps CELP Version 3.1," submitted to *Proceedings of Speech Tech*, 1990.

[21] Welch, V., T. Tremain and J. Campbell, "A Comparison of U.S. Government Standard Voice Coders," *Proceedings of the IEEE Military Communications Conference (MILCOM)*, 1989, p. 269-273.

13

ANALYSIS-BY-SYNTHESIS LINEAR PREDICTIVE SPEECH CODING AT 4.8 kBIT/S AND BELOW

F. F. Tzeng

COMSAT Laboratories
Clarksburg, Maryland 20871-9475

INTRODUCTION

Linear predictive speech coders (LPC) have dominated speech coding applications for the past two decades. A common characteristic of these coders is that open-loop methods are used for the analysis of the spectrum filter and the excitation signal. With these open-loop methods, no performance measure is defined directly between the original speech and the reconstructed speech.

The analysis-by-synthesis method [1], or closed-loop analysis method, has long been used in areas other than speech coding. In recent years, this method has also been successfully applied to several speech coding techniques, such as multipulse-excited LPC [2] and code-excited LPC (CELP) [3]. With a perceptually meaningful distortion measure, the analysis part of a speech coding scheme can be optimized to minimize the chosen distortion measure between the original speech and the reconstructed speech.

In this paper, the application of the analysis-by-synthesis method to speech coders operating at 4.8 kbit/s and below is investigated. For mathematical tractability, a perceptually-weighted mean-squared-error (WMSE) is used as the distortion measure. Coders that employ WMSE are basically waveform coders. In this sense, the resultant coder should be less speaker dependent and more robust against background acoustic noise as the synthesized speech is reconstructed to mimic the individual or the corrupted speech waveform (in a perceptually weighted sense). However, in terms of speech quality, a better distortion measure for narrowband speech coding applications could be used.

The decoder (or the synthesis part) of an analysis-by-synthesis LPC (A-by-S LPC) is identical to that of LPC. To enhance the perceived speech quality, an adaptive post-filter is used. The spectrum filter is typically a tenth-order all-pole filter. The excitation signal can assume a wide range of different models depending on the available data rate. The transfer function of the adaptive post-filter is given in [4]

$$Q(z) = \frac{(1 - \mu z^{-1})\, A(z/a)}{A(z/b)} \qquad (1)$$

where $A(z) = 1 - \sum_{i=1}^{10} a_i z^{-i}$ is the transfer function of the spectrum filter; $0 < a < b < 1$ are design parameters; and $\mu = ck_1$, where $0 < c < 1$ is a constant, and k_1 is the first reflection coefficient.

The perceptual weighting filter, $W(z)$, used in the WMSE distortion measure is defined as

$$W(z) = \frac{A(z)}{A(z/\gamma)} \qquad (2)$$

where $0 < \gamma < 1$ is a constant controlling the amount of spectral weighting.

SPECTRUM FILTER

The spectrum filter coefficients can be represented by several different parameter sets for efficient quantization. Examples are the log-area-ratios (LAR), the line spectrum frequencies (LSF) [5], and so on. Several efficient spectrum filter coding schemes have been developed in recent years. Examples are the 41-bit LAR coding scheme and the 32-bit LSF coding scheme [6]. Both use scalar quantization methods.

In this paper, a 26-bit spectrum filter coding scheme is proposed. This scheme is based on LSF interframe prediction with two-stage vector quantization. The interframe predictor can be formulated as follows [7]. Given the LSF set of the current frame (with mean values removed),

$$F_n = (f_n^{(1)}, f_n^{(2)}, ..., f_n^{(10)})^T$$

the predicted LSF set based on the previous frame is

$$\hat{F}_n = MF_{n-1} \qquad (3)$$

where the optimal prediction matrix, M, which minimizes the mean-squared prediction error, is given by

$$M = [E(F_n F_{n-1}^T)] \, [E(F_{n-1} F_{n-1}^T)]^{-1} \qquad (4)$$

where E is the expectation operator.

For interframe prediction, a 6-bit codebook of predictor matrices (which is precomputed using a large speech database) is exhaustively searched to find the predictor matrix, M, that minimizes the mean-squared prediction error. The predicted LSF vector for the current frame $\hat{F}_n$, is then computed according to (3). The residual LSF vector, which is the difference vector between the current LSF vector (F_n) and the predicted LSF vector ($\hat{F}_n$), is then quantized by a two-stage vector quantizer. Each vector quantizer contains 1,024 (10-bit) vectors.

To improve coding performance, a perceptual weighting factor is included in the distortion measure used for the two-stage vector quantizer. The distortion measure is defined as

$$D = \sum_{i=1}^{10} w_i \, (x_i - y_i)^2 \qquad (5)$$

where x_i, y_i denotes the component of the LSF vector to be quantized, and the corresponding component of each codeword in the codebook, respectively. The corresponding perceptual weighting factor, w_i, is defined as [8]

$$w_i = \begin{cases} u(f_i) \sqrt{D_i/D_{max}}, & 1.375 \leq D_i \leq D_{max} \\ u(f_i) D_i/\sqrt{1.375 D_{max}}, & D_i < 1.375 \end{cases} \qquad (6)$$

where

$$u(f_i) = \begin{cases} 1, & f_i < 1,000 \text{ Hz} \\ \dfrac{-0.5}{3,000} (f_i - 1,000) + 1, & 1,000 \leq f_i \leq 4,000 \text{ Hz} \end{cases} \qquad (7)$$

The factor $u(f_i)$ accounts for the human ear insensitivity to the high-frequency quantization inaccuracy; f_i denotes the i-th component of the LSFs for the current frame; D_i denotes the group delay for f_i in milliseconds; and D_{max} is the maximum group delay, which has been found experimentally to be around 20 ms. The group delay (D_i) accounts for the specific spectral sensitivity of each frequency (f_i), and is well related to the formant structure of the speech spectrum. At frequencies near the formant region, the group delays are larger. Since those frequencies should be more accurately quantized, the weighting factors should be larger. Hence, the group delays provide a convenient way to adapt w_i.

The group delays (D_i) can be computed as the gradient of the phase angles of the ratio filter (defined in [8]) at $-n\pi$ ($n = 1, 2, ..., 10$). These phase angles are computed in the process of transforming the predictor coefficients of the spectrum filter to the corresponding LSFs. With perceptual weighting, the spectrum filter quantization effect is imperceptible.

EXCITATION MODELS

A.   CELP

Figure 1 is a schematic diagram of the CELP speech coder. The excitation signal is formed by filtering the selected random sequence through the selected pitch synthesizer. For the closed-loop excitation analysis, a suboptimum sequential procedure is used. This procedure first assumes zero input to the pitch synthesizer and employs the closed-loop pitch synthesizer analysis method to compute the pitch and the pitch synthesizer coefficients. Pitch synthesizer fixed, a closed-loop method is then used to find the best excitation random sequence, C_i, and compute the corresponding gain, G. To save in computation, a first-order pitch synthesizer is used.

138

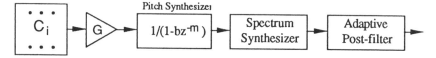

Figure 1. CELP

For speech coding at 4.8 kbit/s, the system parameters are as follows [9]. The sampling rate is 8 kHz. The frame size is 180 samples. The spectrum filter uses the 26-bit coding scheme described in the previous section. Ten bits are used to specify 1,024 random sequences for the excitation codebook. Seven bits are allocated for the pitch period, m, with a range from 16 to 143 samples. Five bits each are allocated for the gain G and the pitch synthesizer coefficient, b, respectively. The excitation information is updated three times per frame.

Due to the block processing nature in the computation of the spectrum filter, the spectrum filter parameters can have abrupt change in neighboring frames. To smooth out the abrupt change and to synchronize with the excitation, each excitation subframe uses a different set of LSF parameters. These LSF parameters are obtained through interpolation using the quantized LSF parameters of the current frame and the previous frame.

The procedures for the closed-loop pitch synthesizer analysis and the selection of the best random excitation sequence are identical. For every excitation subframe, each codeword is used as the input signal to the synthesizing filter. Codeword C_i, together with its corresponding gain G, which minimizes the WMSE between the original speech and the synthesized speech, is selected as the best excitation. The minimization step can be formulated as

$$\text{Minimize } E_W(G, C_i) = \sum_{n=1}^{N} [S_W(n) - GY_W(n)]^2 \qquad (8)$$

where N is the total number of samples in a subframe; $S_W(n)$ denotes the weighted residual signal after the memory of the synthesizing filter has been subtracted from the speech signal; and $Y_W(n)$ denotes the combined response of the synthesizing filter and $W(z)$ to the input signal C_i. The optimum value of the gain term, G, can be derived as

$$G = \sum_{n=1}^{N} S_W(n) \ Y_W(n) \ / \ \sum_{n=1}^{N} Y_W^2(n) \qquad (9)$$

The excitation codeword (C_i) which maximizes the following term is selected as the best excitation codeword:

$$E_W(C_i) = \left[\sum_{n=1}^{N} S_W(n) \ Y_W(n) \right]^2 \ / \ \sum_{n=1}^{N} Y_W^2(n) \qquad (10)$$

For the closed-loop analysis of the random excitation, the synthesizing filter is the combination of the spectrum synthesizer and the pitch synthesizer. For the closed-loop analysis of the pitch synthesizer, the codeword C_i corresponds to the different pitch synthesizer memory due to different pitch periods. The gain term G corresponds to the pitch synthesizer coefficient b.

The closed-loop analysis of the random excitation and the pitch synthesizer require extremely high computational complexity. For real-time implementation using current DSP chips, substantial reduction of the computational complexity is essential. For the pitch synthesizer, it is obvious that the different pitch synthesizer memory as the excitation signal form a set of overlapped excitation sequence. The computation of $Y_w(n)$ in (9) and (10) can thus be greatly simplified. For the random excitation, the same technique can be employed by using a codebook with overlapped random sequences. An even more efficient method for the random excitation search was proposed in [10]. By using these complexity reduction techniques, real-time implementation of the CELP coder becomes practical.

Informal listening tests were conducted for the 4.8-kbit/s coder. The test material consisted of nine Harvard sentences [11], spoken by five female and five male speakers. It was found that the 4.8-kbit/s CELP speech coder is able to produce good quality speech with very high intelligibility and excellent speaker recognizability. Relative to the original speech, however, the 4.8-kbit/s CELP speech is less smooth, and lacks some of the "fullness" of the original speech.

B.   A-BY-S LPC-10

For speech coding at 2.4 kbit/s and below, the LPC-10 model shown in Fig. 2 is usually used. Based on this binary excitation model, a decision is made on each speech frame whether it is a voiced or an unvoiced frame. For an unvoiced frame, a white Gaussian sequence is used as the excitation signal. For a voiced frame, a pitch value is estimated, and a pulse train with its period equal to the pitch value is used as the excitation signal.

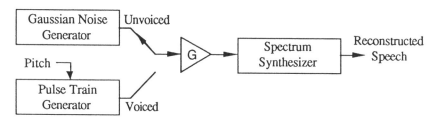

Figure 2.   LPC-10

Previously, open-loop methods were used for V/UV decision-making and pitch estimation/tracking [12],[13]. In [14], an analysis-by-synthesis (A-by-S) LPC-10 based on the closed-loop method for the excitation analysis has been proposed. Several excitation models have been presented and their performance compared. It has been found that the excitation model (Fig. 3) which selects the excitation signal from either a random sequence codebook or a pitch synthesizer produces the best perceived quality speech.

140

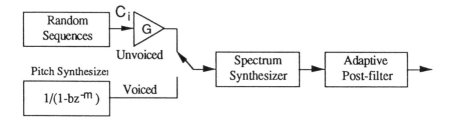

Figure 3. A-by-S LPC-10

The distinctive features of this scheme as compared to the LPC-10 are as follows:

a. The V/UV decision and the pitch estimation/tracking are implicitly performed and are optimum in terms of minimizing the WMSE distortion measure for the current speech frame, be it clean speech or noisy speech.

b. The perceptual weighting effect, which is important in low-data-rate speech coding, is easily introduced.

c. Speech coder performance is improved by using an unvoiced codebook, instead of a single random sequence.

For speech coding at 2.4 kbit/s, the system parameters are as follows. The sampling rate is 8 kHz. The frame size is 180 samples. The spectrum filter uses the 26-bit coding scheme described previously. For the excitation, a single bit is used to indicate which codebook (unvoiced or voiced) is the source of the best excitation codeword. The unvoiced codebook uses 8 bits to specify a total of 256 random sequences, and 5 bits to encode the power gain. For the voiced codebook, a first-order pitch synthesizer is used. Seven bits are used to specify a total of 128 different pitch values. Six bits are used to specify the pitch synthesizer coefficient. The excitation information is updated twice per frame. Five bits usually are enough to encode the coefficient of a first-order pitch synthesizer. With 6 bits assigned, it is possible to extend the first-order pitch synthesizer to a third-order synthesizer. The three coefficients are then treated as a vector and quantized using a 6-bit vector quantizer.

The closed-loop analysis procedure to find the best excitation codeword for both the unvoiced codebook and the voiced codebook is exactly the same as described in the previous section.

Informal listening tests were conducted for the 2.4-kbit/s coder using the same test material as in the 4.8-kbit/s case. The speech produced by the 2.4- kbit/s coder has high intelligibility and good speaker recognizability. In terms of the speech quality, the speech produced was tested against that produced by the 4.8-kbit/s speech coder. It was found that the speech is close in quality to that of 4.8-kbit/s speech, especially when listened to through a telephone handset. The obvious difference is that the 4.8-kbit/s speech sounds smoother. The 2.4-kbit/s speech has more of the audible quantization noises.

The 2.4-kbit/s coder seems to be quite speaker independent due to the fact that WMSE is used as the distortion measure. For the ten male and female speakers in the test speech database, the performance of the coder has been very consistent.

SPECTRUM FILTER ANALYSIS-BY-SYNTHESIS

The analysis of the spectrum filter involves the open-loop computation of the filter coefficients. In [9], the concept of closed-loop spectrum filter analysis was proposed. An implementation method based on the stochastic gradient search was outlined. However, the spectrum filter instability poses a serious problem for that approach. In [15], a simple method is proposed to avoid the filter instability problem. The method is described below.

Given a computed excitation signal, and a set of candidate spectrum filters, the "optimum" spectrum filter is selected as the one filter which generates the minimum WMSE between the original speech and the reconstructed speech. The set of candidate spectrum filters is finite since the number of data bits assigned to specify the information is finite. Depending on the system complexity allowed, an exhaustive search or a partial search can be performed.

To illustrate the analysis method, an example is given here. it is assumed that the 41-bit LAR scalar quantization method is used to encode the 10 spectrum filter coefficients (Other parameter sets with other bit allocation schemes can be easily generalized.)

a. First use the open-loop method to compute the spectrum filter coefficients. Encode them by the spectrum filter coding scheme. Denote the coded filter coefficients as $\{a_1, ..., a_{10}\}$.

b. Use the closed-loop method to compute the excitation signal.

c. From the 41-bit $\{a_1, ..., a_{10}\}$ quantization table (where the bit allocation is 5, 5, 5, 5, 4, 4, 4, 4, 3, 2 bits for the 10 coefficients, respectively), select a set of candidate spectrum filters $C = C_1 \times C_2 \times ... \times C_{10} = \{\hat{a}_1, ..., \hat{a}_{10}\}$ where $\hat{a}_i \in C_i$. Set C_i contains n_i elements which represent n_i quantizing levels closest in the Euclidean distance to a_i.

d. Use the computed excitation signal as the input signal to each candidate spectrum filter, and compute the WMSE between the original speech and the reconstructed speech. The one spectrum filter which results in the minimum WMSE is selected as the best filter to represent the current frame of speech.

It is noted here that using this approach, the stability of the spectrum filter is easily guaranteed. For example, if $\{a_1, ..., a_{10}\}$ are in terms of the reflection coefficients, then the stability of each candidate filter is guaranteed as long as the magnitude of each $\hat{a}_i$ is less than unity. If $\{a_1, ..., a_{10}\}$ are in terms of the line spectrum frequencies, then the stability of each candidate filter is guaranteed as long as the natural ordering of $\{\hat{a}_1, ..., \hat{a}_{10}\}$ is ensured.

The closed-loop analysis method for the spectrum filter, coupled with the closed-loop analysis method for the excitation signal, can be extended to achieve a near optimum design under the constraint of a given speech coder structure and an available data rate. The near optimum performance, in terms of minimizing the WMSE between the original speech and the reconstructed speech, can be realized by iteratively recomputing the spectrum filter and the excitation until a performance threshold is met. The formidable computational requirement involved when using a joint parameter optimization approach can be avoided.

CONCLUSION

Analysis-by-synthesis LPC models with WMSE distortion measure have been proposed for narrowband speech coding applications. Two speech coders operating at 4.8 and 2.4 kbit/s based on the new models are presented. Extension of the closed-loop analysis method to the spectrum filter has also been described. With a better distortion measure (possibly a combination of several distortion measures), these models provide alternative methods to achieve good quality speech at very low data rates.

REFERENCES

[1]   C. G. Bell et al., "Reduction of Speech Spectra by Analysis-by-Synthesis Techniques," *J Acoust Soc Am*, Vol. 33, Dec. 1961, pp. 1725–1736.

[2]   B. S. Atal and J. R. Remde, "A New Model of LPC Excitation for Producing Natural-Sounding Speech at Low Bit Rates," International Conference on Acoustics, Speech, and Signal Processing, 1982, *Proc*, pp. 614–617.

[3]   M. R. Schroeder and B. S. Atal, "Code-Excited Linear Prediction (CELP): High Quality Speech at Very Low Bit Rates," International Conference on Acoustics, Speech, and Signal Processing, 1985, *Proc*, pp. 937–940.

[4]   J. H. Chen and A. Gersho, "Real-Time Vector APC Speech Coding at 4800 BPS with Adaptive Postfiltering," International Conference on Acoustics, Speech, and Signal Processing, 1987, *Proc*, pp. 51.3.1-51.3.4.

[5]   F. Itakura, "Line Spectrum Representation of Linear Predictive Coefficients of Speech Signals," *J Acoust Soc Am*, Vol. 57, Supplement No. 1, 1975, p. 535.

[6]   F. K. Soong and B. H. Juang, "Optimal Quantization of LSP Parameters," International Conference on Acoustics, Speech, and Signal Processing, 1988, *Proc*, pp. 394-397.

[7]   M. Yong, G. Davidson, and A. Gersho, "Encoding of LPC Spectral Parameters Using Switched-Adaptive Interframe Vector Prediction," International Conference on Acoustics, Speech, and Signal Processing, 1988, *Proc*, pp. 402–405.

[8]   G. S. Kang and L. J. Fransen, "Low-Bit-Rate Speech Encoders Based on Line-Spectrum Frequencies (LSFs)," Naval Research Laboratory Report No. 8857, Nov. 1984.

[9]   F. F. Tzeng, "Near-Toll Quality Real-Time Speech Coding at 4.8 kbit/s for Mobile Satellite Communications," 8th International Conference on Digital Satellite Communications, 1989, *Proc*, pp. 93–98.

[10]   I. A. Gerson and M. A. Jasiuk, "Vector Sum Excited Linear Prediction (VSELP)," IEEE Workshop on Speech Coding for Telecommunications, 1989, Abstracts, pp. 66-68.

[11]   "IEEE Recommended Practice for Speech Quality Measurements," *IEEE Transactions on Audio and Electroacoustics*, Vol. AU-17, No. 3, Sept. 1969, pp. 225–246.

[12]   J. P. Campbell, Jr., and T. E. Tremain, "Voiced/Unvoiced Classification of Speech With Applications to the U.S. Government LPC-10E Algorithm," International Conference on Acoustics, Speech, and Signal Processing, 1986, *Proc*, pp. 473–476.

[13]   L. R. Rabiner et al., "A Comparative Performance Study of Several Pitch Detection Algorithms," *IEEE Transactions on Acoustics, Speech, and Signal Processing*, Vol. ASSP-24, Oct. 1976, pp. 399–417.

[14]   F. F. Tzeng, "Analysis-By-Synthesis Linear Predictive Speech Coding at 2.4 kbit/s," Global Telecommunications Conference, 1989, *Proc*, pp. 34.4.1-34.4.5.

[15]   F. F. Tzeng, "An Analysis-by-Synthesis Linear Predictive Model for Narrowband Speech Coding," International Conference on Acoustics, Speech, and Signal Processing, 1990, *Proc*, to be published.

14
BINARY PULSE EXCITATION: A NOVEL APPROACH TO LOW COMPLEXITY CELP CODING

Redwan A. Salami

Department of Electronics and Computer Science
University of Southampton
Southampton SO9 5NH, U.K.

INTRODUCTION

The last decade has witnessed an acceleration in the evolution of speech coding. Near-toll quality speech coders are now available at bit rates from 4.8 to 8 kbit/s. These low bit rate speech coders are becoming increasingly needed for many future applications such as digital mobile radio telephony, mobile satellite links, and the emerging ISDN service. High quality speech at these low bit rates has become possible with the introduction of a new generation of speech coding techniques known as *analysis-by-synthesis predictive coding*. The structure of this new generation of speech coders first appeared with the introduction of *multi-pulse excited linear prediction coder (MPE-LPC)* by Atal in 1982 [1]. Since its introduction, the MPE has received much attention from researchers, and as a result, several analysis-by-synthesis coders have been developed in the bit rate range from 4.8 to 16 kbit/s with different levels of complexity [2]. These systems include the *regular-pulse excited LPC (RPE-LPC)* [3], the *code-excited LPC (CELP)* [4], and the *self-excited LPC* [5] (or *Backward Excitation Recovery (BER)* [6]). All the above mentioned coders exhibit the same structure as the originally proposed MPE-LPC [1] in which the excitation signal is optimized by minimizing the perceptually weighted error between the original and synthesized speech. They differ only in the way the excitation signal is defined and coded.

The MPE-LPC and RPE-LPC have fairly moderate complexity but fail to produce high quality speech below 8 kbit/s due to the fact that the excitation pulses have to be encoded with more than 7 kbit/s to maintain high speech quality. The CELP has proven to be the most promising candidate for producing high quality speech at bit rates as low as 4.8 kbit/s, where bit rate reduction is achieved by vector quantizing the excitation sequence using a large stochastic codebook. The CELP principle is based on the observation that the residual signal, after short-term and long-term prediction, is a noise-like signal and it is assumed, therefore, that the residual can be modelled by a zero-mean Gaussian process with slowly-varying power spectrum. A large excitation codebook (usually 1024 entries) is used and the optimum innovation sequence is determined by exhaustively searching the codebook for the address (and the corresponding gain) which minimizes the mean-squared weighted error criterion.

The need to exhaustively search the CELP excitation codebook has resulted in a computationally demanding algorithm which, until recently, hindered the real time implementation of CELP systems. The CELP complexity has been reduced by using efficient codebook structures such as ternary codebooks [7, 8], overlapping sparse codebooks [7, 9], algebraic codebooks [10], and vector sum excitation [11]. Despite the reduction in the complexity offered by the above mentioned efficient codebook structures, the exhaustive search of the excitation codebook has still to be performed.

We describe here a new approach for representing the stochastic excitation sequence, called *Transformed Binary Pulse Excitation (TBPE)* [12], which significantly reduces the excessive complexity associated with CELP coders. We will describe the coder structure and the excitation definition, before deriving an efficient excitation determination procedure which eliminates the need for the exhaustive search of an excitation codebook.

TRANSFORMED BINARY PULSE EXCITATION

Coder Structure

The Transformed Binary Pulse Excited LPC coder structure is shown in the block diagram of Figure 1. The synthesized speech $\hat{s}(n)$ is found by filtering the excitation signal $v(n)$ through the pitch synthesis filter $1/P(z)$ and the LPC synthesis filter $1/A(z)$. The error between the original speech $s(n)$ and the synthesized speech $\hat{s}(n)$ is then weighted by the perceptual weighting filter $W(z)$ and the excitation signal is determined by minimizing the mean square of the weighted error $e_w(n)$. The LPC synthesis filter is an all-pole time-varying filter of the form

$$H(z) = \frac{1}{A(z)} = \frac{1}{1 - \sum_{i=1}^{p} a_i z^{-i}}, \tag{1}$$

where $\{a_i\}$ are the LPC parameters (short-term predictor coefficients) and p is the predictor order. The filter parameters are determined outside the analysis-by-synthesis loop by minimizing the mean squared short-term prediction residual, and they are updated every 20-30 ms. Short-term prediction removes the correlation between successive speech samples, while pitch prediction removes the long-term correlation in the speech (corresponding to the pitch periodicity). Assuming one-tap long-term prediction (LTP), the pitch synthesis filter is given by

$$F(z) = \frac{1}{P(z)} = \frac{1}{1 - Gz^{-\alpha}}, \tag{2}$$

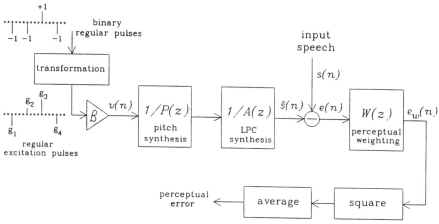

Figure 1: **Block diagram of the TBPE coder**

where G is the LTP gain and α is the LTP delay. Although the pitch synthesis filter parameters can be determined outside the closed optimization loop, a significant speech quality improvement is obtained when the long-term prediction parameters are determined inside the analysis-by-synthesis loop. The selection of the error weighting filter is perceptually motivated. Its role is to shape the spectrum of the quantization noise such that it is concentrated in the frequency regions where the speech has high energy (the formant regions), thereby masking the noise by the speech signal. The error weighting filter is given by [13]

$$W(z) = \frac{A(z)}{A(z/\gamma)} = \frac{1 - \sum_{i=1}^{p} a_i z^{-i}}{1 - \sum_{i=1}^{p} a_i \gamma^i z^{-i}}, \qquad (3)$$

where γ is a fraction between 0 and 1 which determines the degree by which the error spectrum is deemphasized in the format regions. A commonly used value of γ is 0.8.

Excitation Definition

In the TBPE approach, the excitation signal consists of a number of pseudo-stochastic pulses with predefined pulse positions. In an excitation frame of length N, we suppose that there are M nonzero pulses separated by $D - 1$ zeros, where $M = N \operatorname{DIV} D$, and DIV denotes integer division. The excitation vector is given by

$$v(n) = \beta \sum_{i=1}^{M} g_i \delta(n - m_i), \qquad n = 0, \ldots, N - 1 \qquad (4)$$

where $\delta(n)$ is the Kronecker delta, g_i are the pulse amplitudes, m_i are the pulse positions, and β is a scalar gain similar to that which appears in the CELP. As in the RPE approach [3], there are D sets of pulse positions given by

$$m_i^{(k)} = k + (i-1)D, \qquad \begin{aligned} & i = 1, \ldots, M \\ & k = 0, \ldots, D-1 \end{aligned} \qquad (5)$$

where D is the pulse spacing, and k is the position of the first pulse. In RPE coders, the optimum pulse amplitudes and first pulse position are determined by minimizing the mean-squared weighted error between the original and synthesized speech, and this requires solving a set of $M \times M$ equations D times. Further, the pulse amplitudes in RPE are each quantized with 3 bits after scaling by the maximum pulse, or the rms value of the pulses, which is quantized with 5 or 6 bits. The large number of bits needed to quantize the pulse amplitudes in RPE makes it difficult to achieve high quality speech below 9.6 kbit/s. In our TBPE approach, the pulses are pseudo-stochastic random variables, similar to the CELP concept, and they are quantized only with one bit per pulse, in addition to the scaling gain β.

The pulse amplitudes g_i, $i = 1, \ldots, M$, are not obtained from a large stochastic codebook as in the CELP approach. Instead, they are determined by the transformation of a binary vector. That is

$$\mathbf{g} = \mathbf{Ab}, \qquad (6)$$

where $\mathbf{b}$ is an $M \times 1$ binary vector with elements -1 or 1, $\mathbf{A}$ is an $M \times M$ transformation matrix, and $\mathbf{g}$ is the excitation vector containing the pulse amplitudes. The vector $\mathbf{b}$ could be one of 2^M possible binary patterns, which means 2^M different excitation vectors can be obtained using the transformation in Equation (6). Thus, this transformation is equivalent to a 2^M sized codebook with the need to only store an $M \times M$ matrix. The equivalent of smaller codebook sizes can be obtained by setting some of the binary pulses to fixed values, or by omitting some of the columns of the matrix $\mathbf{A}$. If the hypothetic codebook size is to be reduced by a factor m, either m pulses in the binary vector are made fixed (say -1), or m columns are omitted from the matrix $\mathbf{A}$ resulting into an $M \times Q$ transformation matrix and $Q \times 1$ binary vector, where $Q = M - m$. On the other hand, the equivalent of larger codebooks is obtained by utilizing several transformation matrices. Using m different transformation matrices is equivalent to a book of size 2^{M+m}.

For the special case where the transformation matrix is equal to the identity matrix $\mathbf{I}$, the excitation pulses are binary with values -1 or 1 [14] (a sparse version of the codebook proposed in [15]). In this case the excitation vectors can be viewed as 2^M points regularly distributed over the surface of a sphere in N-dimensional space. When the matrix $\mathbf{A}$ is orthogonal (i.e. $\mathbf{A}^T\mathbf{A} = \mathbf{I}$), the transformation results into a vector containing Gaussian random variables. Generating binary pulses at random and examining the distribution of the pulses g_i resulting from the orthogonal transformation reveals that the variables g_i follow a Gaussian distribution with zero mean and unit variance. Applying an orthogonal transformation to the binary vectors rotates the vectors without changing their distribution in the N-dimensional space. Both identity and orthogonal transformations exhibited similar performances.

A slight speech quality improvement was achieved when the transformation matrix was derived from a training set of RPE vectors (which contain the optimum pulse amplitudes). Any other iterative algorithm which minimizes the expectation of the perceptually weighted error between the original and synthesized speech can be used to derive the transformation matrix. In general, the transformation matrix can be chosen to obtain any desired codebook properties. Laflamme et al. [16] have recently proposed an elegant approach for defining the transformation, or the shaping matrix where the matrix is a function of the LPC filter $A(z)$, resulting in a codebook which is dynamically frequency-shaped.

EXCITATION DETERMINATION

The weighted error between the original and synthesized speech is given by

$$e_w(n) = s_w(n) - \hat{s}_w(n), \tag{7}$$

where $s_w(n)$ is the weighted input speech and $\hat{s}_w(n)$ is the weighted synthesized speech. The weighted synthesized speech can be written as

$$\hat{s}_w(n) = \sum_{i=0}^{n} v(i) h_c(n - i) + \hat{s}_0(n), \tag{8}$$

where $h_c(n)$ is the impulse response of the filter $C(z)$, which is the combination of the pitch synthesis, LPC synthesis and error weighting filters, given by

$$C(z) = F(z) H(z) W(z) = \frac{1}{P(z)} \frac{1}{A(z/\gamma)}, \tag{9}$$

and $\hat{s}_0(n)$ is the zero-input response of the combined filter $C(z)$. $\hat{s}_0(n)$ is found by setting the excitation input $v(n)$ to zero and computing the output of the combined filter $C(z)$ due to its initial states.

The pitch synthesis filter, $1/P(z)$, contributes to the impulse response $h_c(n)$ of the combined filter only after the αth sample, where α is the LTP delay. If the pitch delay is restricted to values larger than the excitation frame length N, $h_c(n)$ becomes the impulse response of the filter $1/A(z/\gamma)$. For large excitation frames (e.g. $N = 60$ as in 4.8 kbit/s coding), it is preferred not to restrict the pitch delay α to values larger than N. The LTP delay range from 20 to 147 is commonly used. For delays $\alpha < N$ the adaptive codebook concept, or the virtual search procedure described in [9] can be used.

From Equations (7) and (8), the weighted error can be written as

$$e_w(n) = x(n) - \sum_{i=0}^{n} v(i) h_c(n - i), \tag{10}$$

where

$$x(n) = s_w(n) - \hat{s}_0(n) \tag{11}$$

is the weighted input speech after subtracting the zero-input response of the combined filter $C(z)$. Now, substituting the excitation signal $v(n)$ from Equation (4) into Equation (10) gives

$$
\begin{aligned}
e_w(n) &= x(n) - \sum_{i=0}^{n} \beta \sum_{k=1}^{M} g_k \delta(n - m_k) h_c(n - i), \\
&= x(n) - \beta \sum_{k=1}^{M} g_k h_c(n - m_k), \qquad n = 0, \ldots, N-1, \tag{12}
\end{aligned}
$$

where $h(n - m_k) = 0$ for $n < m_k$. The excitation parameters are determined by minimizing the mean square of the weighted error $e_w(n)$ which is given by

$$E = \sum_{n=0}^{N-1} \left[x(n) - \beta \sum_{i=1}^{M} g_i h_c(n - m_i) \right]^2. \tag{13}$$

Setting $\partial E / \partial \beta$ to zero leads to

$$\beta = \frac{\sum_{i=1}^{M} g_i \psi(m_i)}{\sum_{i=1}^{M} \sum_{j=1}^{M} g_i g_j \phi(m_i, m_j)}, \tag{14}$$

where ψ is the correlation between $x(n)$ and the impulse response $h_c(n)$, given by

$$\psi(i) = \sum_{n=i}^{N-1} x(n) h_c(n - i), \tag{15}$$

and ϕ is the autocorrelation of the impulse response $h_c(n)$ given by

$$\phi(i, j) = \sum_{n=\max(i,j)}^{N-1} h_c(n - i) h_c(n - j). \tag{16}$$

By substituting Equation (14) into Equation (13), the minimum mean squared weighted error between the original and synthesized speech can be written as

$$
\begin{aligned}
E &= \sum_{n=0}^{N-1} x^2(n) - \beta \sum_{i=1}^{M} g_i \psi(m_i), \\
&= \sum_{n=0}^{N-1} x^2(n) - \frac{\left[\sum_{i=1}^{M} g_i \psi(m_i) \right]^2}{\sum_{i=1}^{M} \sum_{j=1}^{M} g_i g_j \phi(m_i, m_j)}. \tag{17}
\end{aligned}
$$

Equation (17) can be written in matrix form as

$$E = \mathbf{x}^T \mathbf{x} - \frac{(\mathbf{\Psi}^T \mathbf{g})^2}{\mathbf{g}^T \mathbf{\Phi} \mathbf{g}} = \mathbf{x}^T \mathbf{x} - \frac{(\mathbf{\Psi}^T \mathbf{A} \mathbf{b})^2}{\mathbf{b}^T \mathbf{A}^T \mathbf{\Phi} \mathbf{A} \mathbf{b}}, \tag{18}$$

where $\mathbf{x}$ is an $N \times 1$ vector, $\boldsymbol{\Psi}$ and $\mathbf{b}$ are $M \times 1$ vectors, and $\boldsymbol{\Phi}$ is an $M \times M$ symmetric matrix with elements $\phi(m_i, m_j)$, $i, j = 1, \ldots, M$. The autocorrelation approach can be used to express $\phi(m_i, m_j) = \phi(|m_i - m_j|)$. In this case, the matrix $\boldsymbol{\Phi}$ is reduced to a Toeplitz symmetric matrix with diagonal $\phi(0)$ and off-diagonals $\phi(D), \phi(2D), \ldots, \phi([M-1]D)$, respectively. Defining

$$z = A^T \boldsymbol{\Psi}, \tag{19}$$

and

$$\Theta = A^T \boldsymbol{\Phi} A, \tag{20}$$

Equation (18) becomes

$$E = \mathbf{x}^T \mathbf{x} - \frac{(\mathbf{z}^T \mathbf{b})^2}{\mathbf{b}^T \Theta \mathbf{b}}. \tag{21}$$

To determine the optimum innovation sequence, one could exhaustively search through all possible binary patterns and select the pattern which maximizes the second term in Equation (21). This can be easily done using a Gray code counter [11, 12], where the Hamming distance between adjacent binary patterns is 1. Using a Gray code counter, the value of

$$\mathcal{N} = \mathbf{z}^T \mathbf{b} \tag{22}$$

in the numerator of the second term of Equation (21) is updated by

$$\mathcal{N}_k = \mathcal{N}_{k-1} + 2z_j b_j^{(k)}, \tag{23}$$

where k is the index of the Gray code and j is the index of the pulse which has been toggled. The energy of the filtered excitation

$$\mathcal{D} = \mathbf{b}^T \Theta \mathbf{b} \tag{24}$$

can be similarly updated by

$$\mathcal{D}_k = \mathcal{D}_k + 4b_j^{(k)} \sum_{\substack{i=1 \\ i \neq j}}^{M} b_i^{(k)} \theta(i, j). \tag{25}$$

Equations (23) and (25) offer a very efficient method to exhaustively search for the best binary excitation pattern. For every new pattern, about $M + 3$ operations are needed to update the second term in Equation (21). However, inspection of Equation (21) leads to an even simpler excitation determination procedure in which the exhaustive search is completely ruled out.

A closer look at the autocorrelation matrix $\boldsymbol{\Phi}$ in Equation (18) suggests that it is strongly diagonal, because the magnitude of $\phi(nD)$ (D is usually 4) is much less than $\phi(0)$. As $A^T A$ is equal to the identity matrix $\mathbf{I}$, Θ of Equation (20) is also strongly diagonal. Therefore, as $\mathbf{b}^T \mathbf{b}$ is constant ($= M$), the denominator

152

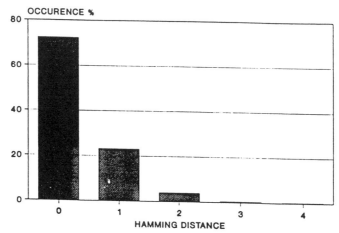

Figure 2: **Hamming distance between the Binary code determined by Equation (24) and the one determined by exhaustive search.**

in Equation (21) can be approximated by a constant equal to $M\phi(0)$. Thus, minimization of the error in Equation (21) can be performed by maximizing the numerator, i.e. maximizing the absolute value of the term $z^T b$, and this can be simply done by choosing the pulses to be equal to the signs of z [14], i.e.

$$b_i = \text{sign}\{z_i\}, \qquad i = 1, \ldots, M. \tag{26}$$

Equation (26) offers an extremely simple excitation determination procedure in which no exhaustive search is needed.

Figure 2 shows the histogram of the Hamming distance between the binary vector b_0 determined using the simple relation of Equation (26) and the optimum binary vector b_{opt} determined by the exhaustive search through all the possible binary vectors for the one which minimizes the mean squared weighted error. The exhaustive search is performed with the joint optimization of the binary vector and excitation gain, where the mean-squared weighted error is expressed by [17]

$$E = x^T x - \hat{\beta}(2\mathcal{N} - \hat{\beta}\mathcal{D}), \tag{27}$$

where $\hat{\beta}$ is the quantized value of the gain $\beta = \mathcal{N}/\mathcal{D}$. We notice that the optimum vector is properly computed 72% of the time since the Hamming distance between b_0 and b_{opt} over that time is zero. For about 23% of the time the Hamming distance is one, which means that whenever Equation (26) fails to determine the optimum binary vector the computed vector differs from the optimum one by only one sign. This observation has led us to the following efficient search procedure. An initial binary vector is first determined using Equation (26), then the second term of Equation (27) is evaluated using the initial vector and the other M vectors which have a Hamming distance of one

Codebook Population	SEGSNR (dB)
Gaussian	14.03
Sparse	14.06
Ternary	13.81
Overlapping sparse	14.09
Binary regular pulses	13.71
Transformed pulses	13.85

Table 1: SEGSNR for the TBPE and different CELP approaches.

from the initial vector. In this efficient procedure the search of a book of size 2^M is reduced to searching a local book of size $M + 1$, yet guaranteeing that 95% of the time the optimum binary vector is identified. Notice that for the $M + 1$ sized local codebook the efficient Gray code procedure is used to update $\mathcal{N}_k$ and $\mathcal{D}_k$ as in equations (23) and (25).

CODER PERFORMANCE

The TBPE coder was evaluated at different bit rates in the range from 4.8 to 8 kbit/s. Good communications quality speech was obtained at 4.8 kbit/s and near-toll quality speech was obtained at 8 kbit/s. Table 1 shows a comparison between the segmental SNRs of the TBPE and different CELP approaches at 7.8 kbit/s. The 20 ms speech frame is divided into 5 subframes of 32 samples length. For the CELP approaches a 9-bit stochastic codebook was used with the gain quantized with 5 bits (4 bits for the magnitude and 1 for the sign), and with the index and gain jointly optimized [17]. The sparse and ternary codebooks were obtained by center-clipping the unit variance Gaussian sequences at 1.2 (77% sparsity). The overlapping codebook is a shift by 2 ternary codebook. In case of TBPE, 8 binary pulses are used (decimation factor of 4) with the first pulse position quantized with 2 bits and the gain with 4 bits (the BPE gain is always positive as the sign information is carried by the pulses themselves). It is clear from the SNR figures in Table 1 that the objective quality of TBPE is very close to CELP. In fact, subjective listening tests did not show any difference in speech quality in either case. When the excitation vectors were not transformed (binary regular pulses), there was slight degradation in speech quality compared to the transformed case. Using untransformed binary pulses reduces the complexity since computing $\mathbf{z}$ and Θ in Equations (19) and (20)

154

is not needed. In 8 kbit/s coding, better performances were obtained using two transformation matrices, where one of the matrices was set to the identity matrix (no transformation) to reduce the complexity and storage requirement.

The TBPE coder has several advantages over the CELP. The main advantage is the significant reduction in the computational complexity. As shown in previous sections, the search of an excitation codebook of size 2^M is reduced to searching a local book of size $M+1$. In case of untransformed vectors, computing the second term of Equation (21) requires about $M^2 + M$ instructions and for the next M vectors in the local codebook, about $M(M+3)$ instructions are needed to update the term. This is repeated D times (for all possible first pulse positions) which results in a total of $2M+4$ instructions per speech sample. Using the transformation requires the computation of z and Θ in Equations (19) and (20). z needs M^2 instructions and it is computed D times, which results in M instructions per speech sample. Θ is computed only once since it is independent of the first pulse position, and it requires about $(3M^2 + M)/(2D)$ instructions per speech sample. For $M = 8$ and $D = 4$ and with no transformation, about 20 instructions per speech samples are required to jointly optimize the binary vector and the first pulse position, and this number rises to 50 when a transformation is used.

The second advantage of the TBPE is the ability to improve the speech quality by utilizing several transformation matrices, which is equivalent to using a very large excitation codebook; a task which becomes impractical with the CELP when the codebook address exceeds 10 bits. Another advantage is the reduction in the storage requirements of the excitation codebook. The equivalent of a 2^M sized codebook is obtained by storing an $M \times M$ matrix, and the storage is eliminated in case of untransformed pulses. Finally, the TBPE possesses an inherent robustness against transmission errors. As the excitation pulses are directly derived from the transmitted binary vector, a transmission error in the binary vector will cause little change in the transformed excitation vector. However, in CELP coders a transmission error in the codebook address will cause the receiver to use an entirely different excitation vector.

Embedding error-correcting coding into the TBPE coder has been investigated [18]. A coder at a gross bit rate of 11.4 kbit/s was studied over mobile radio channels, where a 12 dB channel SNR was sufficient to maintain good speech quality.

References

[1] B.S.Atal and J.R.Remde, "A new model for LPC excitation for producing natural-sounding speech at low bit rates," Proc. ICASSP'82, pp. 614-617.

[2] P.Kroon and E.F.Deprettere, "A class of analysis-by-synthesis predictive coders for high quality speech coding at rates between 4.8 and 16 kbits/s,"

IEEE J. Selected Areas in Commun., vol. 6, no. 2, pp. 353-363, Feb. 1988.

[3] P.Kroon, E.F.Deprettere, and R.J.Sluyter, "Regular-pulse excitation - A novel approach to effective efficient multipulse coding of speech," IEEE Trans. ASSP, vol. 34, no. 5, pp. 1054-1063, Oct. 1986.

[4] M.R.Schroeder and B.S.Atal, "Code-excited linear prediction (CELP): high-quality speech at very low bit rates," Proc. ICASSP'85, pp. 937-940.

[5] R.C.Rose and T.B.Barnwell III, "Quality comparison of low complexity 4800 bps self excited and code excited vocoders," Proc. ICASSP'86, pp. 2375-2378.

[6] N.Gouvianakis and C.Xydeas, "Advances in analysis by synthesis LPC speech coders," J. IERE, vol. 57, no. 6 (supplement), pp. S272-S278, Nov./Dec. 1987.

[7] D.Lin, "New approaches to stochastic coding of speech sources at very low bit rates," Signal Processing III: Theories and Applications (Proceedings of EUSIPCO-86), pp. 445-448, 1986.

[8] C.S.Xydeas, M.A.Ireton and D.K.Baghbadrani, "Theory and real time implementation of a CELP coder at 4.8 and 6.0 Kbits/sec using ternary code excitation," Proc. IERE 5th Int. Conf. on Digital processing of signals in Commun., Univ. of Loughborough, pp. 167-174, Sept. 20-23 1988.

[9] W.B.Kleijn, D.J.Krasinsky, and R.II.Ketchum, "An efficient stochastically excited linear predictive coding algorithm for high quality low bit rate transmission of speech," Speech Communication, vol. 7, no. 3, pp. 305-316, Oct. 1988.

[10] J-P.Adoul et al, "Fast CELP coding based on algebraic codes," Proc. ICASSP'87, pp. 1957-1960.

[11] I.A.Gerson and M.A.Jasiuk, "Vector Sum Excited Linear Prediction (VSELP)," IEEE Workshop on Speech Coding for Telecomm., Sept. 5-8 1989, Vancouver, Canada.

[12] R.A.Salami and D.G.Appleby, "A new approach to low bit rate speech coding with low complexity using binary pulse excitation (BPE)," IEEE Workshop on Speech Coding for Telecomm., Sept. 5-8 1989, Vancouver, Canada.

[13] B.S.Atal and M.R.Schroeder, "Predictive coding of speech signals and subjective error criteria," IEEE Trans. ASSP, vol. 27, pp. 247-254, June 1979.

[14] M.Lever and M.Delprat, "RPCELP: A high quality and low complexity scheme for narrow band coding of speech," Proc. EUROCON, pp. 24-27, June 1988.

[15] A.Le Guyader, D.Massaloux, and F.Zurcher, "A robust and fast CELP coder at 16 kbit/s," Speech Communication, Vol. 7, 217-226, 1988.

[16] C.Laflamme, J-P.Adoul, H.Y.Su, and S.Morissette, "On reducing computational complexity of codebook search in CELP coder through the use of algebraic codes," in ICASSP'90, April 3-6 1990, Albuquerque, New Mexico, U.S.A.

[17] J.P.Campbell, Jr., T.E.Tremain, and V.C.Welch, "The DoD 4.8 kbps standard (proposed federal standard 1016)," To appear in this book.

[18] R.A.Salami, K.H.H.Wong, R.Steele and D.G.Appleby, "Performance of error protected binary pulse excitation coders at 11.4 kb/s over mobile radio channels," in ICASSP'90, April 3-6 1990, Albuquerque, New Mexico, U.S.A.

15

SPEECH CODING WITH DYNAMIC BIT ALLOCATION (MULTIMODE CODING)

Tomohiko Taniguchi †, Yoshinori Tanaka †, Robert M. Gray ††

† Fujitsu Laboratories Ltd.,
1015 Kamikodanaka, Nakahara-ku
Kawasaki 211, Japan

†† Information Systems Laboratory,
Department of Electrical Engineering,
Stanford University, Stanford, CA 94305

INTRODUCTION

Recent research on narrow to medium band speech coding has concentrated on the efficient transmission of the excitation parameters. Using model based excitation analysis techniques, coders with Analysis-by-Synthesis configuration, such as Multi-Pulse Excited LPC (MPLPC) [1] or Code Excited LPC (CELP) [2], can successfully reduce the transmission bit rate for excitation parameters and can provide good speech quality at bit rates around 8 kb/s.

However, below this bit rate, the transmission bit rate for spectral parameters (such as LPC parameters) becomes a large portion of the total transmission bit rate, and it is difficult to spare enough bit rate for transmitting excitation parameters. Thus, a careful consideration on the transmission bit allocation between excitation and spectral parameters and an efficient transmission of spectral parameters are necessary to maintain reproduced speech quality.

Because speech characteristics vary with time, applying time-varying bit allocations to the transmission parameters is an efficient way of transmitting speech through a fixed-rate channel. As a way of achieving the optimum bit allocation with time, we proposed a multimode coding scheme [3] which has several coders with different transmission bit allocations, and dynamically selects the optimum coding mode based on an evaluation of reproduced speech quality.

The basic principle of multimode coding is the same as that of the multiquantizer (MQ) coding scheme [4,5] which one of the authors formerly proposed as a scheme to improve the performance of low bit rate ADPCM. In the ADPCM-MQ system, several ADPCM coders with different quantization characteristics are operated in parallel, and the coder which provides the best speech quality is selected after evaluating the reproduced speech quality of each coder. In this sense, both multiquantizer and multimode coding schemes feature delayed decision coding, and can be considered examples of universal coding.

As is well-known, the characteristic of spectral parameters is slowly varying compared with that of excitation parameters, and there is a strong correlation between the spectral parameters of consecutive frames. For the efficient transmission of spectral parameters, it is meaningful to utilize this inter-frame correlation of LPC parameters. Thus, we exploit variable frame rate transmission of LPC parameters [6] jointly with multimode coding. The idea of variable frame rate coding of LPC parameters is that LPC parameters are updated only when perceivable changes between the current and the previous frame occur. Otherwise, the previous frame parameters are repeated in the current frame. This scheme is similar to delta coding [7] used in packetized speech transmission.

Although multimode coding itself contributes very much to improving the performance of low bit rate CELP coders [8], there is still another important thing on which special consideration is necessary. The subjective quality of reproduced speech is quite dependent on the characteristics of the distortion in the reproduced speech. The distortion caused by reducing the transmission rate of spectral parameters can be characterized as spectral distortion, and that of excitation parameters can be characterized as waveform distortion. Thus, in multimode coding, not only the waveform distortion (perceptually weighted mean square error) but also spectral distortion should be evaluated to select the optimum bit assignment (coding mode). We use the SNR of LAR (log-area-ratio) parameters as a criterion to evaluate spectral distortion, and introduce combined mode decision by both waveform and spectral distortion measure [8,9].

First we will describe the multimode coding algorithm and its application to CELP coders. Then we will give simulation results for the performance of 4.8 and 8.0 kb/s multimode CELP coders. We will describe the distortion characteristics in multimode coding and introduce a combined mode decision using a spectral distortion measure in addition to the waveform distortion measure to improve the subjective quality of the reproduced speech.

CODING ALGORITHM

As shown in Fig. 1, multimode coding has several speech coders, each of which has a different bit assignment for the excitation and spectral parameters. In each frame, these coders process the speech signal in parallel, and the coder which provides the best reproduced speech quality is selected. The assignment of the transmission information is dynamically controlled by switching between coders based on an evaluation of the reproduced speech quality. The index of the selected coder, which specifies the bit assignment of the frame, is transmitted with the coded information. At the decoder, the transmitted information is decoded while switching the decoder according to the coder

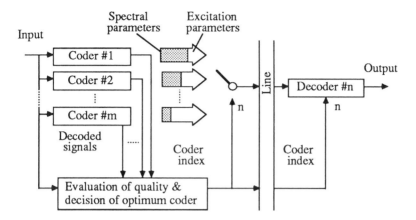

Figure 1. Multimode coding

index. In this algorithm, the larger the number of coders operating in parallel, the better the bit assignment. Having too many coders, however, will increase the additional information specifying the optimum coding mode and decrease transmission efficiency. The increased computational complexity also becomes a problem in implementation.

Considering the computational complexity, we adopted multimode coding to CELP in its simplest form, that is, only two coding modes were used. Here, variable frame rate encoding of LPC parameters was also applied for an efficient transmission of spectral parameters. (This variable frame rate transmission of LPC parameters is especially effective in low bit rate speech coders with coarse LPC quantizers in which quantized LPC parameters often take the same value as previous frame.) As bit assignments of these two coding modes, the bit assignments with and without the transmission of LPC parameters were used respectively, and one additional information bit was added per frame to specify the coding mode. The block diagram of the CELP coder with multimode coding is shown in Fig. 2. In this figure, the A-mode coder is just the same as a conventional CELP coder which carries out LPC analysis and transmits the parameters every frame. On the other hand, the B-mode coder avoids transmitting LPC parameters by using the same coefficients as the previous frame, and increases the bits allocated to the excitation instead. In each frame, the mode selection takes place based on an evaluation of the reproduced speech quality, and the assignment of transmission information is dynamically controlled by switching between the two modes. The coefficient memory is updated only when the A-mode is selected, and the stored coefficients are unchanged during B-mode coding.

There are two dominant factors which determine the reproduced speech quality in predictive coding. One is a prediction gain of the predictor, and the other is a quantization gain of the quantizer. The signal-to-noise ratio of the reproduced speech is assumed to be the sum of these two factors. In our multimode CELP coder, the prediction gain degradation is caused by not updating the LPC parameters in B-mode, while the quantization gain improvement is achieved by allocating the extra bits to the excitation. Thus, if the quantization gain improvement exceeds the prediction gain degradation, the

160

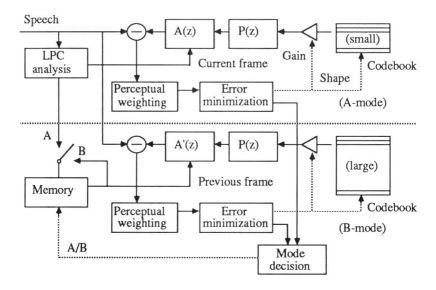

Figure 2. Multimode CELP coder
(A(z): Short-term predictor, P(z): Long-term predictor)

resultant SNR in B-mode is greater than that of A-mode. To decide the coding mode, the SNR of each coding mode should be evaluated in each frame. If the SNR of A-mode is greater than that of B-mode, the A-mode coder is selected. Otherwise the B-mode coder is selected, and the transmission bits are all allocated to the residual signal.

Similar ideas to our multimode coding were independently developed and reported by Yong and Gersho [10] and Jayant and Chen [11] recently. In [10], the likelihood ratio was used as a distortion criterion for mode decision and the optimum coding mode was selected in a forward fashion, whereas the mode decision in our system is made in backward fashion using a criterion based on the perceptually weighted error between original and reconstructed speech. As stated before, we tested a two-mode CELP coder in which the transmission of spectral parameters was drastically switched completely on or completely off, but an example of a four-mode CELP coder using intermediate sets of bit assignments has been presented in [11].

PERFORMANCE OF MULTIMODE CELP CODERS

The performances of multimode CELP coders at bit rates of 4.8 and 8.0 kb/s were evaluated by computer simulation. The bit assignments used in the coders are listed in Table 1. The bit assignment for the A-mode is simply the bit assignment of a conventional CELP coder. In both coders, 10th order LPC analysis was carried out every 20 ms (using a 25 ms Hamming window of the input speech), and extracted LPC parameters were coded as log-area-ratio (LAR) parameters. In the 8.0 kb/s system, the extra available bits achieved by not transmitting LPC coefficients are mainly used to

Table 1.   Transmission bit assignments

	A-mode		B-mode	
	bits / ms	kb/s	bits / ms	kb/s
LPC parameters	28 / 20	1.4	0 / 20	0.0
Long-term predictor Delay	7 / 5	1.4	7 / 5	1.4
Gain	4 / 5	0.8	5 / 5	1.0
Stochastic codebook Shape	6 / 2.5	2.4	9 / 2.5	3.6
Gain	5 / 2.5	2.0	5 / 2.5	2.0
Total	8.0 kb/s		8.0 kb/s	

	A-mode		B-mode	
	bits / ms	kb/s	bits / ms	kb/s
LPC parameters	26 / 20	1.3	0 / 20	0.0
Long-term predictor Delay	7 / 10	0.7	7 / 5	1.4
Gain	4 / 10	0.4	4 / 5	0.8
Stochastic codebook Shape	7 / 5	1.4	8 / 5	1.6
Gain	5 / 5	1.0	5 / 5	1.0
Total	4.8 kb/s		4.8 kb/s	

increase the size of excitation codebook. On the other hand, the extra available bits are mainly allocated to the long-term predictor parameters in the 4.8 kb/s system.

The results of the computer simulation are summarized in Table 2. Applying the multimode coding to CELP, the average transmission rate of the LPC parameters was reduced to as little as 13% of that of the conventional CELP coder. This introduces a significant improvement in segmental SNR (SNRseg) by increasing the bits for excitation parameters instead of transmitting LPC parameters. Compared with the conventional CELP coder, approximately 2 dB of SNRseg improvement was achieved. On the other hand, however, the LPC Cepstrum Distance (LPC-CD) calculated between original and reproduced speech was degraded 1.2-1.7 dB by introducing multimode coding.

The characteristics of the SNRseg and LPC-CD with time are shown in Fig. 3. As shown in this figure, the SNRseg characteristic of the multimode CELP coder (solid line) is superior to that of the conventional CELP coder (dotted line) while B-mode is selected. However, the LPC-CD characteristic is inferior to that of conventional CELP. (While A-mode is selected, these two characteristics takes the same value in both coders, because the A-mode bit assignment is the same as that of conventional CELP.) Apparently, the mode decision based on SNR evaluation sometimes selects the B-mode even when the degradation of the prediction gain is significant, because even this large degradation doesn't exceed the B-mode improvement of the quantization gain. This causes spectral distortion and degrades the subjective quality, even though the waveform distortion is minimized in the mean square error sense. In informal listening tests, the speech repro-

162

duced by the multimode CELP coder sounds somewhat 'synthetic', although it contains less audible quantization noise.

Table 2. Performance of multimode CELP coder

8.0 kb/s	SNRseg (dB)	SNR (dB)	CD (dB)	LPC bit rate (bits/s)	B-mode (%)
CELP	15.06	15.01	2.8	1,400	0.0
CELP-MMC	17.21	16.80	4.0	195	86.1

4.8 kb/s	SNRseg (dB)	SNR (dB)	CD (dB)	LPC bit rate (bits/s)	B-mode (%)
CELP	10.88	10.59	2.9	1,300	0.0
CELP-MMC	13.06	12.93	4.6	168	87.1

MMC : Multimode coding

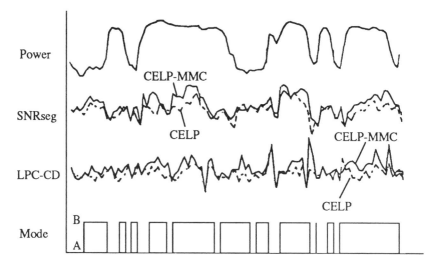

Figure 3. SNRseg and LPC cepstrum distance characteristics

DISTORTION CHARACTERISTICS IN MULTIMODE CODING

To solve this problem and to achieve subjectively better reproduced speech with the multimode CELP coder, we investigated the distortion characteristics in multimode coding. Figure 4 shows the distortion characteristics of the multimode CELP coder as a function of the frequency of B-mode selection. In this figure, B-mode ratio of 0 %

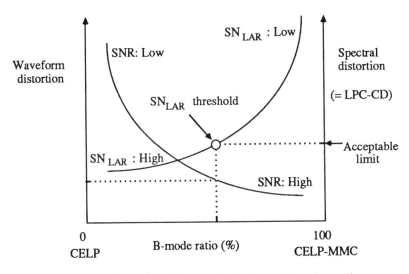

Figure 4. Distortion characteristics in multimode coding

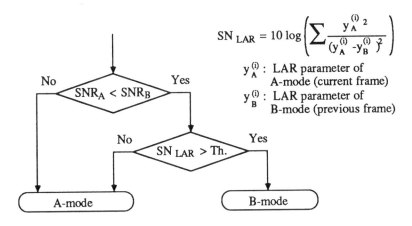

$$SN_{LAR} = 10 \log \left(\sum \frac{y_A^{(i)\,2}}{(y_A^{(i)} - y_B^{(i)})^2} \right)$$

$y_A^{(i)}$: LAR parameter of A-mode (current frame)

$y_B^{(i)}$: LAR parameter of B-mode (previous frame)

Figure 5. Combined mode decision

corresponds to the conventional CELP coder. Introducing multimode coding, the waveform distortion decreases and the SNR is improved as the B-mode ratio increases. However, the spectral distortion increases as the B-mode ratio increases, because LPC parameters are not transmitted and not updated during B-mode coding.

To evaluate the spectral distortion and suppress it to an acceptable level, we introduced an SNR of LAR parameters (SNLAR) as a second criterion for the mode decision. Our new mode decision scheme is a combined mode decision by SNR and SNLAR as shown in Fig. 5. (SNLAR is known as a good measure of log-spectral distortion [12]. Our multimode CELP coder evaluates it according to the formula in Fig. 5.) In this scheme, the SNR of each mode is evaluated first, and if the SNR of A-mode is greater

164

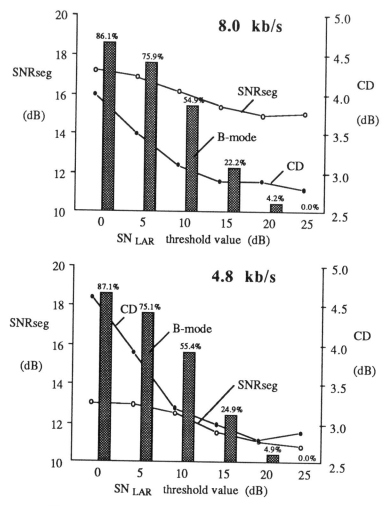

Figure 6. Performace of the multimode CELP coder
with combined mode decision

than that of B-mode, then A-mode is selected. Otherwise, the SNLAR is evaluated, and
only when the value exceeds the predetermined threshold level, B-mode is selected.

Simulation results for 4.8 and 8.0 kb/s multimode CELP coders using this
combined mode decision are summarized in Fig. 6. In the simulation, we varied the
threshold level of SNLAR from 0 dB to 25 dB. The threshold level of 0 dB corresponds
to the multimode CELP using the mode decision by SNR. At the threshold value of 25
dB, B-mode ratio is 0 %, and this case corresponds to the conventional CELP coder. The
best SNRseg value is achieved when the threshold value is 0 dB (multimode CELP coder
without SNLAR evaluation), and the SNRseg is gradually degraded as the threshold value
becomes higher. On the other hand, LPC-CD is rapidly degraded when the threshold
value is less than 10 dB. Considering these aspects, the threshold value should be set at

10-15 dB, so that we can achieve reproduced speech with high SNRseg and low spectral distortion. Informal listening tests also show that the subjective quality of the reproduced speech at the threshold value of 10 dB is the best in both 4.8 and 8.0 kb/s multimode CELP coders.

CONCLUSION

We proposed a new coding algorithm for narrow and medium band speech coding called multimode coding. Multimode coding dynamically balances the transmission bit rate between the excitation parameters and the spectral parameters. In this coding algorithm, dynamic bit allocation is carried out by selecting the optimum bit assignments frame by frame among several coders with different bit assignments based on an evaluation of the reproduced speech quality. When we applied this algorithm to 4.8 and 8.0 kb/s CELP coders, we achieved approximately 2 dB of SNRseg improvement over the conventional CELP coders. Furthermore, the SNR of the LAR parameters was introduced as an additional criterion for the mode decision, and the improvement of the reproduced speech quality was confirmed by informal listening tests.

ACKNOWLEDGEMENT

We would like to express our thanks to M. A. Johnson of M.I.T. for his thorough review and constructive comments. We would also like to thank K. Murano and S. Unagami of Fujitsu Laboratories Ltd. for their valuable suggestions and consistent encouragement.

REFERENCES

[1] B. S. Atal and J. R. Remde, "A New Model of LPC Excitation for Producing Natural-Sounding Speech at Low Bit Rates," Proc. ICASSP'82, pp. 614-617, May 1982.

[2] M. R. Schroeder and B. S. Atal, "Code-Excited Linear Prediction (CELP): High Quality Speech at Very Low Bit Rates," Proc. ICASSP'85, pp. 937-940, Mar. 1985.

[3] T. Taniguchi, S. Unagami, and R. M. Gray, "Multimode Coding: A Novel Approach to Narrow and Medium Band Speech Coding," J. Acoust. Soc. Am., Supplement 1, Vol. 84, pp. S12, Nov. 1988.

[4] T. Taniguchi, K. Iseda, S. Unagami, and S. Tominaga, "A High Efficiency Speech Coding Algorithm based on ADPCM with Multiquantizer," Proc. ICASSP'86, pp. 1721-1724, Apr. 1986

[5] T. Taniguchi, S. Unagami, K. Iseda, and S. Tominaga, "ADPCM with a Multiquantizer for Speech Coding," IEEE, J. Select. Areas Commun. vol. SAC-6, pp. 410-424, Feb. 1988.

[6] R. Viswanathan, J. Makhoul, R. Schwartz, and A. W. F. Huggins, "Variable Frame

Rate Transmission: A Review of Methodology and Application to Narrow-Band LPC Speech Coding," IEEE, Trans. Commun. vol. COM-30, pp. 674-686, Apr. 1982.

[7] D. T. Magill, "Adaptive Speech Compression for Packet Communication Systems," Proc. Nat. Telecommun. Conf., pp. 29D1-29D5, Nov. 1973.

[8] T. Taniguchi, S. Unagami, and R. M. Gray, "Multimode Coding: Application to CELP," Proc. ICASSP'89, pp. 156-159, May 1989.

[9] T. Taniguchi, S. Unagami, Y. Tanaka, F. Amano, and Y. Ohta, "Distortion Measure for Mode Decision of Multimode CELP Coder (Dynamics of Waveform Distortion and Spectral Distortion)," IEEE Speech Coding Workshop, pp. 45-48, Sep. 1989.

[10] M. Yong and A. Gersho, "Vector Excitation Coding with Dynamic Bit Allocation," Proc. GLOBECOM'88, pp. 290-294, Dec. 1988.

[11] N. S. Jayant and J.-H. Chen, "Speech Coding with Time-Varing Bit Allocations to Excitation and LPC Parameters," Proc. ICASSP'89, pp. 65-68, May 1989.

[12] B. S. Atal, "Stochastic Gaussian Model for Low-bit-rate Coding of LPC Area Parameters," Proc. ICASSP'87, pp. 2404-2407, Apr. 1987.

16

7 KBPS - 7 MIPS - HIGH QUALITY ACELP
FOR CELLULAR RADIO

C.Galand

IBM Thomas J.Watson Research Center
P.O. Box 704
Yorktown Heights, NY 10598

J.Menez

LASSY UA CNRS 1376
41 Bd Napoleon III
06041 Nice Cedex - France

M.Rosso, F.Bottau, B.Pucci

IBM Laboratory
06610 La Gaude - France

INTRODUCTION

In this chapter, we propose a modification of the classical CELP (Code Excited Linear Predictive) algorithm in order to reduce its computational complexity and required memory size, while preserving the quality of the reconstructed speech.

Rather than performing the individual weighting of each candidate sequence, we suggest a global implementation of the vocal tract weighting function at the code-book level, thanks to the use of an adaptive code-book. As a result, the analysis-by-synthesis procedure does not require the processing of all the candidate sequences through the synthesis and weighting filters, and therefore the complexity requirement of the algorithm is much reduced.

The chapter is organized as follows. In the first section, we briefly remind the classical CELP algorithm, as well as the main directions of the re-

168

search works which have been largely developed in the past years to overcome the CELP native complexity, and we introduce our own approach to this problem. In the second section, we describe the basic technique which can be used to implement the weighting at the code-book level. We also give an alternate approach to this implementation, and we report the results of the application of this technique to two architectures: the CELP and the SELP (Stochastically Excited Linear Predictive coder). In the last section, we describe an implementation of a 7.2 kbps adaptive code-book CELP (ACELP) on a fixed point DSP, which requires a processing capability of 7 MIPS. The speech quality is reported from formal quality tests which show that the 7.2 kbps ACELP algorithm presents the same level of quality as the GSM algorithm which operates at 13 kbps with a processing requirement of 4 MIPS.

CELP DEVELOPMENTS

Originally proposed by B.S.Atal and M.R.Schroeder [1,2], the Code Excited Linear Prediction (CELP) algorithm is a method using vector quantization for high quality speech coding at low bit rates.

Fig.1. Basic CELP (Atal and Schroeder, 1984)

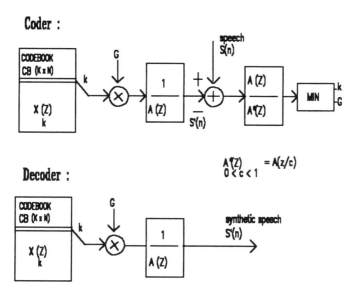

Basic CELP

In CELP (Fig.1), the speech is analyzed by blocks of N samples. The best approximation to each block s(n),(n = 1,...,N) of speech signal is searched through stored code-book sequences x(k,n) (n = 1,...,N; k = 1,...,K). The search through the code-book is performed by an iterative analysis-by-synthesis procedure in which an adaptive noise spectral shaping is achieved by using a perceptual weighting filter $A(Z)/A^*(Z)$ where $A^*(Z) = A(Z/c)$, $0 < c < 1$ [1]. The criterion which is used to select the candidate sequence of the code-book is the weighted mean squared error between the original speech signal s(n) and the synthetic speech signal s'(n) (Fig.1). The sequence giving the minimum weighted error is selected, and a gain G is computed. Both the index k and the gain G are transmitted to the receiver, along with the coefficients of the vocal tract filter A(Z). The full search requires every sequence of the code-book to be processed through the short-term and weighting filters, thus involving a high computational load.

CELP Research

Although the original CELP algorithm has been shown to achieve high quality speech coding at low bit rates, its huge computational complexity and a large code-book storage represented a drawback for applications either requiring a low power consumption (portable handsets in cellular radio), or having a high cost sensitivity. For example, an 8 kbps rate could be achieved assuming 5 ms blocks (N = 40), 1 K code-book (K = 1024), and an updating rate of 10 ms for the vocal tract and pitch coefficients. In this case, the required complexity would be about 40 MFLOPS and the code-book storage would take 40 kbytes of fast access and therefore costly memory. This is why, in the past few years, many studies have mainly concentrated on the reduction of the implementation requirements of the CELP concept in terms of computational complexity [3,5,7,10]. These studies have dealt with various aspects of the basic CELP, and resulted in valuable simplifications and original techniques like self excitation, linear code-book, algebraic coding of gains. Our own approach is outlined hereafter.

Our approach to the CELP problem

In our research to reduce the computational complexity and the storage memory size of CELP while preserving the quality of the reconstructed speech, we have focused on the elaboration of a method which could avoid the filtering of each candidate sequence. We finally found out that one can globally consider the perceptual weighting of the code-book instead of individually weighting each candidate sequence, as shown in next section.

PROPOSED CELP TECHNIQUE

Basic technique
    We have represented in Fig.2 two coder architectures which use the same decoder of Fig.1. In the following discussion, we show that, under some assumptions, these architectures are equivalent to the basic CELP (Fig.1).

Fig.2. Proposed CELP

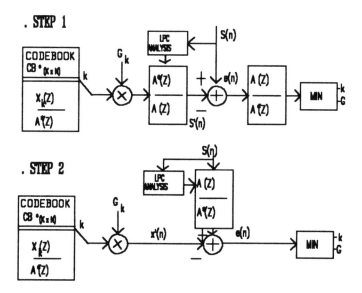

The coder shown in step 1 is strictly equivalent to the basic CELP shown on Fig.1, if one assumes that the code-book CB* is a spectrally shaped version of CB used in the basic CELP. If X(k,Z) denotes the Z-transform of the kth candidate sequence of the stochastic code-book CB, then the kth sequence of the spectrally shaped code-book CB* of step 1 has a Z-transform equal to X(k,Z)/A*(Z). The coder shown in step 2 actually represents the proposed architecture. It is obtained from step 1, by removing the filters from the paths of the signals s'(n) and e(n), and by introducing a filter in the path of the signal s(n). This shows that the search of the best sequence does not require the filtering of each candidate sequence by the vocal tract filter and the filtering of each error sequence by the weighting filter.

The problem is now to have the sequences of the code-book properly weighted. This can be approximated by considering an adaptive linear code-book (ALCB). Such a code-book CB^* is a long sequence $(K >> N)$ of samples $CB^*(i)(i = 1,...,K + N-1)$. Each candidate sequence $x(k,n)$ is given by:

$$x(k,n) = CB^*(k+n-1) \qquad n = 1,...,N \;\; ; \;\; k = 1,...,K$$

The code-book is updated at each block with the selected sequence:

$$CB^*(i) = CB^*(i+N) \qquad i = 1,...,K\text{-}1$$
$$CB^*(K+n-1) = x(k,n) \qquad n = 1,...,N$$

Each selected sequence is an approximation of the signal $s(n)$ filtered by the filter $A(Z)/A^*(Z)$, and therefore has the same spectral behavior as $1/A^*(Z)$. Since the selected sequences with spectral behavior $1/A^*(Z)$ are used to adapt the code-book, one can consider that the code-book has the same spectral behavior and therefore meets the assumption stated in the previous discussion of the coder shown on Fig.2 (step 1). Thus, the analysis of each segment of speech is made with a code-book which is shaped according to past segments of speech. In practice this approximation holds for quasi-stationary segments of speech since the code-book usually comprises 128 to 256 sequences, corresponding to a short delay (16 to 32 ms).

Alternate implementation

We now show that one can ensure an adaptive spectral shaping of the code-book CB^* by using a simpler method where the pole-zero inverse filter $A(Z)/A^*(Z)$ is replaced by a filter $A'(Z)$. This method (Fig.3) will be considered in the subsequent sections reporting our implementation.

Fig.3. Adaptive pre-emphasis

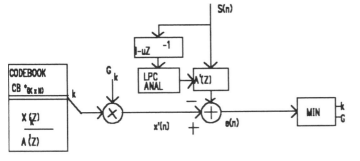

172

The filter $A'(Z)$ is computed from the pre-emphasized speech signal, with $\mu = R(1)/R(0)$, where $R(i)$ represents the short-term auto-correlation coefficients of $s(n)$. After filtering by $A'(Z)$, the signal $x(n)$ is vector quantized into $x'(n) = x(n) + e(n)$ with an error $e(n)$. At the receiver, there is no post-emphasis: the signal $x'(n)$ is just filtered by the filter $1/A'(Z)$ to give the synthetic signal $s'(n)$. We now show that the signal $s'(n)$ has the same spectral behavior as $s(n)$, and that the coding noise is spectrally shaped accordingly.

The pre-emphasis filter $(1 - \mu Z^{-1})$ removes the average slope of the speech signal $s(n)$. The LPC analysis which is performed on the first-order spectrum flattened signal thus results in a filter $1/A'(Z)$ the spectrum of which has approximately a null average slope and presents formants at about the same frequencies as $1/A(Z)$. In addition, the spectrum envelope of $x(n)$ presents the same behavior as $(1 - \mu Z^{-1})$. In the Z domain, one can write:

(1) $X(Z) = S(Z).A'(Z)$
(2) $X'(Z) = X(Z) + E(Z)$
(3) $S'(Z) = X'(Z)/A'(Z) = S(Z) + E(Z)/A'(Z)$

Assuming a high signal to noise ratio, relation (3) shows that the synthesized speech $s'(n)$ has the same spectral behavior as the original speech $s(n)$. Under the additional assumption that the error $e(n)$ is a white noise, relation (3) shows that the coding noise has the same spectral behavior as $1/A'(Z)$, and thus presents maxima at the formant frequencies. Fig.4 shows that the spectral behavior of $S'(Z)$, given by $1/(1 - \mu Z^{-1})A'(Z)$ approximates the spectral behavior of $S(Z)$, given by $1/A(Z)$.

Fig.4. Spectral noise shaping by pre-emphasis

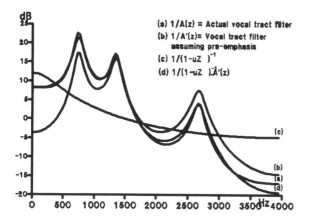

Application to CELP and SELP

We have applied our weighting by pre-emphasis technique to two coders of the CELP family: the ACELP [9] and the SELP [7] coders. Each coder (Fig.5) incorporates a system to remove the long-term redundancy of the speech signal. In the ACELP structure, there is a single adaptive linear code-book (ALCB) associated with a LTP (Long Term Prediction) loop, while the SELP structure exhibits two cascaded code-books (the first one being completely adaptive, the second one fixed). The LTP loop of the ACELP structure plays the role of the first code-book [4,6] of the SELP structure while the code-book of ACELP plays the role of the second code-book of SELP, and in addition includes an adaptive part. Note that, on Fig.5, CBM and CBR respectively denote the signal to code-book matching operation and the reconstruction of the signal from the code-book.

Fig.5. ACELP and SELP with pre-emphasis based weighting

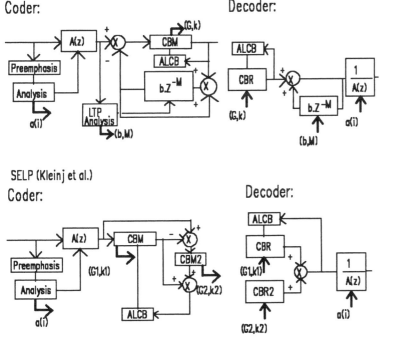

Each of the coders has been simulated at various bit rates, using either the classical weighting technique [1,2,7], or the proposed weighting by pre-emphasis technique. In all cases, we experimentally noted that when using the pre-emphasis, the speech quality is preserved while the processing requirement is reduced.

We have also compared the ACELP and SELP coders with pre-emphasis. The coders were tested for bit rates in the range (4-13 kbps). We found that both coders provide the same level of quality when operated at the same global bit rate. At 13 kbps, the quality is transparent, and gracefully degrades as the bit rate decreases. We determined that the coders could still give high quality speech around 7 kbps, so we decided to implement and thoroughly test a 7.2 kbps ACELP, as detailed in next section.

IMPLEMENTATION OF A 7.2 KBPS - 7 MIPS CODER

In this section, we describe the application of the proposed technique to a real-time operating ACELP coder, with a bit rate of 7.2 kbps.

7.2 kbps ACELP characteristics

The ACELP coder shown on Fig.5 has been implemented for an operation at 7.2 kbps, according to the bit allocation shown on Table I. The LPC analysis is performed by using the auto-correlation method on 20 ms segments, with adaptive pre-emphasis. The parcor coefficients are extracted through the LeRoux-Gueguen algorithm, coded by a piece-wise linear quantizer, decoded and converted to the direct-form coefficients $a(i)$ by using a step-up algorithm. The $a(i)$ coefficients are then time interpolated to ensure smooth block transitions. The long-term predictor parameters (b,M) are extracted every 5 ms by peak picking of the cross-correlation function between the input short-term residual signal and its reconstructed waveform. The sequence matching algorithm uses a linear code-book of 256 samples, and is executed every 2.5 ms (20 sample sub-blocks). The search is achieved through a cross-correlation computation between the input sequence and the code-book sequence, and result in a gain G and an index k. The linear code-book is normalized after each updating, thus enabling to use a fast sequence matching algorithm.

In fact, as detailed in [9], the code-book is composed of two parts. The first one is fixed, and the second one is adaptive. The first part (115 samples) is populated with a long sequence of a long-term prediction residual with typical stochastic characteristics. The stochastic sequences are essentially selected in the analysis of transient portions of the speech signal.

Table I. Bit allocation of the 7.2 kbps ACELP (20 ms blocks)

Short-term predictor (8th order)		28
Long term predictor lag M	2 x 7 =	14
Long term predictor gain B	2 x 2 =	4
Vector quantizer index	8 x 8 =	64
Vector quantizer gain	8 x 4 =	32
Control bits		2
Total		144

Bit rate $144/20 = 7.2$ kbps

Fig.6 shows the time variations of the input and decoded speech signals. Also shown are the signals at the input and at the output of the vector quantizer, as well as the quantizing error (amplified by 16 for display clarity). The binary signal, denoted CB INDEX, shows the selection of the fixed or variable portion of the adaptive code-book.

Fig.6. Example of signals in the 7.2 kbps ACELP

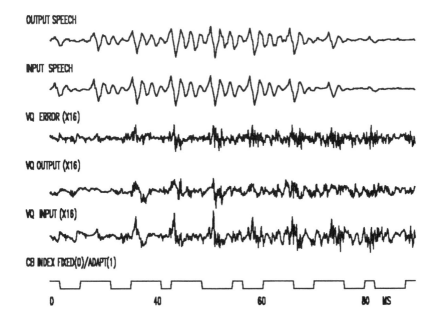

Fixed point DSP implementation

The proposed ACELP algorithm has been implemented on a programmable 16 bit fixed point DSP operating at a cycle time of 100 ns (IBM proprietary DSP) [8]. The implementation required about 70% of the DSP cycles, a program storage of less than 4 K instructions, and a data storage of 1.5 K half-words. Table II shows the measured processing load in MIPS (millions of instructions per second) for each major signal processing function. The LPC analysis includes the pre-emphasis, the auto-correlation, the parcor determination and coding-decoding, and the step-up algorithm. The STP (short term prediction) filtering includes the actual filtering operations as well as the interpolation of the a(i) coefficients. The VQ block includes the search, the code-book updating and normalization, and all the coding, decoding, scaling, move operations necessary for these functions.

Table II. Processing load (MIPS) of the 7.2 kbps ACELP

CODER:	LPC analysis	0.56
	STP filtering	0.22
	LTP analysis	0.61
	VQ block	4.33
	LTP filtering	0.18
DECODER:	VQ block	0.47
	LTP filtering	0.11
	STP filtering	0.34
I/O, MPX, DMPX		0.24
TOTAL		7.06

Speech quality

The speech quality of the 7.2 kbps ACELP coder has been evaluated as follows. 32 French sentences (16 male voices and 16 female voices) were processed through the 7.2 kbps ACELP coder, and through the 13 kbps GSM coder which was taken as a reference coder [11]. Then the 32 pairs of sentences were evaluated by standard pair-to-pair comparative tests. The pairs of sentences were randomly played to 10 listeners who were asked to mark their preference. It appeared that 43% of the 320 answers indicated a preference for the GSM, 35% indicated a preference for the ACELP, and 22 % of the answers indicated no preference.

From this evaluation test, we conclude that the ACELP coder operating at 7.2 kbps provides a level of quality approximately equivalent to that of the 13 kbps GSM coder, with a complexity only increased from 4 to 7 MIPS.

CONCLUSION

In this chapter, we have proposed an alternate way to implement the perceptual weighting function which is a mandatory function of CELP coders, though being very cost effective when implemented in the classical way. Our proposal basically consists in a global and adaptive weighting of the code-book instead of an individual weighting of every candidate sequence, and allows to drastically reduce the MIPS requirements of the basic CELP. The method has been evaluated on ACELP and SELP coders with good results.

We have demonstrated the method on a 7.2 kbps ACELP operating with the same level of quality as the 13 kbps GSM coder which has been normalized for the speech coding in the European cellular radio network. The coder has been implemented on a fixed-point DSP and required a processing capability of 7 MIPS. This low requirement is of high interest in applications like radio mobile which are sensitive to cost and power consumption.

REFERENCES

[1] B.S.Atal and M.R.Schroeder, "Stochastic coding of speech signals at very low bit rates". ICC Amsterdam pp.1610-1613, May 1984.

[2] M.R.Schroeder and B.S.Atal, "Code-excited linear prediction, high quality speech coding at very low bit rates", pp937-940 ICASSP 1985

[3] G.Davidson and A.Gersho, "Complexity reduction methods for vector excitation coding", pp.3055-3058, ICASSP 1986

[4] R.C.Rose and T.P.Barnwell III, "Quality comparison of low complexity 4800 bps self excited and code excited vocoders", ICASSP 1987.

[5] I.M.Trancoso and B.S.Atal, "Efficient procedures for finding the optimum innovation in stochastic coders", pp.2379-2382, ICASSP 1986

[6] R.C.Rose, T.B.Barnwell III,"The self excited vocoder, an alternate approach to toll quality at 4800 bps", ICASSP 1986, pp.453-456

[7] W.B.Kleijn, D.J.Krasinski, R.H.Ketchum, "Improved speech quality and efficient vector quantization in SELP", ICASSP 1988, New-York.

[8] J.P.Beraud, "Signal processor chip implementation", IBM Journal of Research and Development, Vol.29, N.2, April 1985.

[9] J.Menez, C.Galand, M.Rosso, "Adaptive Code Excited Linear Predictive Coder (ACELP)". ICASSP 89, Glasgow.

[10] J-P.Adoul, P.Mabilleau, M.Delprat, S.Morissette, "Fast CELP coding based on algebraic codes", ICASSP 87 Vol.4 Dallas, pp.1957-1960

[11] K.Hellwig, P.Vary, D.Massaloux, JP.Petit, C.Galand, M.Rosso, "Speech coder for the European mobile radio system", p.1065, GLOBECOM 1989

17

A 6 KBPS REGULAR PULSE CELP CODER FOR MOBILE RADIO COMMUNICATIONS

M. Delprat, M. Lever and C. Gruet

MATRA Communication
Rue J.P. Timbaud, B.P. 26
78392 Bois d'Arcy Cedex
FRANCE

INTRODUCTION

Low bit rate coding techniques are currently an important topic in speech research, because of the wide range of emerging applications such as narrow band digital speech transmission or voice messaging. In designing a low bit rate speech coder, the main issue is to achieve good quality with low enough complexity to allow for real time implementation. Moreover, in mobile telephony, the coder must also be robust to adverse transmission conditions (background noise, transmission errors...) [1].

In the past years, much has been done to improve the quality of speech coders at low bit rate. In that field, Code-Excited Linear Prediction (CELP) is undoubtly the most popular technique [2]. In CELP coding, the speech signal is modelled as a random process with a slowly varying power spectrum. Synthetic speech is produced by filtering successive innovation sequences through long and short term predictors. For each block, the optimum innovation sequence is selected from a codebook of vectors using an analysis by synthesis procedure with a perceptual error criterion.

This technique ensures that the reconstructed speech is close (in the subjective sense) to the original one, but it results in huge complexity and the basic CELP performances (quality and robustness) must still be improved for general public applications. More recently, several related schemes with a reduced computational load have been proposed [3,4,5]. On the other hand, improved quality CELP coders have also been presented [6] but their complexity remains very high. For the realization of low cost and low power consumption mobiles, the speech coder must be implemented on low cost and widespread DSP chips. Thus, efforts are still necessary to further reduce complexity and to improve quality at low bit rate.

This chapter deals with the design of a very low complexity CELP coder, suitable for mobile radio communications. First, several error minimization criteria are compared; the role of the "memory" of linear prediction (LP) filters is pointed out and a convenient perceptual weighting filter is introduced. Then the excitation model is discussed, presenting a new Regular Pulse (RP) innovation codebook. A wide range of fast codebook search algorithms are described, relying on modified error criteria and/or codebook structure. Finally the design of a robust and very low complexity 6 kbps Regular Pulse CELP coder is addressed.

ERROR MINIMIZATION CRITERION

Selection of the Optimum Innovation Sequence

In the original CELP [2], the optimum innovation sequence is selected by filtering each possible codeword c_k, scaled by a gain factor G_k, through both long and short term predictors, respectively $1/B(z)$ and $1/A(z)$. The resulting synthetic signal is compared with the original one and the difference signal is processed through the perceptual weighting filter $W(z)=A(z)/A(z/\gamma)$ with γ around 0.8. The codeword that minimizes the weighted error signal energy is then selected for the current block. An equivalent structure for the codebook search [3,5] appears in Fig. 1.

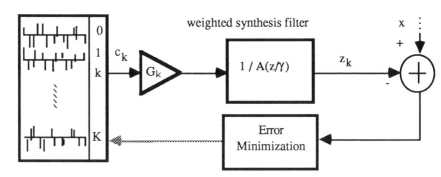

Figure 1: **Modified CELP Analysis Procedure.**

The weighted error signal energy is now expressed as

$$E(k) = \|x - z_k\|^2 = \|x - G_k.H.c_k\|^2 \tag{1}$$

where x represents the perceptually weighted original signal with the contribution from past excitations subtracted and H is the impulse response matrix of the weighted synthesis filter $1/A(z/\gamma)$. The optimum innovation sequence (c_{k_0}, G_{k_0}) which minimizes $E(k)$ in equation (1) is then determined in two steps:

1) Find the index k_0 which maximizes the weighted inner product $P_w(k)$:

$$P_w(k) = (x^t.H.c_k) / \|H.c_k\| \tag{2}$$

2) Compute the related gain G_{k_o}:

$$G_{k_o} = P_w(k_o) \, / \, \| H.c_{k_o} \| \tag{3}$$

Covariance Versus Autocorrelation Approach

The minimization of E(k) described above is traditionally carried out over the current block, of length L. It is referred to as a "covariance" method and in this case the impulse response matrix H is a LxL lower triangular Toeplitz matrix built on the impulse response h(i) of $1/A(z/\gamma)$. Another approach [7,8,9,10] is to consider longer sequences of residual and excitation signals, completing the L actual samples with J zeros, where J is chosen in such a way that h(i) is practically zero for i >J. In this "autocorrelation" approach, H becomes a (L+J)xL Toeplitz matrix:

$$
H = \begin{bmatrix}
h(0) & 0 & 0 & \cdots\cdots\cdots\cdots\cdots & 0 \\
h(1) & h(0) & 0 & \cdots\cdots\cdots\cdots & 0 \\
h(2) & h(1) & h(0) & \cdots\cdots\cdots\cdots & 0 \\
\vdots & \vdots & \vdots & \ddots & \vdots \\
h(J-1) & h(J-2) & h(J-3) & \ddots & \vdots \\
0 & h(J-1) & h(J-2) & \ddots & \\
\vdots & \vdots & \vdots & & h(0) \\
& & & & \vdots \\
0 & \cdots\cdots\cdots\cdots\cdots\cdots & 0 & h(J-1)
\end{bmatrix} \tag{4}
$$

The terminology used here for the error minimization approach refers to LPC analysis. In the autocorrelation method the covariance terms R_{ij} of the impulse response h involved in the error minimization procedure become autocorrelation terms $R_{|i-j|}$ (stationarity assumption), leading to interesting symmetry properties [9], that will be exploited for the design of fast search algorithms. The covariance method results in a more accurate matching at the beginning of the block than at the end, while in the autocorrelation method all the excitation samples are evenly weighted.

In the autocorrelation method, when a rectangular window is applied over the block [13], signals are supposed to be zero outside of the minimization interval and then, strictly speaking, the memory terms of the LP filters should be discarded [10]. Indeed, in our experiments, we noticed that the outputs of $1/A(z/\gamma)$ with a zero excitation are quite similar for the original and synthetic signals. Though, the minimization procedure can take these memory terms into account by applying an extended rectangular window that begins J samples before the current block [4].

On the contrary, since the covariance method does not evaluate the influence of current excitation on future blocks, the memory terms of the weighted synthesis filter play an important role in the analysis process. We have compared both approaches and, as shown in Table 1, their performances are very similar. However, as expected, the difference is more obvious when the memory terms are neglected.

Table 1: **Covariance vs Autocorrelation Method Performances.**

	Covariance		Autocorrelation	
	With memory	Without	With memory	Without
SNR	12.1	11.2	12.1	11.6
SNR-Seg	11.0	10.0	11.0	10.5

These results have been obtained with a block size L=20 and a 7 bits Regular Pulse codebook. SNR values are averaged over 50 seconds of clean speech from 4 speakers (2 males, 2 females). In informal listening tests, no degradation could be heard when suppressing the memory terms in the autocorrelation method.

Perceptual Weighting Filter

Perceptual weighting is known to significantly improve the subjective quality of low bit rate coders but the widely used form $W(z)=A(z)/A(z/\gamma)$ is not the only expression of the perceptual weighting filter that leads to high subjective quality. We have studied the perceptual weighting performed by a convenient filter given by

$$W'(z) = A(z) / C(z/\gamma) \tag{5}$$

where $1/C(z)$ is an average low order linear short term speech predictor. The weighted synthesis filter is then modified in $1/C(z/\gamma)$ whose coefficients are time invariant, a property that will be exploited for complexity reduction. Such a filter has already been used in Multi Pulse coding of speech [10] and has proved its remarkable ability to provide almost equivalent subjective results. When applied to a CELP coder [7,8], it produces very small (if any) audible distorsion in spite of a relatively lower SNR (up to 2.5 dB). An interpretation of the high performances obtained with $W'(z)$ is that perceptual weighting is essentially efficient for voiced sounds and for these segments $W'(z)$ is close to $W(z)$ (since $1/C(z)$ has the form of a smooth low pass filter).

EXCITATION MODEL

Recent studies have shown that choosing the innovation codebook is not that critical for the quality of the coder. Comparable performances have been obtained with a wide range of codebooks such as random codebooks with different statistical distributions [2], sparse codebooks [5] and binary or ternary codebooks that may be derived from algebraic codes [3,4,11]. On the other hand, as will be shown next, coder complexity may greatly benefit from a strong codebook structure.

Good results achieved with codebooks including pulse sequences suggest using excitation sequences of length L that have a regular structure consisting in q equidistant pulses separated by D-1 zeros. The first pulse (initial phase p) is at one of the locations 0 to D-1. The Regular Pulse (RP-) codebook [7,8], populated in a stochastic or deterministic manner, may be constituted of K independent sequences or of the D possible shifts of a basic set of K/D RP-sequences with initial phase zero.

A "binary" RP-codebook, as represented in Fig. 2, built from the 2^q binary words of length q (0 becoming -1), is particularly efficient to reduce both computational load and storing requirements (see next section).

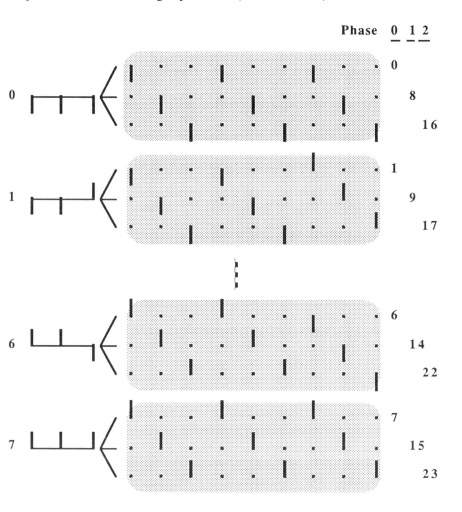

Figure 2: **Example of a Binary RP-Codebook (D=3, q=3 and K=24).**

Furthermore, RP-codebooks achieve comparatively high performances [7] because they allow for a better representation of the phase information in the excitation signal. Regular Pulse excitation has already been used at higher rates in RPE coders [10] and such a low complexity coder has recently been chosen as a standard for the Pan-European mobile radio system [12]. Introducing a RP-codebook in a CELP coder can be seen as a RPE technique in which the pulse amplitudes are optimally vector quantized and in that sense may be considered as a Base Band CELP coding technique.

FAST CODEBOOK SEARCH ALGORITHMS

A high complexity remains in the modified CELP structure described before. It mainly comes from the filtering of all codewords to find the one which maximizes the inner product $P_w(k)$ in (2). Major ways to reduce this amount of computations are as follows: to exploit the structure of particular codebooks (e.g. algebraic [3] or simply binary [4, 11]), to choose a suitable perceptual weighting filter [7, 8], to modify the error criterion [4, 7, 8]. We present below different approaches that may be combined to speed up the search procedure in the codebook. A fast and efficient scheme called RPCELP is then derived.

Suitable Error Criterion

The complexity considerably decreases when the weighted synthesis filter is fixed, as proposed above. The filtered vectors $H.c_k$ and the weighting factors $\|H.c_k\|$ involved in the inner product computation may all be precomputed and stored. A very simple but more memory consuming method (M1) is then obtained.

In the autocorrelation approach defined previously, signal x is expressed as $x=H.r$ where r represents the LP residual vector with the contribution of long term prediction (LTP) subtracted. The weighted inner product becomes:

$$P_w(k) = y^t.c_k \ / \ \|H.c_k\| \qquad (6)$$

where $y^t = r^t.H^t.H$ can be obtained as the result of a particular filtering operation (with time varying coefficients in general case) of each residual vector r before the codebook search [7,8], as illustrated in Fig. 3.

The remaining weighting term $\|H.c_k\|$ may be neglected in a first approximation. This leads to a very fast but suboptimal method (M2) in which the unnormalized inner product $P(k) = y^t.c_k$ is to be maximized. This approximation has a small impact on speech quality if the excitation sequences are such that all the $\|H.c_k\|$ have almost the same value. This is the case for the RP-codebook defined above.

An optimum scheme requires the computation and storage of the norms of the filtered vectors once per frame [5], (method M3). One way to efficiently perform this filtering operation is to convolve the excitation sequences with the (truncated) impulse response h of the time-varying weighted synthesis filter. This product may be obtained as the sum of the contributions of each excitation pulse [11]. The number of operations varies with the number of pulses and this method (M4) is rather fast in the case of sparse codewords (e.g. RP-codebook). Moreover, the norms may be jointly computed using the fact that sequences with the same values up to sample n lead to identical partial sums up to index n.

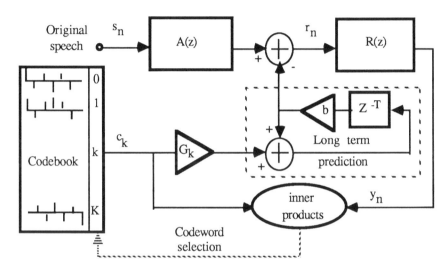

Figure 3: **Fast CELP Structure with Modified Error Criterion.**

The RPCELP Approach

When introducing a regular pulse excitation, together with the autocorrelation approach, the codebook structure can be exploited to further accelerate the search procedure. The normalization factors become:

$$\|H.c_k\|^2 = c_k{}^t.H^t.H.c_k = d_m{}^t.R_D.d_m \tag{7}$$

where R_D is a qxq symetrical Toeplitz matrix whose i^{th} diagonal term is the autocorrelation $R((i-1)D)$ and d_m is a q-dimensional vector. Note that R_D is independent of the phase p as a result of the autocorrelation method defined above.

Moreover, R_D can be forced to a diagonal matrix, using a reasonable approximation on the weighted synthesis filter: for instance, its impulse response can be shortened in order that $h(n) = 0$ for $n \geq D$, as described in [10]. With the normalization by $R(0)$, the matrix R_D becomes the identity matrix and (7) leads to

$$\|H.c_k\|^2 = \|d_m\|^2 \tag{8}$$

Assuming that the codewords are normalized, the search procedure comes down to maximize the inner product $P(k) = y^t.c_k$, which represents a small amount of computations (all the more because the codewords are sparse). This RPCELP method was first introduced in [7].

Besides, with a binary RP codebook the optimum codeword is efficiently determined in a two-steps procedure [7]:

1) Find the phase p_0 which maximizes $M(p)$ with

$$M(p) = \Sigma_i \mid y(p+iD) \mid , \text{ sum over } i= 0,...,q-1 \tag{9}$$

2) Choose the vector d_{m_0} such that $y^t.c_{k_0} = M(p_0)$:

$$d_{m_0}(i) = \text{sign of } y(p_0+iD) , \text{ for } i =0,...,q-1. \tag{10}$$

The related gain is then given by $G_{k_0} = M(p_0)/q$, since $\|H.c_k\|^2 = q$ for any k.

The perceived speech quality obtained with this fast RPCELP algorithm has been found to be equivalent to that of the original CELP.

Complexity Evaluation

We have evaluated the computational requirements of the codebook search for the different methods described above. Table 2 gives both approximated analytical expressions and estimated values for a particular configuration: codebook size K=128, number of non-zero samples q=5, LP order M=8, number of blocks per frame B=8, decimation factor D=4.

Table 2: **Codebook Search Complexity Comparisons (in Number of Operations per Sample)**

Original CELP	M1,M2	M3	M4	RPCELP	Fast RPCELP
$K(2M+4)$	K	$K(\frac{M+1}{B}+1)$	$K(\frac{1+q/4}{B}+1)$	K/D	5
2560	128	270	162	32	5

Note that the complexity of the fast RPCELP algorithm is practically independent of the codebook size K. This scheme is especially suitable for real time implementation since it requires less than 20 operations per sample for the entire excitation determination (filtering of the residual signal, index and gain computation).

DESIGN OF A 6 KBPS RPCELP CODER

Quantization of side information is a major issue in CELP coding of speech. Though the intrinsic quality of the CELP algorithm is high, its performances may drop significantly with successive quantizations, especially for low bit rate coders (below 8 kbps). Thus great care has been taken in designing the quantizers.

Coefficients of the 8th order LP filter are non-linearly scalar quantized at a rate of 1.5 kbps. In spite of its efficiency at low rates, vector quantization of LP filters has not been considered for the present coder because of its complexity and of its sensitivity to transmission conditions [1].

The single tap long term predictor is updated twice per frame and quantized with only 0.8 kbps. The delay is represented on 6 bits and the gain is efficiently quantized with 2 bits.

With a block length of 20 samples, the 7 bits binary RP-codeword index is transmitted at a 2.8 kbps bit rate. The excitation gains are quantized relatively to the maximum value of each frame, which is first logarithmically quantized. Reducing the update rate of the gains (with only one value for two blocks) gave unsatisfactory performances. A better solution consists of vector quantizing the successive couples of relative gain values using a small codebook derived from their statistical distribution. It can be easily shown that for the proposed coder, the optimal distance measure (in the sense of the minimization of the weighted error signal energy) associated to the vector quantizer is the euclidean distance. The overall bit rate devoted to the excitation gains is 0.9 kbps.

Fixed point simulations have been carried out to prepare real time implementation on a 16 bits DSP chip. Scaling factors have been optimized to achieve maximum precision, using dynamic scaling and 32 bits computations when necessary. The output speech quality of the fully quantized 6 kbps fixed point coder is close to that of the unquantized floating point version. Real time implementation is under way and the complexity of the whole coder/decoder scheme is around 4 Mips.

The robustness of the coder to transmission errors has been evaluated and the specific sensitivity of each parameter has been tested. We found that protecting the most significant bits of the first LP filter coefficients and of the energy term (maximum gain) is sufficient to preserve good intelligibility, even with a high error rate (more than 10^{-2}), but it does not maintain a high quality output speech. As a matter of fact, errors on LTP parameters or on excitation parameters result in uniform degradation that gradually increases with error rate. However, it should be noted that the binary RP-codebook is especially robust against transmission errors since one error on a given bit of a codeword index (except for the phase information) produces only a single wrong pulse in the innovation sequence [4, 7].

CONCLUSION

A class of CELP speech coding methods, based on a regular pulse excitation model, has been presented. We focused on the perceptual error criterion to derive several low complexity methods. The least complex of them, called fast RPCELP, which has been fully quantized at 6 Kbps and implemented in 16 bits fixed point arithmetics requires only 4 MIPS. The reconstructed speech quality is equivalent to that of the original CELP and the coder is intrinsically robust against transmission errors. Therefore, it appears to be an efficient solution for mobile radio applications.

REFERENCES

[1] I. Lecomte, M. Lever, L. Lelièvre, M. Delprat, A. Tassy, "Medium band speech coding for mobile radio communications," in Proc. ICASSP, Apr. 1988.
[2] M. R. Schroeder and B. S. Atal, "Code-excited linear prediction (CELP): high-quality speech at very low bit rates," in Proc. ICASSP, Mar. 1985.
[3] J. P. Adoul, P. Mabilleau, M. Delprat and S. Morissette, "Fast CELP coding based on algebraic codes," in Proc. ICASSP, Apr. 1987.
[4] A. Le Guyader, D. Massaloux, J. P. Petit, "Robust and fast code-excited linear predictive coding of speech signals," in Proc. ICASSP, May 1989.
[5] G. Davidson, M. Yong and A. Gersho, "Real-time vector excitation coding of speech at 4800 bps," in Proc. ICASSP, Apr. 1987.
[6] J. P. Campbell, V. C. Welch, T. E. Tremain, "An expandable error-protected 4800 bps CELP coder (U.S. Federal Standard 4800 bps voice coder)," in Proc. ICASSP, May 1989.
[7] M. Lever and M. Delprat, "RPCELP: A high quality and low complexity scheme for narrow band coding of speech," in Proc. EUROCON, June 1988.
[8] M. Delprat, M. Lever and C. Gruet, "Efficient excitation model and fast selection in CELP coding of speech," in Proc. EUROSPEECH, Sep. 1989.
[9] W. B. Kleijn, D. J. Krasinski and R. H. Ketchum, "Improved speech quality and efficient vector quantization in SELP," in Proc. ICASSP, Apr. 1988.
[10] P. Kroon, E. F. Deprettere, R. J. Sluyter, "Regular pulse excitation: a novel approach to effective and efficient multi-pulse coding of speech," IEEE Trans. ASSP, Vol. 34, Oct. 1986.
[11] L. Cellario, G. Ferraris and D. Sereno, "A 2 ms delay CELP coder," in Proc. ICASSP, May 1989.
[12] P. Vary, K. Hellwig, R. Hofmann, R. J. Sluyter, C. Galand, M. Rosso, "Speech codec for the European mobile radio system," in Proc. ICASSP, Apr. 1988.
[13] M. Berouti, H. Garten, P. Kabal, P. Mermelstein, "Efficient computation and encoding of the multipulse excitation for LPC," in Proc. ICASSP, Mar. 1984.

PART V

ALTERNATIVE TECHNIQUES
FOR LOW RATE SPEECH CODING

In spite of the major advances in speech coding during the past decade, researchers still face the challenge of achieving high quality speech at low bit rates. A considerable effort is being directed at the region of 2.4 to 4 kb/s. The paradigm of using analysis-by-synthesis to determine an excitation signal as input to a speech production filter model has offered a popular and valuable approach to attain reasonable quality at 4.8 kb/s and above. Nevertheless, this paradigm is not the "ultimate" solution and the quality achieved with existing coders at lower rates does not yet meet the requirements for most applications Some of the alternative approaches being considered consitute an entirely different approach while others attempt to build on this paradigm in combination with new ideas. Also of interest is the very low bit rate region of 300 to 600 b/s coding where the paradigm does not appear to be applicable.

This section reports on some of the alternate techniques currently being investigated. A fresh look at the kind of excitation signal really needed is described by Atal and Caspers. Recent advances in sinusoidal modeling and coding of speech are described by McAulay et al. A coding scheme based on a novel model of the excitation spectrum is presented by Brandstein, Hardwick, and Lim. The use of phonetic segmentation of speech followed by distinctive encoding of segments according to their phonetic classes is reported by Wang and Gersho. A coding technique based on a homomorphic vocoder framework that includes the use of analysis-by-synthesis is presented by Chung and Schafer. Finally, a 400 b/s vocoder is described by Liu which is based on efficient postprocessing of the parameters of the U.S. government standard LPC-10 algorithm.

18
BEYOND MULTIPULSE AND CELP
TOWARDS HIGH QUALITY SPEECH AT 4 KB/S

Bishnu S. Atal and Barbara E. Caspers

Acoustics Research Department
AT&T Bell Laboratories
Murray Hill, NJ 07974

INTRODUCTION

In recent years, rapid progress has been made in producing high quality speech at low bit rates [1-10]. The introduction of Multi-Pulse Linear Predictive Coding (MPLPC) in 1982 [1] and Code-Excited Linear Prediction (CELP) in 1984 [2,3] led to research in a new class of analysis-by-synthesis speech coders [4-6] that enabled us to encode high quality speech at bit rates as low as 4.8 kb/s [7,8]. The success of these coders has stimulated much interest in using the low bit rate speech coding technology for practical applications. Already, a new speech coding standard [11-12] has been adopted for cellular telephone service in Europe, and similar efforts are under way in North America. A standard based on CELP coding has been proposed by the United States Government for secure telephone, mobile satellite, and land mobile applications [8]. The rapidly increasing demand for cellular telephone services is providing the motivation to decrease the bit rate even further - down to 4 kb/s and lower.

Although both multi-pulse and stochastic (CELP) excitation models represent very important steps in synthesizing natural-sounding speech, these models are not able to reduce the bit rate for high quality speech down to 4 kb/s or lower. The performance of CELP coders at present degrades rapidly below 4.8 kb/s. We provide in this paper a critical review of the multi-pulse and code-excited coders and discuss their major limitations in providing high speech quality at lower bit rates.

Synthesis of high quality speech at low bit rates using linear predictive coding techniques requires efficient encoding of both the parameters of the LPC filter and its input excitation. Important progress has been made in recent years in reducing the bit rate for encoding the LPC parameters. But the bit rate for encoding the excitation continues to remain high. For example, in the present 4.8 kb/s code-excited linear predictive (CELP) coders, the excitation for the LPC filter is encoded at approximately 3.2 kb/s - using up 66% of the total number of bits. Significant reduction in the excitation bit rate is needed to achieve transmission of

192

high quality digital speech at 4 kb/s. At low bit rates, there are not enough bits to produce a signal that matches the original speech waveform. At best, one can produce a signal that is perceptually close to the natural speech signal. We present in this paper some results from our recent research on determining the perceptually important components of the excitation.

An important step in encoding the excitation at low bit rates is to create a parametric representation that provides a compact and accurate representation of the excitation for speech synthesis. Lack of an efficient representation of the excitation remains a major obstacle at present in synthesizing high quality speech at low bit rates. We present in this paper a method for creating an efficient and accurate representation of the LPC excitation for reducing its bit rate. In this method, the technique of singular-value decomposition (SVD) is used to represent the waveform of the LPC excitation as a weighted linear combination of orthogonal components and to transform the convolution operation of the LPC filter to a diagonal matrix form. The orthogonality of SVD components ensures that a change in the amplitude of any one component in the excitation produces a change at the filter output only in that component. Such a representation allows us to study the effect of changes in the filter excitation on the speech quality in a systematic manner and to determine which components of the excitation are perceptually important and which are not.

MULTI-PULSE AND CELP CODERS

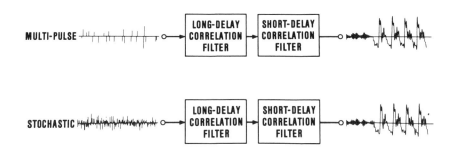

Fig. 1. Speech Synthesis Models Used in Multipulse and Stochastic (CELP) Coders.

Let us compare the multi-pulse and CELP coders by looking at the structure of the synthesizers in these coders [13]. Figure 1 shows the speech synthesis models used in multi-pulse and CELP coders. In both coders, the excitation is filtered through two filters, a long-delay predictive filter to introduce pitch periodicity and a short-delay LPC filter to add spectral characteristics to the signal. The major difference between the two coders is in the method of coding the excitation. In a multi-pulse coder, the location and amplitude of each pulse is coded separately, one pulse at a time - a relatively inefficient procedure. The excitation in the CELP coder is selected from a codebook of stored sequences and thus the entire sequence is coded as one vector providing an efficient representation of the excitation. In general, the multi-pulse coders need a much higher bit rate than the CELP coders to produce speech of comparable quality.

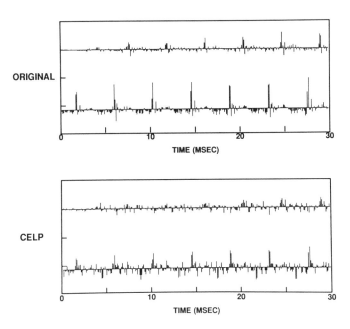

Fig. 2. Waveforms of the Original LPC Residual and the CELP Excitation for a Short Voiced Speech Segment.

Let us look carefully at the excitation for the LPC filter that is produced in a CELP coder. Figure 2 shows an example of the original LPC residual together with the excitation generated at the output of the long-delay filter in the CELP coder. We observe that, at the onset of voiced speech, the CELP coder has difficulty in generating the periodic pulse behavior exhibited by the LPC residual. This is due to the restricted codebook sizes typically used in CELP coders at bit rates of 8 kb/s and below. The waveform of the LPC residual in each pitch period consists of a large major pulse surrounded by a number of small pulses. The major pulse in the LPC excitation of the CELP coder is not very clearly defined and is barely noticeable in the first 30 ms of the waveform. What is the importance of the large major pulse in each period for the synthesis of high quality voiced speech?

COMBINED SINGLE-PULSE AND CODE-EXCITED SYNTHESIS MODEL

We have investigated the importance of major pitch pulses by developing a new coder in which the periodic segments are synthesized by exciting the LPC filter with a single pulse every pitch period and the nonperiodic segments are synthesized using random sequences selected from a codebook. Figure 3 shows a block diagram of a combined single-pulse and code-excited synthesis model.

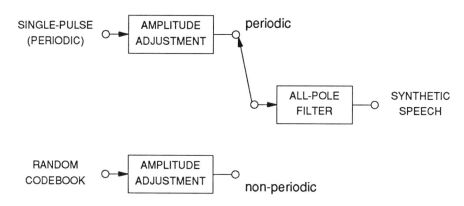

Fig. 3. Combined Single-Pulse and Code-Excited LPC Synthesis Model.

The locations of the major pitch pulses for this synthesis model were determined automatically by a pitch tracking algorithm using a pulse search procedure similar to one used in multi-pulse analysis. Codebooks were created using vector quantization training procedures to quantize the locations and amplitudes of pulse locations and their amplitudes. It was found that 14 bit codebooks, each with 16384 entries, provided sufficient accuracy to represent various combinations of pulse locations and amplitudes for 25 ms speech segments, resulting in a bit rate of 560 b/s for each one of these parameters. An additional 160 b/s were used to indicate mixed frames containing both periodic and nonperiodic excitations yielding a total bit rate of $560 + 560 + 160 = 1280$ b/s for the periodic segments of the excitation. A codebook with 1024 random white sequences was used to select nonperiodic excitation vectors for 5 ms frames. The quality of synthetic speech using the combined model was judged to be very good in informal listening tests for both male and female speakers. In careful listening on headphones, the synthetic speech was found to have a slightly muffled sound quality. The reasons for the muffled sound quality are not exactly clear but they could be attributed to either the large pole bandwidths often found in the LPC filters or to the idealization of the LPC excitation by a single pulse in a pitch period. Further studies are needed to determine exactly which one of these factors is responsible for the slightly muffled sound quality. But, it is clear that a single pulse in each pitch period can, by itself, bring the speech quality to a high level. Yet, the pitch loop in the CELP coder has great difficulty in producing such a pulse with an excitation signal consisting of randomly located pulses.

SINGULAR VALUE REPRESENTATION OF LPC EXCITATION

Next, we discuss the filtering properties of the LPC filter and describe a new method of representing the LPC filter such that the filtering reduces to a diagonal matrix operation [14]. Consider the input-output relationship for the LPC filter as shown in Fig. 4. The LPC filter is assumed to include the effect of perceptual noise weighting (shifting the poles towards the origin in the z-plane) used in CELP or multipulse coders. The input excitation $x(n)$ and the output noise-weighted speech signal $y(n)$ at the nth sampling instant are related by the convolution expression,

$$y(n) = \sum_{k=0}^{n} h(n-k)x(k), \qquad n = 0, 1, 2, \cdots , \qquad (1)$$

where $h(n)$ is the impulse response of the noise-weighted LPC filter. It is obvious from (1) that the changes in any one sample of the input excitation $x(n)$ are propagated to every sample of the output speech signal $y(n)$. This makes it very difficult to study the effect on the output of changing a particular sample of the input.

Consider now a frame of N speech samples. The output noise-weighted speech signal $y(n)$ can be written as

$$y(n) = \mu(n) + w(n), \qquad n = 1, 2, \cdots , N, \qquad (2)$$

where $\mu(n)$ is the signal contributed by the excitation from the previous frames and $w(n)$ is the signal contributed by the excitation in the current frame. Then, $w(n)$ can be expressed as

$$w(n) = \sum_{k=1}^{n} h(n-k)x(k), \qquad n = 1, 2, \cdots , N. \qquad (3)$$

We can write (3) in matrix notations as

$$w = Hx \qquad (4)$$

where w and x are column vectors with their nth components represented by $w(n)$ and $x(n)$, respectively, and H is a $N{\times}N$ matrix with its (i,j) term given by $h(i-j)$.

The singular value decomposition (SVD) is a well known method for diagonalizing rectangular matrices [15]. We apply SVD to the $N{\times}N$ matrix H and write it as $H = UDV^t$, where U and V are unitary matrices and D is a diagonal matrix with non-negative elements. The diagonal elements $\{d_i\}$ of D are called the singular values of H.

Let us go back to (4) and rewrite it using the singular value decomposition. Then, $w = UDV^tx$ and on multiplying both sides of this equation by U^t, one obtains $U^tw = DV^tx$. By representing U^tw and V^tx by new vectors θ and ξ,

Fig. 4. Time-Domain Input-Output Relationship for the LPC Filter. The Noise-Weighted LPC Filter Includes the Effect of Perceptual Noise Weighting.

respectively, (4) is expressed in the SV (Singular Value) domain by $\theta = D \xi$ and in scalar notations as

$$\theta_n = d_n \xi_n, \qquad n = 1, 2, \cdots, N. \tag{5}$$

Equation (5) provides a complete representation of the input and output of the LPC filter in terms of SV components. By using SVD, the convolution in (3) is transformed to a simple multiplication providing important advantages in interpreting the role of each SV component in the excitation. Figure 5 shows the input-output relationship for the LPC filter in the SV domain.

Fig. 5. Singular Value Domain Input-Output Relationship for the LPC Filter. The Noise-Weighted LPC Filter Includes the Effect of Perceptual Noise Weighting.

The procedure for producing any desired output is as follows: Let w be the desired output signal in the time domain. The output is then expressed in the SV domain by $\theta = U^t w$. The excitation in the SV domain is given by $\xi = D^{-1}\theta$ in vector notation and $\xi_n = \theta_n / d_n$ in scalar notation, assuming that $d_n > 0$ for all n. The SV components corresponding to zero singular values cannot be produced at the output of the LPC filter. Finally, the excitation in the time domain is given by $V\xi$.

PROPERTIES OF SINGULAR VALUE REPRESENTATION

As can be seen from (5), the amplitude of each SV component of the filter output θ_n is equal to the amplitude of the corresponding SV component ξ_n at the input multiplied by the singular value d_n. Thus any change in a particular component of the filter input (for example, produced by quantization) appears only in the same component at the filter output. The singular value representation allows a systematic study of the effect of changes in the filter excitation on the speech output. Moreover, changes in excitation components associated with smaller singular values produce relatively small changes in the output speech signal.

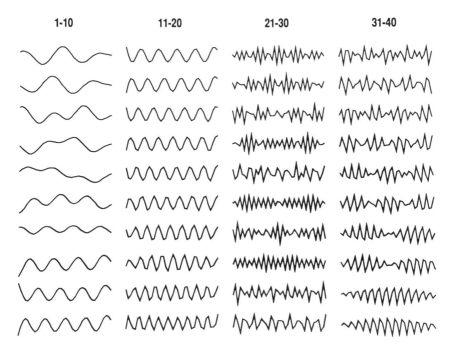

Fig. 6. Waveforms of the Right Singular Vectors for a Typical LPC Filter for Voiced Speech Arranged in order of Decreasing Singular Values.

Figure 6 shows the waveforms of eigenvectors of $H^t H$ (also the right singular vectors of H) arranged in order of decreasing singular values for an LPC filter typical of voiced speech. The duration of each eigenvector is 5 ms. Since the low frequency components of voiced speech have high amplitudes, the eigenvectors corresponding to higher singular values have predominant low frequency components. In contrast, the eigenvectors corresponding to lower singular values have predominant high frequency components. Some, but not all, of these singular vectors are approximately sinusoidal in shape and have narrowband spectra. Although the singular vectors from the SV decomposition appear similar to sine and cosine functions used in Fourier analysis, there are important differences between the two. The SV decomposition not only provides an orthogonal set of basis functions but also the simple input-output relationship expressed in (5) over a finite time interval.

QUANTIZATION OF SV COMPONENTS

We have determined the precision that is necessary in representing various SV components without producing audible distortion in the synthetic speech signal. It is obvious from (5) that, in general, SV components with larger singular values need much higher precision as compared to SV components with smaller singular values. The singular values determine the importance of each SV component only partially. Extensive listening tests are necessary to determine exactly the role of each SV component in determining the speech quality.

Each SV component was quantized using nonuniform quantizers with different number of quantization levels. The nonuniform quantizers were designed using a one-dimensional clustering procedure similar to the one used for training of vector quantizers. The number of quantization levels were varied from a minimum of 2 to a maximum of 32 for each of the SV components. It was found that no distortion was audible in synthetic speech if the number of quantization levels for each SV component was greater that 12. Furthermore, the minimum number of quantization levels required for no audible distortion varied with the index of the SV component – varying from a maximum of 12 for the first few components to a minimum of 2 for the last few components. Using a procedure that allocated different number of bits to different SV components depending on the singular values revealed that a maximum of 80 bits in each 5-ms excitation frame (2 bits/sample) was sufficient to keep the synthetic signal perceptually indistinguishable from the original.

The quantization at 2 bits/sample represents the maximum accuracy in any frame and is necessary only when each 5-ms frame is quantized independently of other frames. In general, the voiced speech frames required much higher precision in representing the excitation as compared to the unvoiced speech frames. Using vector quantization techniques, it was found that no distortion was audible when the unvoiced speech frames were quantized using 20 bits for each 5-ms frame. For voiced speech, segments with higher periodicity needed more precision than segments with lower periodicity. With differential quantization, an average of 1 bit/sample provided sufficient precision in the excitation to keep the quantization noise inaudible.

SELECTIVE MODIFICATION OF SV COMPONENTS

The singular value representation allows a systematic study of the effect of changes in the filter excitation on the speech output. We have studied the effect on speech quality of modifying various SV components.

The singular values can be ordered according to their values - from the largest to the smallest. Is it possible to set some of the SV components associated with smaller singular values equal to zero? Our experience, based on informal listening tests, is that elimination of even the smallest 20% of the components introduces audible distortion in synthetic speech. In another experiment, we replaced the smallest 20% of the SV components with properly scaled random numbers and found that the synthetic speech had audible distortions similar to ones heard in a 4.8 kb/s CELP coder.

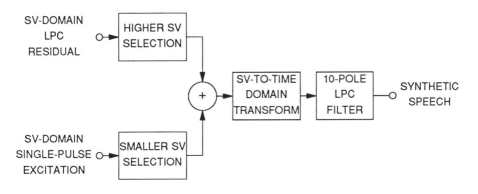

Fig. 7. LPC Speech Synthesis Using the Excitation Obtained by Mixing Higher Amplitude SV Components of the LPC Residual with Smaller Amplitude SV Components of the Single-Pulse Excitation.

In another experiment, the smaller SV components were replaced by the corresponding components from the single-pulse excitation. Figure 7 shows the method used to synthesize speech using a combination of single-pulse and LPC residual excitations. No distortion was audible, even when 70% of the smallest SV components were replaced by the corresponding SV components of the single-pulse excitation. This finding suggests that reproduction of the smaller SV components does not require additional bits over what is needed to produce the single-pulse excitation.

CONCLUSIONS

The present CELP coder is quite capable of synthesizing high quality speech at 4 kbit/s for both unvoiced and transient segments but not for voiced segments of speech. Our studies indicate strongly that the major deficiency in a CELP coder is its inability to reproduce the major pitch pulse in voiced speech. An important reason for the success of multipulse and CELP coders is that these coders eliminate the need for pitch and voiced-unvoiced analysis that is performed in vocoders. New research is needed to develop speech coding algorithms that add the single-pulse component to the LPC excitation in CELP coders and maintain robustness of the speech coding procedure. The lower amplitude components (spectral or singular-value) of the LPC excitation in CELP coders are mostly reproduced with random amplitudes. These random components are audible as noise in CELP coders at low bit rates. The use of single-pulse excitation for the low amplitude components can eliminate these noise components in CELP coders.

REFERENCES

[1] B. S. Atal and J. R. Remde, "A new model of LPC excitation for producing natural-sounding speech at low bit rates," in *Proc. Int. Conf. on Acoustics, Speech and Signal Proc.*, vol. 1, pp. 614-617, May 1982.

[2] B. S. Atal and M. R. Schroeder, "Stochastic coding of speech signals at very low bit rates," in *Proc. Int. Conf. Commun.*, vol. ICC84, part 2, pp. 1610-1613, May 1984.

[3] M. R. Schroeder and B. S. Atal, "Code-excited linear prediction (CELP): high quality speech at very low bit rates," in *Proc. Int. Conf. on Acoustics, Speech and Signal Proc.*, vol. 1, pp. 937-940, March 1985.

[4] P. Kroon, E. F. Deprettere and R. J. Sluyter, "Regular-pulse excitation: a novel approach to effective and efficient multi-pulse coding of speech," *IEEE Trans. ASSP*, vol. ASSP-34, pp. 1054-1063 (1986).

[5] G. Davidson and A. Gersho, "Complexity reduction method for vector excitation coding," in *Proc. Int. Conf. on Acoustics, Speech and Signal Proc.*, vol. 4, pp. 3055-3058, April 1986.

[6] P. Kroon and E. F. Deprettere, "A class of analysis-by-synthesis predictive coders for high quality speech coding at rates between 4.8 and 16 kbit/s," *IEEE Jour. on Selected Areas in Commun.*, vol. 6, no. 2, pp. 353-363, Feb. 1988.

[7] D. P. Kemp, R. A. Sueda, and T. E. Tremain, "An evaluation of 4800 bps voice coders," *Proc. Int. Conf. on Acoustics, Speech and Signal Proc.*, vol. 1, pp. 200-203, March 1989.

[8] J. P. Campbell, Jr., V. C. Welch, and T. E. Tremain, "An expandable error-protected 4800 bps CELP coder (U.S. Federal Standard 4800 bps voice coder)", *Proc. Int. Conf. on Acoustics, Speech and Signal Proc.*, vol. 2, pp. 735-738, March 1989.

[9] K. Ozawa and T. Araseki, "High quality multi-pulse speech coder with pitch prediction," *Proc. Int. Conf. on Acoustics, Speech and Signal Proc.*, vol. 3, pp. 1689-1692, April 1986.

[10] S. Singhal and B. S. Atal, "Amplitude optimization and pitch prediction in multipulse coders," *IEEE Trans. ASSP*, vol. 37, pp. 317-327 (1989).

[11] J. E. Natvig, "Pan-European speech coding standard for digital mobile," *Speech Communication*, vol. 7, no. 2, pp. 113-123, July 1988.

[12] C. Galand et al., "MPE/LTP coder for mobile radio application," *Speech Communication*, vol. 7, no. 2, pp. 167-178, July 1988.

[13] B. S. Atal, "High-quality speech at low bit rates: multi-pulse and stochastically excited linear predictive coders," in *Proc. Int. Conf. on Acoustics, Speech and Signal Proc.*, vol. 3, pp. 1681-1684, April 1986.

[14] B. S. Atal, "A model of LPC excitation in terms of eigenvalues of the autocorrelation matrix of the impulse response of the LPC filter," *Proc. Int. Conf. on Acoustics, Speech and Signal Proc.*, vol. 1, pp. 45-48, March 1989.

[15] G. H. Golub and C. F. Van Loan, *Matrix Computations*, Baltimore, MD: Johns Hopkins University Press, 1983.

19

SINE-WAVE AMPLITUDE CODING AT LOW DATA RATES[1]

Robert Mcaulay †, Thomas Parks †, Thomas Quatieri †, Michael Sabin ‡

† Lincoln Laboratory, MIT
Lexington, MA 02173-9108

‡ CYLINK, Inc.
Sunnyvale, CA 94086

INTRODUCTION

An analysis/synthesis system based on the sinusoidal speech model has been developed [1]. In that system, the sine-wave amplitudes and frequencies are located by searching for the peaks of the magnitude of the short-time Fourier transform (STFT) of the input speech. The phases are computed from the real and imaginary parts of the STFT at the measured frequencies. The frequencies on successive frames are matched, used in a cubic phase interpolator and applied to a sine-wave generator. Each sine wave is amplitude-modulated by the linear interpolation of the matched sine-wave amplitudes. At a 10 ms frame rate, this system produces speech that is perceptually indistinguishable from the original [1]. Since it is not possible to code all of the sine-wave parameters at low data rates, a system has been developed that codes the sine-wave frequencies by fitting a harmonic set of sine waves to the input waveform using a modified mean-squared error criterion [2], and codes the phase information implicitly using a voicing adaptive transition frequency to provide for a mixed voiced/unvoiced phase excitation model [3]. Provided a postfilter is used at the synthesizer to attenuate the noise in the formant nulls, the speech synthesized by this system is of quite high quality having achieved a DAM score of 63.0 in the uncoded mode. Since the fundamental frequency can be coded using ≈ 7 bits and the voicing measure can be coded using ≈ 3 bits, then the possibility exists for good speech quality at low data rates provided the sine-wave amplitudes can be coded efficiently. In this paper the zero-phase, harmonic analysis/synthesis system and the post-filter design methodology will be described and then the various techniques that have been examined for coding the sine-wave amplitudes will be discussed.

---

[1] This work was sponsored by the Department of the Air Force.

THE ZERO-PHASE HARMONIC ANALYSIS/SYNTHESIS SYSTEM

The sine-wave model that forms the basis of the low-rate vocoder is given by

$$\tilde{s}(n) = \sum_{k=1}^{K(\omega_o)} A(k\omega_o)\cos[(n - n_o)k\omega_o + \theta_k] \tag{1}$$

where $\omega_o = 2\pi f_o/f_s$ is the fundamental frequency, $K(\omega_o)$ is the number of harmonics in the speech bandwidth, $\{A(k\omega_o)\}$ are the sine-wave amplitudes at the harmonics, n_o is the onset time for the current frame and $\{\theta_k\}$ are the sine-wave phases. Since the fundamental frequency is chosen to make $\tilde{s}(n)$ a "best fit" to the input speech waveform, the error in the fit becomes a measure of harmonicity or voicing. If P_v represents the probability of voicing and if a voicing transition frequency is defined by

$$\omega_c(P_v) = \pi P_v \tag{2}$$

then the phases are determined by the rule

$$\theta_k = \begin{cases} 0 & \text{if } k\omega_o < \omega_c(P_v) \\ U[-\pi, \pi] & \text{if } k\omega_o \geq \omega_c(P_v) \end{cases} \tag{3}$$

where $U[-\pi, \pi]$ is a uniformly distributed random variable on $[-\pi, \pi]$. During strong voicing, therefore, the phase of each sine wave is determined by the frame onset time n_o which determines the time at which all sine waves come into phase. It is this ability to maintain phase coherence across sine waves that removes the "reverberant" quality reported in other implementations of the sine-wave system. Although methods have been developed for estimating the onset time explicitly [4], it has been found that simply maintaining the locations of the sequence of pitch pulses (determined from the estimated fundamental frequency) is sufficient to provide the necessary phase coherence [5]. Methods have also been developed for estimating the probability of voicing [2]. One such measure is based on the normalized correlation coefficient $\rho(\omega_o)$ given by

$$\rho(\omega_o) = \frac{\displaystyle\sum_{n=-N/2}^{N/2} s(n)s(n - \tau_o)}{\displaystyle\sum_{n=-N/2}^{N/2} s^2(n)} \tag{4}$$

where $s(n)$ is the input waveform and $\tau_o = 2\pi/\omega_o$ represents the estimated pitch period. If the sine-wave representation is used for $s(n)$ [1], then

$$s(n) = \sum_{\ell=1}^{L} A_\ell \cos(n\omega_\ell + \phi_\ell) \tag{5}$$

where $\{A_\ell, \omega_\ell, \phi_\ell\}$ are the amplitudes, frequencies and phases obtained by picking the peaks of the magnitude of the short-time Fourier Transform (STFT). It is easy to show that

$$\rho(\omega_o) = \frac{\sum_{\ell=1}^{L} A_\ell^2 \cos\left(2\pi \frac{\omega_\ell}{\omega_o}\right)}{\sum_{\ell=1}^{L} A_\ell^2} \qquad (6)$$

and a typical rule for the voicing probability might be to set

$$P_v = \begin{cases} 1 & \text{if} \quad P(\omega_o) \geq .9 \\ 1.8 \quad (\rho - .35) & \text{if} \quad .33 < \rho(\omega_o) < .9 \\ 0 & \text{if} \quad \rho(\omega_o) \leq .33. \end{cases} \qquad (7)$$

A slightly different measure of voicing is derived in [2].

During unvoiced speech, the phase of each sine wave is completely random which imparts the appropriate noise-like quality to the synthetic speech waveform, a property which is insured by defaulting the harmonic spacing to 100 Hz for all frequencies above the voicing transition. Moreover, for those speech frames for which the speech is neither completely voiced, nor completely unvoiced the above model allows for a mixed excitation that has been found to be beneficial for improving speech quality [3].

While the synthetic speech produced by this system is of quite high quality, a slight muffling effect can be detected. Such a quality loss has also been reported in code-excited LPC systems where it has been speculated that the muffling is due to coder noise showing up in the formant nulls [6]. Techniques have been developed for filtering out this noise by using a postfilter. In the sinewave-based system the noise-rejection postfilter is very straightforward to implement, since all of the necessary data is available in the frequency domain. The design of the postfilter is the subject of the next section.

DESIGN OF THE POSTFILTER

The first step in the design of the postfiltering procedure is to remove the spectral tilt from the log-spectral data, using the first two cepstral coefficients (Figure 1a). These coefficients are computed according to the equation

$$c_m = \frac{1}{\pi} \int_o^\pi \log S(\omega) \cos(m\omega) d\omega \quad m = 0, 1 \qquad (8)$$

where $S(\omega)$ is the envelope obtained by applying linear interpolation to successive sine-wave amplitudes. The spectral tilt is then given by

$$\log T(\omega) = c_o + 2\, c_1 \cos \omega \qquad (9)$$

and this is removed from the measured envelope to give the residual envelope

$$log R(\omega) = log S(\omega) - log T(\omega) \qquad (10)$$

which is then normalized to have unity gain, and compressed using a root-γ compression rule ($\gamma \cong .2$). If R_{max} is the maximum value of the residual envelope then the postfilter is taken to be

$$P(\omega) = \left[\frac{R(\omega)}{R_{max}} \right]^{\gamma} \qquad 0 \leq \gamma \leq 1. \qquad (11)$$

The idea is that at the formant peaks the normalized residual envelope will have unity gain and will not be altered by the compressor. In the formant nulls, the compressor will reduce the fractional values further so that overall, a Weiner filter characteristic will result (Figure 1b). The resulting compressed envelope then becomes the postfilter and is applied to the measured envelope to give

$$\hat{S}(\omega) = P(\omega)S(\omega) \qquad (12)$$

which causes the formants to narrow and reduces the depth of the formant nulls thereby reducing the effects of the coder noise (Figure 1c). When applied at the synthesizer of the zero-phase harmonic system the resulting system produces high-quality speech in which the muffled quality has been completely eliminated.

In fact, a floating-point simulation of this system has achieved a DAM score of 63.0. Since the excitation for this system is described in terms of the pitch ($\approx$ 7 bits), voicing ($\approx$ 3 bits), and gain ($\approx$ 5 bits), then high-quality speech at low data rates (2400 b/s) appears to be achievable, provided the sine-wave amplitudes can be coded efficiently. Several approaches to the sine-wave amplitude coding problem will now be explored.

SINE-WAVE AMPLITUDE CODING USING DPCM TECHNIQUES

The first approach used to code the sine-wave amplitudes is an extension of the method used in the Spectral Envelope Estimation Vocoder (SEEVOC) [7]. In the SEEVOC implementation the sine-wave amplitudes are interpolated linearly in frequency and treated as a waveform to be low-pass-filtered, down-sampled and coded using DPCM techniques [8]. In the Sinusoidal Transform Coder (STC) implementation, instead of using linear sampling in the frequency domain, the frequency axis is first warped according to the Bark scale and then the low-pass filtering, downsampling and coding are applied. This nonlinear sampling is intended to exploit the formant broadening which occurs at the higher frequencies, thereby providing a more efficient allocation of the available bits. One problem with the DPCM coding method is the fact that at low data rates its dynamic range at successive channels is limited and a positive slope

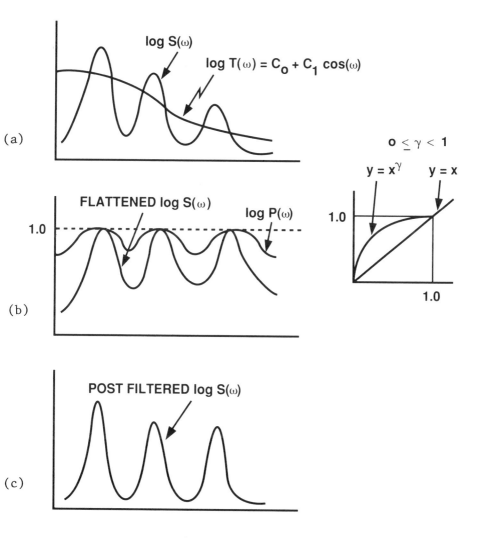

Figure 1: Frequency-domain postfilter design.

overload condition can occur in which the coder fails to track the leading edge of a sharply-peaked formant. This can cause the formant frequencies to shift and distort the synthetic speech. An example of this effect is shown in Figure 2.

Once the parameters of the DPCM coder have been set (i.e., the number of channels, the number of bits per channel, and the step size per bit), it is possible to define a "reachability" condition that insures that positive slope overload will not occur. For example, if the distance to the current amplitude from its decoded neighbor is less than the largest step size that can be taken at that channel, then slope overload will not occur. This condition has been used to develop a recursion for an amplitude threshold which is used to replace the DPCM quantized amplitudes by a set of amplitudes which will not experience the positive slope overload condition (provided the coder parameters were reasonably well designed in the first place) [9]. Therefore, as the DPCM coding procedure proceeds, if an amplitude to be coded is below this threshold, it is replaced by the value at the threshold. Although this will result in some formant broadening, the formant shift will not occur, and this seems to be the more deleterious effect on synthetic speech quality. An example of the application of this algorithm to the data in Figure 2 is shown in Figure 3, which demonstrates the protection against the slope-overload condition at the expense of the formant broadening. The postfilter is, of course, applied after the amplitude decoding, and this offsets some of the effect of the formant broadening. This method has been incorporated into the STC DPCM-based systems at 4800 b/s and 2400 b/s. The real-time fixed-point implementation of these algorithms using three ADSP-2100 digital signal processing chips, has been tested by Dynastat, using the Diagnostic Acceptability Measure (DAM) and the Diagnostic Rhyme Test (DRT), and the results are given in Table 1.

Table 1: **STC DAM & DRT TEST RESULTS**

Diagnostic Acceptability Measure (DAM)

Rate	6-Speaker	Male	Female
4800 b/s	62.0 (1.7)	64.0	59.8
2400 b/s	56.8 (2.0)	58.2	55.3

Diagnostic Rhyme Test (DRT)

Rate	6-Speaker	Male	Female
4800 b/s	92.7 (.78)	94.4	91.0
2400 b/s	90.1 (.77)	92.3	88.0

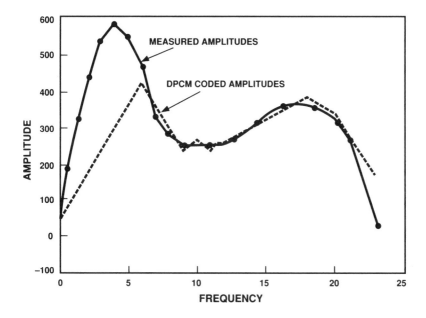

Figure 2: DPCM amplitude coding.

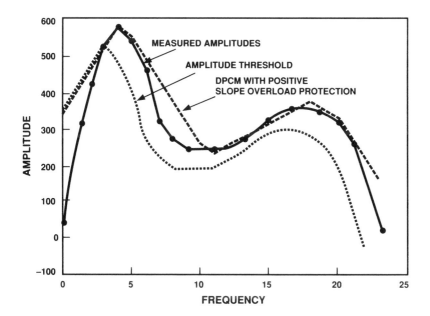

Figure 3: DPCM amplitude coding with positive slope overload protection.

OPTIMUM DPCM CODING

It is well known that DPCM coders can lead to improved performance by delaying the coder decisions to account for the future states that the coder can reach. In the present context one can think of generating all possible amplitude states that can be reached by the coder over the entire speech bandwidth and then selecting the set of states that provides the best fit to the measured channel amplitudes. While an exhaustive evaluation of all possible states would be computationally prohibitive, the use of Dynamic Programming can render the search problem more tractable. The analysis/synthesis simulation program has been generalized to use the Dynamic Programming technique using a metric based on the sum of the logarithmic errors between the measured channels and their DPCM approximations evaluated over the entire speech bandwidth. The "best" coded sequence could then be determined for each input sequence. Since the input to the algorithm is the set of sine-wave amplitudes, the squared log-error and the Itakura-Saito spectral matching criterion [10] could be computed directly on the measured data. The latter metric was found to be beneficial in reducing the slope-overload problem in the optimum coder. Examples of this coding technique are shown in Figure 4. It is interesting to note that the Itakura-Saito criterion avoids the slope-overload condition at the expense of some formant broadening. Unfortunately there were insufficient resources available to perform a thorough evaluation of this approach, and it is not possible to say at this time whether or not overall benefits in speech quality could be obtained.

SINE-WAVE AMPLITUDE CODING USING ALL-POLE MODELING

Since the above results have demonstrated that the envelope of the sine-wave amplitudes is an adequate basis for good quality sinusoidal speech synthesis, then it seems reasonable that an all-pole model could be fitted to this envelope, and that its parameters might be more efficiently coded than those obtained using the SEEVOC downsampling technique. By using the Itakura-Saito spectral matching criterion [10], it follows that the all-pole envelope depends on a set of reflection coefficients computed from the normal equations (similar to those obtained in LPC analysis) that are based on a set of correlation coefficients corresponding to the inverse Fourier transform of the sine-wave amplitude envelope. An example in which a 10th order all-pole model is fitted to the SEEVOC envelope is shown in Figure 5. The advantage of this frequency-domain approach is that it is independent of pitch, since its effect has been properly normalized by the sine-wave analysis system. Moreover, the proposed scheme should be robust, since the sinusoidal analysis system provides considerable coherent integration against noise. Finally, it is worth noting that the potential exists for a 2400 b/s version of the STC system that is interoperable with the Government standard LPC-10. Work is currently underway to develop

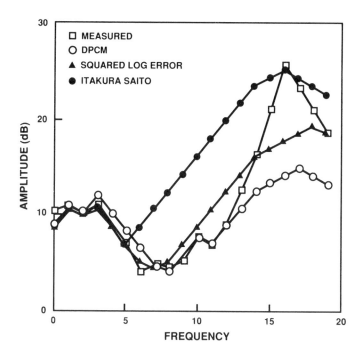

Figure 4: DPCM amplitude coding with delayed decision dynamic programming.

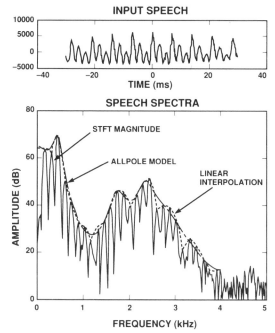

Figure 5: Sine-wave amplitude coding in terms of an all-pole model.

and evaluate the all-pole modeling approach in the STC/LPC context [11].

DISCUSSION OF THE IMPORTANCE OF PHASE

In this paper it was shown that the harmonic sine-wave system that used a voicing-adaptive phase model resulted in synthetic speech that was "muffled". It was also shown that the "muffling" could be eliminated by using a postfilter which narrowed the formant bandwidths and increased the depths of the formant nulls. It is interesting to note that when the modelled phases were replaced by the harmonic samples of the STFT phase, then the synthetic speech, while not perceptually equivalent to the original, was of very high quality and, moreover, that this high quality was obtained without the use of a postfilter. This suggests that the effect of the postfilter is to attenuate those sine waves for which the zero-phase model is replicating the phases incorrectly.

It was conjectured that perhaps the muffling phenomenon was due to structural phase components that the zero-phase system failed to model. For example, due to the phase of the glottal pulse, perhaps due to the phase of the vocal tract system function, or perhaps due to an incorrect estimate of the voiced/unvoiced frequency transition. In an attempt to answer some of these questions various modifications were made to the phases of the zero-phase system, and the resulting speech was compared to a baseline system that used the harmonic frequencies and the measured phases obtained by sampling the STFT phase at the harmonics.

In the first experiment the zero-phase system was modified such that the phases above 3000 Hz were always randomized regardless of the voicing-adaptive transition frequency and the modeled phases were used below 3000 Hz. In this case the muffling was reduced at the expense of an additional "shrillness", but for two all-voiced sentences ("Why were you away a year ago Roy?" and "Nanny may know my meaning?"), the muffling was completely eliminated and the synthetic speech was equivalent to that of the baseline system. This result suggests that the muffling may have been due to too much temporal correlation in phases.

In another experiment, the zero-phase system was altered to include a minimum phase component which was computed using the envelope of the sine-wave amplitudes. Remarkably, the muffling was essentially eliminated. It was speculated that perhaps the time and frequency variation of the system phase imparted the correct degree of phase decorrelation. In any case it can be stated that it is now possible to recreate high-quality synthetic speech without having to use any measured phases.

References

[1] R.J. McAulay and T.F. Quatieri, "Speech Analysis/Synthesis Based on a Sinusoidal Representation," IEEE Trans. Acoust., Speech and Signal Proc., Vol. ASSP-34, No. 4, August 1986, p. 744.

[2] R.J. McAulay and T.F. Quatieri, "Pitch Estimation and Voicing Detection Based on a Sinusoidal Speech Model," IEEE Int. Conf. Acoust., Speech and Signal Proc. (ICASSP'90), Albuquerque, NM, April 1990.

[3] J. Makhoul, R. Viswanathan, R. Schwartz, and A.W.F. Huggins, "A Mixed-Source Model for Speech Compression and Synthesis," Proc. IEEE Int. Conf. Acoust., Speech and Signal Proc. (ICASSP'78), Tulsa, OK, p. 163, April 1978.

[4] R.J. McAulay and T.F. Quatieri, "Phase Modeling and Its Application to Sinusoidal Transform Coding," Proc. IEEE Int. Conf. Acoust., Speech and Signal Proc. (ICASSP'86), Tokyo, Japan, p. 1713, April 1986.

[5] R.J. McAulay and T.F. Quatieri, "Phase Coherence in Speech Reconstruction for Enhancement and Coding Applications," Proc. IEEE International Conf. Acoust., Speech and Signal Proc. (ICASSP'89), Glasgow, Scotland, p. 207, May 1989.

[6] J.-H. Chen and A. Gersho, "Real-Time Vector APC Speech Coding at 4800 b/s with Adaptive Postfiltering," IEEE Int. Conf. Acoust., Speech and Signal Proc. (ICASSP'87), Dallas, TX, p. 51.3.1, April 1987.

[7] D.B. Paul, "The Spectral Envelope Estimation Vocoder," IEEE Trans. Acoust., Speech and Signal Proc., Vol. ASSP-29, No. 4, p. 786, August 1981.

[8] J.N. Holmes, "The JSRU Channel Vocoder," in Proc. Inst. Elect. Eng., 127, Pt. F, February 1980.

[9] M.J. Sabin, "DPCM Coding of Spectral Amplitudes without Positive Slope Overload," IEEE Trans. Acoust., Speech and Signal Proc. (to appear).

[10] F. Itakura and S. Saito, "A Statistical Method for Estimation of Speech Spectral Density and Formant Frequencies," *Electron. Commun. Japan,* Vol. 53-A, p. 36, 1970.

[11] R.J. McAulay and T. Champion, "Improved Interoperable 2.4 kb/s LPC Using Sinusoidal Transform Coder Techniques," IEEE Int. Conf. Acoust., Speech and Signal Proc. (ICASSP'90), Albuquerque, NM, April 1990.

20

THE MULTI-BAND EXCITATION SPEECH CODER

Michael Brandstein, John Hardwick, and Jae Lim

Research Laboratory of Electronics
Massachusetts Institute of Technology
Cambridge, MA 02139

INTRODUCTION

There has been considerable interest in the development of low bit rate, high quality speech analysis/synthesis systems. Applications for such systems include voice mail, low bit rate digital communications, and high security telephony. One class of speech analysis/synthesis systems (vocoders) which has been studied extensively and used widely in practice is based on an underlying model of speech. For this class, segments of speech are represented as the product of excitation and system spectra. The excitation parameters generally consist of a pitch period and a voiced/unvoiced (V/UV) decision. The system parameters are typically the spectral envelope or impulse response of the vocal tract. Speech is generated in the vocoder by exciting the system with a periodic impulse train in the case of voiced speech or random noise in the case of unvoiced speech. While vocoders of this type are capable of producing intelligible speech, they have not been successful in synthesizing high quality speech. In addition, the performance of these vocoders is known to degrade rapidly in the presence of back-ground noise. Considerable attention has been devoted to improving these systems. These improvements have focused primarily on the specification and quantization of the excitation signal after removal of the pitch structure. While these techniques have improved the quality, they have significantly increased algorithm complexity, which has precluded the real-time implementation of these systems on low cost architectures.

In the Multi-Band Excitation (MBE) Speech Model a different approach is taken toward representing the excitation signal [2]. The MBE speech model replaces the binary voiced/unvoiced classification with a series of such decisions over harmonic intervals. This added degree of freedom allows each speech segment to be partially voiced and partially unvoiced. The result is a speech analysis/synthesis system which is capable of generating high quality speech in a wide range of environments without a marked increase in computational complexity.

Previous work has shown that the MBE speech model produces high quality speech at 4.8 kbps [4,11]. An improved version of the MBE model (IMBE) has recently been developed [6] which includes higher speech quality along with reduced computational requirements. An additional effort has produced a 2.4 kbps speech coding system based on the IMBE speech model. Informal listening tests have shown the new 2.4 kbps system to be substantially better than the government standard LPC-10e speech coder. The computational simplicity of the IMBE algorithm makes it particularly well suited to real-time implementation at a significantly lower cost than LPC based systems producing similar speech quality. The current system performs the full-duplex IMBE algorithm on a single WE DSP32 processor.

MULTI-BAND EXCITATION SPEECH MODEL

The Fourier Transform, $S_w(\omega)$, of a windowed speech segment, $s_w(n)$, can be modeled as the product of an excitation spectrum $E_w(\omega)$ and a spectral envelope $H_w(\omega)$. The primary difference between the MBE speech model and previous ones lies in the form of the excitation spectrum. In previous models the excitation is entirely specified by the fundamental frequency ω_o and a voiced/unvoiced decision for the entire speech segment. For voiced speech $E_w(\omega)$ is equal to $P_w(\omega)$, the Fourier Transform of a windowed impulse train with spacing equal to $M = 2\pi/\omega_o$. If the effects of aliasing are ignored, $P_w(\omega)$ corresponds to the Fourier Transform of the window sequence centered at each harmonic of ω_o. Speech segments which do not exhibit this periodic property are declared unvoiced. $E_w(\omega)$ for these segments is modeled as the spectrum of a windowed white noise sequence. This approach allows only two different types of excitation as shown in Figure 1a. Consequently a conventional model's ability to represent the full range of speech signals is severely limited.

The extension which has resulted in the MBE speech model is to replace the binary voiced/unvoiced distinction with a more continuous division as shown in Figure 1b. Instead of making a single voiced/unvoiced decision for each speech segment, the new model divides the spectrum into a number of regions. Within each of these regions the speech spectrum is analyzed, and a voiced/unvoiced decision is made. The resulting excitation spectrum is a frequency dependent mixture of voiced and unvoiced energy. In each band which is declared voiced $P_w(\omega)$ is used in the excitation spectrum. The remaining frequency bands correspond to unvoiced regions, and they are filled with noise energy. The use of many voiced/unvoiced decisions allows the MBE speech model to have fine control over the makeup of the excitation spectrum.

This approach was motivated by the observation that many speech segments have some frequency regions which are dominated by noise energy while other regions are filled with periodic, voiced energy. This is especially the case in mixed voicing segments of clean speech and in voiced segments of noisy speech. We hypothesize that humans can discriminate between frequency regions dominated by harmonics of the fundamental and those dominated by noise-like energy and employ this information in the process of separating voiced speech from random noise. Elimination of this

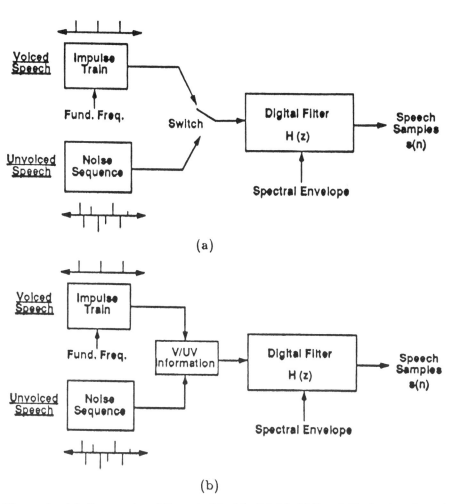

Figure 1: (a) Conventional Speech Model, (b) Multi-Band Excitation Speech Model.

acoustic cue in vocoders based on simple excitation models may help to explain the significant intelligibility decrease observed with these systems in noise [4].

The spectrum of each speech segment is only partially determined by the excitation spectrum. In addition the spectral envelope $H_w(\omega)$ determines the relative amplitude of each frequency component. Since the excitation spectrum is assumed to have a constant amplitude, the spectral envelope provides the scaling between $E_w(\omega)$ and the actual speech spectrum. In this manner $H_w(\omega)$ can be viewed as the frequency response which will map $E_w(\omega)$ into $S_w(\omega)$. Since $H_w(\omega)$ is generally slowly varying, it is often assumed to be a smoothed version of the actual speech spectrum $S_w(\omega)$.

In Figure 2a the spectrum of a typical speech segment is shown. This was

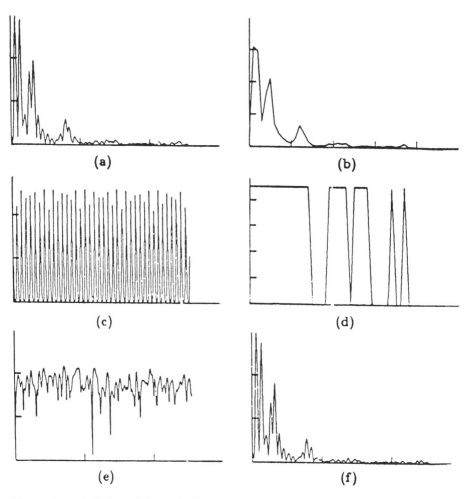

Figure 2: (a) Original Speech Spectrum, (b) Spectral Envelope, (c) Periodic Spectrum, (d) V/UV Information, (e) Noise Spectrum, (f) Synthetic Speech Spectrum.

obtained by windowing the speech signal with a 256 point Hamming window and then calculating the Discrete Fourier Transform of the windowed sequence. Figure 2b shows the spectral envelope which has been found for the segment. One can see that this is a smooth contour containing the general shape of the spectrum shown in 2a. The pitch period which has been estimated for this segment is 78 samples at a 10 kHz sampling rate. $P_w(\omega)$ corresponding to this pitch period is shown in Figure 2c. The voiced/unvoiced information is displayed in Figure 2d. A high value on this graph corresponds to a voiced region of the spectrum where $P_w(\omega)$ would be used in the excitation spectrum. Frequency regions having a low value in Figure 2d are unvoiced and noise energy as shown in Figure 2e is used in the excitation spectrum. This combination of voiced and unvoiced spectra is then multiplied by the spectral envelope to create the synthetic speech. This product is shown in Figure 2f.

SPEECH ANALYSIS AND SYNTHESIS

Figure 3 is an outline of the MBE algorithm. The parameters of the MBE speech model consist of the fundamental frequency, the V/UV information, and the spectral envelope. Our approach to estimating these parameters is similar to the one presented by Griffin in [2]. This approach attempts to estimate the excitation and system parameters which minimize the difference between the original and synthetic speech spectra. In general the error between the original and synthetic speech spectra can be expressed as:

$$E = \frac{1}{2\pi} \int_{w=-\pi}^{\pi} G(\omega)|S_w(\omega) - \hat{S}_w(\omega)|^2 d\omega \qquad (1)$$

where $\hat{S}_w(\omega)$ is the synthetic speech spectrum and $G(\omega)$ is a frequency dependent weighting function. In order to find the parameter set which achieves the minimum error it is necessary to solve a highly non-linear optimization problem. For this reason we use a different approach in which we first minimize over the spectral envelope and the fundamental frequency assuming that the speech is voiced, and then we determine the V/UV information.

The resulting method can be viewed as an analysis-by-synthesis system. For a given fundamental frequency the spectral envelope can be represented by a set of complex harmonic coefficients, which correspond to the value of the spectral envelope at the harmonics of the fundamental frequency. The harmonic coefficients which minimize the error for a given fundamental frequency are found through the solution of a set of uncoupled linear equations. This combination of fundamental frequency and harmonic coefficients can then be used to generate a synthetic spectrum which is used to evaluate (1). The resulting error is the minimum attainable for that particular fundamental frequency. By calculating this error function versus all fundamental frequencies of interest, a global minimum can be found. The V/UV decisions are made based upon the spectrum of the minimum error. We first obtain the error spectrum which is the difference between $S_w(\omega)$ and the synthetic spectrum with the minimum error. The average value of the magnitude error spectrum is then found over the region corresponding to each group of three harmonics. If this average exceeds a threshold then the region is declared unvoiced, otherwise the region is declared voiced.

To synthesize speech from the estimated model parameters we use separate techniques for the voiced and unvoiced portion of the speech signal. The voiced speech is synthesized using a bank of tuned oscillators. For a particular speech segment, an oscillator is assigned to each harmonic which has been declared voiced. The amplitude, phase and frequency of each oscillator are varied over the length of each segment. Once the oscillator parameters have been calculated for each harmonic, the voiced portion of the speech signal is formed by summing the contribution from each harmonic oscillator.

In order to complete the synthesis procedure the unvoiced speech must be reconstructed. This is accomplished by calculating the spectrum of a windowed noise

220

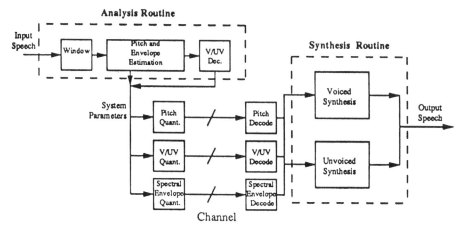

Figure 3: Outline of the MBE Algorithm

sequence and weighting the magnitude according to the estimated harmonic coefficients. The regions corresponding to voiced harmonics are zeroed out, so that they do not contribute any energy. The inverse transform of this spectrum is then taken and used with the weighted overlap-add procedure [7] to generate the unvoiced speech.

PARAMETER QUANTIZATION

The model parameters which are estimated for each frame include the fundamental frequency, the voiced/unvoiced decisions, and the spectral envelope information. The number of bits available for encoding these values is a function of the bit rate and analysis frame rate. For each vocoder system implemented, 2.4, 4.8, and 8.0 kbps, the model analysis and synthesis routines are identical. With the exception of the bit allocation, the coding schemes are also similar. The fundamental frequency needs accuracy of about 1 Hz. The present encoding scheme is straightforward and requires about 9 bits per frame. The V/UV decisions are encoded with one bit per decision. The remaining bits are allocated to error control and the spectral envelope information. Experimental results demonstrate that substantial interdependencies of these parameters exist in both time and frequency. We adopted a transform coding approach to exploit this [5]. Through adaptive bit allocation and uniform quantization of the Discrete Cosine Transform (DCT) coefficients, a high degree of coding efficiency is obtained.

REAL-TIME HARDWARE IMPLEMENTATION

The computaional simplicity of the MBE algorithm has facilitated the development of a real-time hardware implementation [8]. A single WE DSP32 processor is responsible for all the signal processing while a Mac II provides the user interface, a means of downloading software, and data transfer. Figure 4 is a block diagram of the

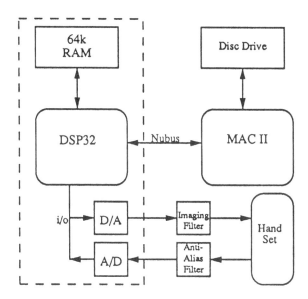

Figure 4: Real-Time Development System

current development system. Real-time results were achieved with no compromises in algorithm complexity, but did require extensive hand coding of the C code into DSP32 assembly language. We estimate that only 2.5 MIPS are necessary for algorithm implementation while the remainder of the 6.25 MIPS available are being expended on overhead and compiler inefficiency.

Table 1 is offered as a means for comparison of the instruction counts of various 4.8 kbps speech coders. These figures are all self reported computational estimates [9] and should not be confused with DSP chip MIPS ratings. It has been found that DSP chips require two to three times these estimated MIPS numbers due to program overhead. This figure agrees with our experiments. Note that none of these other systems have the potential to be implemented on the 6.25 MIPS DSP32 or an equivalent processor without some modification. Of particular interest is DoD's proposed federal standard 4.8 kbps voice coding system. This system has an upper bound of 12.5 MIPS when limited to integer pitch delays and a lower bound of 5.4 MIPS when the stochastic code book is reduced by a factor of eight [10].

PERFORMANCE RESULTS

Table 2 is a comparison of Diagnostic Rhyme Test (DRT) scores for an early version of the 8 kbps MBE vocoder and a 7.45 kbps Single Band Excitation (SBE) Vocoder [2]. The vocoders were tested with both clean speech and speech corrupted by additive white Gaussian noise. In the case of clean speech the results were similar for both systems. For noisy speech the 8 kbps MBE system clearly outperformed the conventional model based system. These results demonstrate the utility of the extra

Algorithm	MIPS
MBE	2.5
Motorola VSELP	6.4
DoD CELP	9.3
AT&T SELP	9.4
UCSB VAPC	3.9
Motorola RELP	6.2
Govt. Standard	5.4-12.5

Table 1: Computational Estimates of Some 4.8 Kbps Coders

System	clean speech	noisy speech
Uncoded mean	97.8	63.1
std	.30	1.8
MBE mean	96.2	58.0
std	.35	1.6
SBE mean	96.0	46.0
std	.44	1.6

Table 2: DRT Scores for the MBE and SBE Systems

voiced/unvoiced bands in the MBE vocoder.

In a recent government study of 4.8 kbps speech coders [9], the DAM and DRT scores for the MBE algorithm were shown to be comparable to other systems, including the proposed government standard CELP system. The IMBE coder has been developed subsequent to this study and possesses a noticeable increase in speech quality over its predecessor. Although we have not performed formal intelligibility and quality tests at this point, informal listening tests by a number of subjects have placed the IMBE algorithm on par, if not superior, to other 4.8 kbps systems. We have also found that the algorithm output quality degrades gracefully with bit rate. The 2.4 kbps coder currently available sounds quite similar to its 4.8 kbps counterpart.

CONCLUSIONS

In this paper, we have presented the MBE speech model. The model has shown to have definite advantages over the traditional single band excitation speech model, particulary in the presence of background noise. The model has been applied to the development of high quality, low bit rate speech coders and has proven amenable to inexpensive, real-time development systems. In addition to speech coding, the MBE Vocoder has various other applications. These include time-scale modification, pitch modification, and speech recognition.

REFERENCES

[1] Daniel W. Griffin and Jae S. Lim, "A New Model-Based Speech Analysis/Synthesis System," *Proc. of IEEE Int. Conf. on Acoustics, Speech and Signal Proc.*, pp. 513-516, Tampa, Florida, March 26-29, 1985.

[2] Daniel W. Griffen and Jae S. Lim, "Multi-Band Excitation Vocoder," *IEEE Trans. on Acoustics, Speech and Signal Proc.*, vol. ASSP-36, pp. 1223-1235, Aug. 1988.

[3] Daniel W. Griffin and Jae S. Lim, "A High Quality 9.6 kbps Speech Coding System," *Proc. of IEEE Int. Conf. on Acoustics, Speech and Signal Proc.*, pp. 125-128, Tokyo, Japan, April 13-20, 1986.

[4] B. Gold and J. Tierney, "Vocoder Analysis Based on Properties of the Human Auditory System," M.I.T. Lincoln Laboratory Technical Report, TR-670, December 1983.

[5] John C. Hardwick and Jae S. Lim, "A 4.8 KBPS Multi-Band Excitation Speech Coder," *Proc. of IEEE Int. Conf. on Acoustics, Speech and Signal Proc.*, pp. 374-377, NY, NY, April 11-14, 1988.

[6] John C. Hardwick and Jae S. Lim, "A 4800 bps Improved Multi-Band Excitation Speech Coder," *IEEE Speech Coding Workshop*, Vancouver, B.C., Canada, Sept. 5-8, 1989.

[7] Daniel W. Griffin and Jae S. Lim, "Signal Estimation From Modified Short-Time Fourier Transform," *IEEE Trans. on Acoustics, Speech and Signal Processing*, vol. ASSP-32, no. 2, pp. 236-243, April 1984.

[8] Michael S. Brandstein, Peter A. Monta, John C. Hardwick, and Jae S. Lim, "A Real-Time Implementation of the Improved MBE Speech Coder," *Proc. of IEEE Int. Conf. on Acoustics, Speech and Signal Proc.*, Albuquerque, NM, April 3-6, 1990.

[9] D.P. Kemp, R.A. Sueda, T.E. Tremain, "An Evaluation of 4800 BPS Coders," *Proc. of the Military and Government Speech Tech '89*, pp. 86-90, Arlington, VA, Nov. 13-15, 1989.

[10] Joseph P. Campbell,Jr.,Vancy C. Welch, and Thomas E. Tremain, "The New 4800 bps Voice Coding Standard," *Proc. of the Military and Government Speech Tech '89*, pp. 64-70, Arlington, VA, Nov. 13-15, 1989.

21
PHONETIC SEGMENTATION
FOR LOW RATE SPEECH CODING

Shihua Wang and Allen Gersho

Center for Information Processing Research
Department of Electrical and Computer Engineering
University of California at Santa Barbara
Santa Barbara, CA 93106

INTRODUCTION

Efforts to bridge the gap between waveform coders and vocoders has led to a new class of hybrid speech coders. These coders perform analysis-by-synthesis encoding of an excitation signal and reconstruct speech from the coded excitation signal and a quantized time-varying filter model of speech production. Most notable of these coders are those which use vector quantization to code the excitation signal as a sequence of vectors. The coding technique is called Code Excited Linear Prediction (CELP) [1], or Vector Excitation Coding (VXC) [2]. VXC coders result in coded speech with a waveform approximating the original and are able to achieve a satisfactory, natural-sounding quality at bit rates as low as 4.8 kb/s. When the bit-rate is reduced below 4.8 kb/s, the quality of VXC coders degrades rapidly and becomes inferior to the synthetic quality of an LPC vocoder operating at 2.4 kb/s. There remains then the challenging problem to find an algorithm that at 2.4 kb/s (or even at 3.6 kb/s) will achieve the quality that VXC offers at 4.8 kb/s

The main deficiency of VXC lies in its rigid coding configuration, wherein the waveform is divided into frames with a fixed frame size, and a fixed coding structure and fixed bit allocation are used for each frame and each vector within the frame, regardless of the local phonetic content of the frame. Consequently, at very low bit rates (e.g., 2.4 to 3.6 kb/s) we no longer have enough bits to adequately and consistently code both the excitation and filter parameters for the wide range of phonetically distinct characteristics that the waveform can take on in different frames. Yet these parameters are perceptually critical and vary in importance from frame to frame.

Recent developments in dynamic bit allocation in VXC offer a useful first step toward adapting the coder to match phonetically distinct frames. Dynamic bit allocation has been proposed, either among excitation vectors [3] or between excitation vectors and speech production filter parameters [4, 5, 6]. For a coding rate of 4.8 kb/s or higher, these coders can attain natural speech quality and offer some modest improvement over fixed bit-allocation methods, but the bit allocation is not closely related to the phonetic character of the frame and the limitation of a fixed analysis frame size is retained.

Some forms of speech segmentation have previously been used in speech coding. Roucos, Schwartz and Makhoul [7] have used segmentation for very low bit rate coding of speech at 150 b/s. Other approaches to speech segmentation for low-rate coding were proposed by Copperi [8] and by Ono and Ozawa [9].

Recently, we introduced a novel approach to the use of phonetic segmentation for speech coding [10] and in this chapter, we present an updated and more comprehensive treatment of the approach which might lead to a speech coder operating at bit rates significantly below 4.8 kb/s yet with a quality comparable to current 4.8 kb/s coders. In this method, speech is segmented into a sequence of contiguous variable-length segments constrained to be an integer multiple of a fixed unit length. The segments are classified into one of a small number of phonetic categories. This provides the front-end to a bank of VXC coders that are individually tailored to the different categories.

The motivation for our work derives from the fact that phonetically distinct speech segments require different coding treatments for preserving what we call *phonetic integrity*. With phonetic segmentation, we can assign the wide variety of possible speech segments into a small number of phonetically distinct groups. In each group, different analysis methods and coding strategies can be used to emphasize the critical parameters corresponding to important perceptual cues. It also becomes easier to identify each individual coding problem in isolated phonetic groups and optimize a multi-mode coding algorithm to suit various phonetic categories.

Some work which resembles our phonetic segmentation approach has been reported recently [11, 12]. In [11], different types of excitation signals are used for various speech segments such as vowels, fricatives, stops, etc. While in [12], speech is divided into seven phonetic categories with a speaker-independent classifier. The LPC parameters of the corresponding categories are quantized by dedicated vector quantizers.

The rationale for phonetic segmentation and the segmentation method is discussed in the next section. Then, the details of exploiting the phonetic features of each class in speech coding are given. Finally, the coding results are presented and compared with the conventional VXC coder.

PHONETIC SEGMENTATION

Phonetic Classification

In most low bit rate speech coding schemes, the speech signal is processed frame by frame with a fixed analysis frame length and boundary. In each frame, the same type of acoustic features are extracted regardless of the local signal contents. Although this analysis strategy is signal-independent and easily implemented, it lacks the ability to adapt a coder to transient parts of speech and distinguish between the parameters with different perceptual meanings. For example, if an analysis frame with a fairly large frame size (often desirable to bring down the bit rate) spans a transient from an unvoiced segment to a voiced one, the resulting speech spectral estimate will represent neither the voiced segment nor the unvoiced one due to the interference of the two adjacent phonemes. This poor spectral estimate produces a large prediction error which puts more burden on the excitation signal. Therefore, the ideal frame boundary should coincide with the transient. Another example is the coding of the voice onset time. When a stop occlusion is released into a voiced sound, the time from stop release (the start of the burst) to the start of vocal fold periodicity is called the *voice onset time*. The exact time location of the burst carries vital information for distinguishing between voiced and unvoiced consonants; they thus need to be well rendered. Furthermore, unvoiced speech segments have a nonstationary character, which suggests that they require a higher update rate on spectral envelope estimation than voiced segments. On the other hand, their spectral resolution (i.e., LPC order) is far less critical than that of steady-state voiced segments.

In voiced speech, the interference of the glottal excitation with the vocal tract imposes more difficulties on the estimation of spectra. Not only does a fast-changing pitch contour – and the corresponding spectral fine structure – have to be tracked accurately (by long-term "pitch" prediction), but for correct vowel perception the formant locations (especially the second formant) must also be correctly reproduced.

Voiced sounds can be subdivided into two basic categories: *full-band*, which contains vowels and voiced fricatives, and *lowpass*, which includes nasals, some glides and the part of voiced stops before air release. The spectra of *lowpass* segments is characterized by: 1) the existence of a very low first formant, well separated from the upper formant structure; 2) relatively high damping of lower formants. For best rendering of these two classes, different coding strategies are desirable: *lowpass* sounds are sensitive to even slight modifications of pole and zero positions, while for *full-band* sounds additional emphasis should be given to reproducing a clear harmonic structure at high frequencies.

Motivated by these considerations, while attempting to avoid excessive complexity, we define six phonetically distinct categories, classified by the tree structure shown in Fig. 1. At the top level, *unvoiced* also includes silence, while the *onset* category is defined as a transition from an unvoiced region into a voiced one. With this definition of *onsets*, 1 bit per frame is enough, on average, to classify segments at the top level.

The voiced class is further subdivided into four categories through two layers of segmentation, leading through the *full-band* and *lowpass* sounds already discussed, to *transient* and *steady-state* sub-categories. Transients occur at the boundary of two adjacent phonemes and after glottal stops, as well as during a fast ascending or descending pitch contour, while steady states dominate sustained vowels and semivowels. Each of these groups requires a distinct coding method.

Segmentation Procedure

We begin by dividing the input speech into *unit frames* of equal length. A binary-valued "raw" voicing parameter is tentatively assigned to each unit frame. The voicing decision is made heuristically by an algorithm using a set of thresholds

228

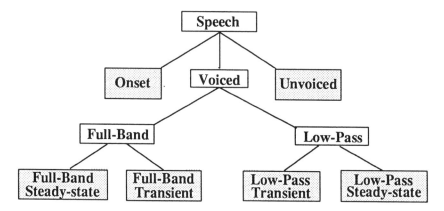

Figure 1. Tree-structured speech segmentation

on six acoustic parameters computed once per unit frame: speech energy, zero-crossing rate, first reflection coefficient, ratio of pre-emphasized signal energy to full band energy, forward and backward pitch prediction gains. The tentative sequence of voicing decisions are then smoothed by a three-point median smoother to yield a revised voicing classification for the unit frame sequence. The revised voicing decision for the current frame is equal to the majority class of the prior, current, and next unit frames. This process introduces a time delay of three unit frames at the transmitter.

Since *unvoiced* and *onset* segments are typically non-stationary, it is necessary to update their parameter sets more frequently. On the other hand, voiced segments are quasi-stationary. We therefore group pairs of unit frames with voiced segments into *coding frames*, while *unvoiced* and *onset* unit frames remain the original (unit) size for coding. A coding frame (later denoted simply as a *frame*) is the smallest piece of the waveform that will be coded as one entity by a particular VXC coding mode.

Voiced segments are first split into two categories by comparing to a preset threshold a combination of three parameters: zero-crossing rate, first reflection coefficient, and pre-emphasized energy ratio. At the second level, the split is determined by the spectral distance between the current and previous frames and the open loop pitch prediction gain, as will be explained later.

Thus, the waveform is partitioned into a sequence of (coding) frames, each frame having a particular class that is one of the six phonetic classes. Any set of contiguous frames belonging to one class can be concatenated into a single block belonging to that phonetic class. A *segment* is then defined as the largest such block of frames so that it is preceded and followed by frames belonging to a different class.

PHONETICALLY SEGMENTED VECTOR EXCITATION CODING

Preliminary research has been done to develop an adaptive coding technique exploiting phonetic features of speech [10]. The structure of this coder evolved from VXC [13], and thus we call it Phonetically Segmented VXC (PS-VXC). The first stage of the PS-VXC coder is speech segmentation. We partition the input speech waveform into a sequence of contiguous segments where each segment is classified into one of six categories. Each segment contains an integer number of frames and each frame is coded in turn by a VXC coder tailored to the particular segment class. The decoder is informed of the sequence of frames and their classes. The actual length of a segment is not explicitly needed by either encoder or decoder. Table 1 summarizes the segment classification and coding structures used for these classes by specifying salient features and coder parameters for each of the six categories. Table 2 lists the bit-allocation for each category in PS-VXC. In this section, we describe the basic structure of PS-VXC and explain how to exploit phonetic features of each class in speech coding.

Table 1. Classification of Segment Types

| Segment Type | Voiced | | | | Unvoiced | Onset |
| | Full-band | | Low-pass | | | |
	Transient	Steady State	Transient	Steady State		
Frame size (ms)	30	30	30	30	15	15
LPC order	10	10	12	12	6	10
LPC-Q type	SIVP-SQ	SIVP-SQ	SIVP-SQ	SIVP-SQ	SQ	VQ
Long-term LPC	Closed	Open	Closed	Open	none	none
Exc. CB type	Gaussian	same	same	same	same	residual
Exc. CB dim.	60	40	60	40	60	30

Table 2. Bit Allocation in PS-VXC

| Segment Type | Voiced | | | | Unvoiced | Onset |
| | Full-band | | Low-pass | | | |
	Transient	Steady State	Transient	Steady State		
Vectors/frame	4	6	4	6	2	4
Short-term LPC	30	19	34	24	20	11
Long-term LPC	26	13	26	13	0	0
Excitation	4×8=32	6×8=48	4×8=32	6×8=48	2×8=16	4×9=36
Gain	17	25	13	20	8	15
Classification	3	3	3	3	1	1
Total	108	108	108	108	45	63

Coding of Unvoiced Segments

Since *unvoiced* segments have a noise-like waveform with fairly weak near-sample correlation, their spectral envelopes can be sufficiently well represented by a 6th-order LPC model. We choose the log area ratios as the parameter set for representing spectral envelopes, and apply to it scalar quantization with a total of 20 bits. The use of a long-term predictor is omitted for coding unvoiced segments. Excitation vectors with dimension 60 samples are selected from an 8-bit Gaussian codebook. We experimented with bit allocation for unvoiced segments to find the minimum bit-rate at which a further increase in rate would cause a barely distinguishable perceptual improvement. This approach was also taken in designing most other components of the PS-VXC algorithm. The result is an overall bit-rate of 45 bits/frame, or 3 kb/s, for unvoiced segments.

Coding of Onset Segments

Onsets are defined as rapid transitions from unvoiced to voiced sounds, in which both the energy and the spectrum change rapidly. The signal is then poorly predictable, and the prediction gain is small. Interestingly, for such a segment, the ear's sensitivity to spectral errors is also small, so that greater spectral quantizing errors can be tolerated during transitions. On the other hand, onsets carry vital phonetic information for distinguishing consonants, with exact timings playing a crucial role. This leads us to place more emphasis on excitation coding, by reducing the dimension of the excitation vectors. We code the LPC spectral parameters in Log Area Ratio format with an 11-bit vector quantizer. Long-term prediction is not performed.

The excitation vectors for the *onset* segments are 30 samples long and are selected from a 9-bit codebook. The codebook is designed from a training set of normalized prediction residual vectors via the LBG algorithm. Since an unvoiced segment coded with only 45 bits always precedes an onset, and 63 bits/frame is allocated to onsets, the pair of frames (unvoiced followed by onset) has an average of 54 bits/frame, corresponding to an average coding rate of a 3.6 kb/s. *Onsets* are thus coded at the rate of 4.2 kb/s.

LPC Analysis and Voiced Segment Classification

For voiced frames both long-term and short-term prediction are needed to extract most of the speech information. At very low bit rates, the quantization of LPC parameters becomes especially critical since large spectral distortions can no longer be corrected by coarsely quantized excitation vectors. It is necessary to reserve enough bits for LPC parameter quantization. On the other hand, during a steady-state voiced segment, the spectral envelope changes slowly and there is a high correlation between consecutive spectral envelopes. This means that extra bits can be saved by interframe LPC parameter coding, thereby improving the accuracy of long-term prediction or excitation coding. We use Switched-adaptive Interframe Vector Prediction (SIVP) [14] for interframe coding using the Log Area Ratios (LAR) as the parameter set for short-term LPC quantization of voiced segments.

As we mentioned before, voiced segments in either the *lowpass* subset or the *full-band* subset are further divided into *steady-state* and *transient* classes. The classification is carried out by measuring the weighted Euclidean distance between the LPC parameter set of the current and previous frames and the open loop pitch prediction gain G_p. The weighted spectral distance measure is given by:

$$D_w = \sum_{i=1}^{P} w(i) \, [L_{x_n}(i) - L_{\hat{x}_{n-1}}(i)]^2 \qquad (1)$$

where $L_{x_n}(i)$ is the ith LAR coefficient of the current frame and $L_{\hat{x}_{n-1}}(i)$ is the quantized i^{th} LAR coefficient of the previous frame. The weighting function $w(i)$ is chosen such that the first four components of the LAR set are more heavily weighted than the others. By setting two empirical threshold values T_s and T_p, we can classify each subset into two categories : when $D_w < T_s$ and $G_p < T_p$, the spectrum of the current frame is considered as not having changed significantly from the previous frame, and the frame is classified as *steady-state*. Otherwise, the current frame is classified as *transient*.

Coding of Full-Band Voiced Segments

The accurate modeling and coding of the short-term and long-term predictors will help to extract the most perceptually significant information from the speech waveform, and thus to reduce the complexity of the residual signal, hence requiring less bits for coding. For example, speech is not exactly periodic, and the pitch may drift irregularly even within an analysis frame. Whenever we cannot perform an adequate pitch prediction during a fast changing pitch contour, the resulting residual signal will have many high peaked pulses and exhibit some of the periodicity of the original speech waveform. It is extremely difficult to correctly render these large pulses, as well as the periodicity, with a codebook constructed of large dimensional, noise-like codevectors. With the large pulses misplaced or altogether missing, and with the periodicity lost, the harmonic structure in the reconstructed spectrum becomes blurred, making the synthetic speech sounds hoarse and "gravelly" [3]. Thus the coding emphasis in this category is placed on the short-term and long-term prediction parameters.

In coding the short-term LPC parameters of *full-band transients*, we employ the SIVP [14] algorithm to predict the current LPC parameter set from the previous quantized set. Each individual component of the resulting prediction error vector is scalar quantized. We use a variable bit allocation for the different components, according to the relative perceptual importance of the different LAR parameters, which in practice means that lower order parameters get more bits. Altogether 30 bits are devoted to quantizing the LPC spectral parameter set.

The long-term prediction of the *full-band transient* segments is performed with a closed-loop method [15]. This method has been shown to greatly improve the synthetic speech quality and remove much of the gravelly character but at the price of an increase in complexity and coding bits. To minimize this effect, we apply the restrictive pitch deviation coding method [16] to code the long-term predictor. The total number of bits for the pitch predictor is 26. The excitation codebook used for this category consists of clipped Gaussian random noise vectors with unit variance and vector dimension of 60.

In *full-band steady-state* segments the spectral envelope and the pitch contour change slowly, and thus we no longer need a fast update rate for filter parameters. Instead, we can allocate more bits to the excitation. The short-term LPC coding for this class differs from that of the *full-band transient* class in which only the first six SIVP prediction error vector components are quantized and the rest are substituted in the receiver by the same components from the previous frame. The long-term prediction is performed with an open-loop method. A 3-tap long-term predictor is determined in the analysis stage from the short-term LPC residual and updated once per frame. Due to the saving of coding bits in the quantization of filter parameters, we can now afford to code the residual signal with a short excitation vector of dimension 40, and thus improve the overall speech quality.

Coding of Lowpass Voiced Segments

Most of the speech spectrum energy in this class is concentrated at low frequencies and the overall prediction gain is fairly high, indicating that the main resonances in the speech waveform can be extracted by the short- and long-term linear predictors. In the original VXC coder we observed two major problems in coding this type of speech segments. One is that the high concentration of spectral energy at low frequencies and the existence of antiformants, introduced by an oral closure, often results in very sharp formant peaks. This increases the sensitivity of the spectral distortion to LPC parameter quantization.

The other problem is the absence of the higher order formants in the reconstruction of nasal sounds. The nasal passage branches towards its middle into two separate tubes, which terminate in the nostrils. When the two nasal passages are appreciably different in their acoustic dimensions, the volume velocity at the nostrils no longer represents the output from a single nasal tube, and the effective nasal tract transfer function may contain additional pole-zero pairs at relatively high frequencies [11]. With only a 10-th order LPC predictor, these pole-zero pairs cannot be represented well. When the residual signal which contains the missing formants is quantized coarsely, these high resonances cannot be recovered in the reconstructed speech.

In order to solve these two problems, we select a 12-th order LPC short-term predictor for *lowpass* segments and add more bits to code the predictor coefficients. The short-term LPC parameters are quantized with the SIVP method and 36 bits are used for the *lowpass transient* class and 25 bits for the *lowpass steady-state* class.

The long-term predictor and the excitation part in *lowpass transient* and *lowpass steady-state* segments are coded as are their counterparts in the *full-band* subset.

CODING RESULTS AND CONCLUSION

A novel speech waveform coder of the VXC (or CELP) family has been presented, in which phonetic analysis precedes coding in order to match the coding algorithm to the phonetic segment type. The three main segment types, if coded individually, would yield following transmission rates: unvoiced – 3 kb/s, unvoiced/onset pairs – 3.6 kb/s, voiced – 3.6 kb/s. For typical speech files, the average rate is 3.4 kb/s, which could be achieved as a fixed rate with buffering of the encoder output. Alternatively, a fixed rate of 3.6 kb/s is readily attainable with some padding of the bit stream.

Informal listening tests indicate that the quality at a fixed 3.6 kb/s rate is roughly comparable to that of conventional VXC at 4.8 kb/s. Nevertheless, there is room for considerable improvement in both the coding algorithm for particular segment categories and in the definition and number of the phonetic classes used in the segmentation process. An end-to-end coding delay of approximately 100 ms (including overhead) is anticipated.

References

1. M. R. Schroeder and B. S. Atal, "Code-Excited Linear Prediction (CELP): High-Quality Speech at Very Low Bit Rates," *Proceedings of IEEE International Conference on Acoustics, Speech, and Signal Processing*, pp. 937-940, Tampa, March 1985.

2. G. Davidson and A. Gersho, "Complexity Reduction Methods for Vector Excitation Coding," *Proceedings of IEEE International Conference on Acoustics, Speech, and Signal Processing*, pp. 3055-3058, Tokyo, Japan, April 1986.

3. P. Kroon and B. S. Atal, "Strategies for Improving the Performance of CELP Coders at Low Bit Rates," *Proceedings of IEEE International Conference on Acoustics, Speech, and Signal Processing*, vol. 1, pp. 151-154, New York City, April 1988.

4. Mei Yong and Allen Gersho, "Vector Excitation Coding with Dynamic Bit Allocation," *Proceedings of IEEE International Conference on Communication*, vol. 1, pp. 0290-0294, Florida, November 1988.

5. N. S. Jayant and J. H. Chen, "Speech Coding with Time-Varying Bit Allocation to Excitation and LPC Parameters," *Proc. IEEE Conf. Acoust., Speech, Sign. Processing*, vol. 1, pp. 65-68, May 1989.

6. T. Taniguchi, S. Unagami, and R. Gray , "Multimode Coding: Application to CELP," *Proc. IEEE Conf. Acoust., Speech, Sign. Processing*, vol. 1, pp. 156-159, May 1989.

7. S. Roucos, R. M. Schwartz, and J. Makhoul, "A Segment Vocoder at 150 b/s," *Proceedings of IEEE International Conference on Acoustics, Speech, and Signal Processing*, pp. 61-64, Boston, April 1983.

234

8. Maurizio Copperi, "Rule-Based Speech Analysis and Application to CELP Coding," *Proceedings of IEEE International Conference on Acoustics, Speech, and Signal Processing*, vol. 1, pp. 143-146, New York City, April 1988.

9. Shigeru Ono and Kazunori Ozawa, "2.4 Kbps Pitch Prediction Multi-pulse Speech Coding," *Proceedings of IEEE International Conference on Acoustics, Speech, and Signal Processing*, vol. 1, pp. 175-178, New York City, April 1988.

10. Shihua Wang and Allen Gersho, "Phonetically-Based Vector Excitation Coding of Speech at 3.6 kbit/s," *Proceedings of IEEE International Conference on Acoustics, Speech, and Signal Processing*, Glasgow, May 1989.

11. Osamu Fujimura and Kazunori Ozawa, "High-quality Speech Coding Using Multiple Types of Excitation Signals at 4.8 kb/s and Below," *Advances in Speech Coding*, Kluwer Academic Publishers, 1990.

12. T. Liu and H. Hoege, "Phonetically-based LPC vector quantization of high quality speech," *Eurospeech 89, section 39.4*, Paris, September 89.

13. G. Davidson, M. Yong, and A. Gersho, "Real-Time Vector Excitation Coding of Speech At 4800 bps," *Proceedings of IEEE International Conference on Acoustics, Speech, and Signal Processing*, vol. 4, pp. 2189-2192, Dallas, April 1987.

14. M. Yong, G. Davidson, and A. Gersho, "Encoding of LPC Spectral Parameters Using Switched-Adaptive Interframe Vector Prediction," *Proceedings of IEEE International Conference on Acoustics, Speech, and Signal Processing*, vol. 1, pp. 402-405, New York City, April 1988.

15. S. Singhal and B.S. Atal, "Improving Performance of Multi-Pulse LPC coders at Low Bit Rates," *Proceedings of IEEE International Conference on Acoustics, Speech, and Signal Processing*, pp. 1.3.1-1.3.4, San Diego, 1984.

16. Mei Yong and Allen Gersho, "Efficient Encoding of the Long-term Predictor in Vector Excitation Coders," *Advances in Speech Coding*, Kluwer Academic Publishers, 1990.

22
VECTOR EXCITATION
HOMOMORPHIC VOCODER

Jae H. Chung and Ronald W. Schafer

School of Electrical Engineering
Georgia Institute of Technology
Atlanta, GA 30332

INTRODUCTION

The use of analysis-by-synthesis in determining the excitation signal has significantly advanced the state of the art in LPC vocoders [1,2]. The same principle has been adopted to obtain the excitation signal in the homomorphic vocoder, i.e., an exhaustive search procedure is used to determine the excitation by minimizing a perceptually weighted difference between the original speech and the output of the vocoder synthesizer [3]. This new homomorphic vocoder, called the *vector excitation homomorphic vocoder*, is a promising low bit rate vocoder, with performance that is far superior to that of a pitch-excited homomorphic vocoder [4] and fully comparable to that of vector-excited LPC vocoders such as code-excited or self-excited vocoders at a bit rate of 4800 bps.

In this paper, the homomorphic vocoder framework is further developed with more efficient quantization of the gains of the excitation signal and more effective modeling of the excitation signal. After a brief review of previous work [3], a new efficient method of coding the two gain parameters of a vector excitation homomorphic vocoder is discussed. The approach involves a time varying gain normalization, which transforms the original uncorrelated gain parameters into highly correlated parameters that can be jointly quantized to achieve a significant reduction in bit rate over independent quantization of the two gain parameters. Then, an improved excitation model is described. This model emphasizes either the pitch dependent periodic nature or the more random nature of the sound sources depending on the state of each speech frame. The cepstrum (which is the basis for the homomorphic vocoder) is also used for classifying a given speech segment into 3 different classes each of which is modeled differently.

THE FUNDAMENTAL VOCODER FRAMEWORK

The homomorphic filtering procedure which provides the vocal tract information in the homomorphic vocoder is depicted in Figure 1. Each block of

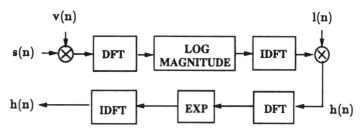

Figure 1: **Homomorphic filtering for estimating the vocal tract impulse response.**

a pre-emphasized speech signal $s(n)$ is first weighted by a Hamming window $v(n)$, and then the corresponding cepstrum $c(n)$ is computed by applying the sequential operations diagrammed in the upper part of Figure 1. The low-time cepstrum $\hat{h}(n)$ representing the vocal tract information is then extracted using a "lifter" $l(n)$; i.e., $\hat{h}(n) = c(n)l(n)$ with

$$l(n) = \begin{cases} 2, & 1 \le n < n_0 \\ 0, & \text{otherwise}, \end{cases} \tag{1}$$

where n_0 is less than the pitch period. Finally, applying the operations depicted at the lower part of Figure 1, the impulse response $h(n)$ is derived from the low-time cepstrum $\hat{h}(n)$. Observe that the impulse response $h(n)$ is normalized automatically because $\hat{h}(n) = 0$. This impulse response represents the vocal tract response during the time interval corresponding to the input block of speech samples.

For the given block of the speech signal $s(n)$, the excitation signal $e(n)$ is determined based on the analysis-by-synthesis excitation algorithm shown in Figure 2. The excitation signal $e(n)$ is composed of the following two parts: $\beta_1 f_{\gamma_1}(n)$, where $f_{\gamma_1}(n)$ is a zero-mean Gaussian codebook sequence corresponding to index γ_1 in the codebook, and $\beta_2 e(n - \gamma_2)$, which represents a short segment of the past (previously computed) excitation beginning γ_2 samples before the present excitation frame, i.e.,

$$e(n) = \beta_1 f_{\gamma_1}(n) + \beta_2 e(n - \gamma_2). \tag{2}$$

(Henceforth, β_1 will be called the codebook gain and β_2 will be called the self-excitation gain.) The perceptually weighted synthetic speech $x(n)$ corresponding to $e(n)$ has the form of

$$x(n) = \beta_1 x_1(n) + \beta_2 x_2(n), \tag{3}$$

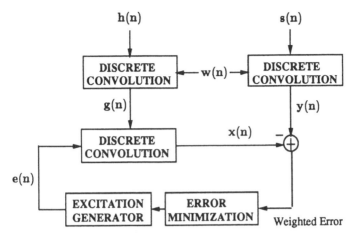

Figure 2: **Analysis-by-synthesis method for obtaining the excitation in a homomorphic vocoder.**

where $x_1(n) = g(n) * f_{\gamma_1}(n)$, $x_2(n) = g(n) * e(n - \gamma_2)$, and $g(n) = w(n) * h(n)$ is the perceptually weighted vocal tract impulse response. The weighting of the speech signal and the impulse response with the weighting filter $w(n)$ before synthesis is to improve subjective speech quality by concentrating the coding noise in the formant regions of the spectral envelope. In the homomorphic vocoder, the weighting filter is derived directly from the cepstrum [3].

The optimum parameters, $\beta_1, \gamma_1, \beta_2$, and γ_2, are determined pairwise in the following manner. First, the parameters γ_2 and β_2 are chosen by minimizing the mean-squared error

$$E_0 = \sum_n |y(n) - \beta_2 x_2(n)|^2. \tag{4}$$

For a given γ_2, the value of β_2 that minimizes the mean squared error in (4) is given by

$$\beta_2 = \frac{\sum_n y(n)x_2(n)}{\sum_n x_2^2(n)}. \tag{5}$$

The optimum values for γ_2 and β_2 are found by an exhaustive search with values of γ_2 restricted to a finite range. Then the residual signal $y_1(n) = y(n) - \beta_2 x_2(n)$ is formed and the parameters γ_1 and β_1 are chosen by an exhaustive search of the Gaussian codebook to minimize

$$E_1 = \sum_n |y_1(n) - \beta_1 x_1(n)|^2. \tag{6}$$

As before, the value of β_1 that minimizes the mean squared error for a given codebook sequence $f_{\gamma_1}(n)$ is

$$\beta_1 = \frac{\sum_n y_1(n)x_1(n)}{\sum_n x_1^2(n)}. \tag{7}$$

238

At the synthesizer, each block of an excitation sequence $e(n)$ is convolved with the corresponding vocal tract impulse response $h(n)$ to produce the synthetic speech output $\tilde{s}(n)$ using the overlap-add method [5].

EFFICIENT QUANTIZATION OF EXCITATION GAINS

Figures 3(a) and 3(b) show $|\beta_1|$ and $|\beta_2|$ as a function of the frame index. In this example, for speech signals sampled at 8 KHz, the vocal tract analysis

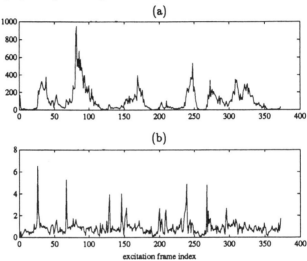

Figure 3: (a) Codebook gain $|\beta_1|$. (b) Self-excitation gain $|\beta_2|$.

was updated every 20 msec (160 samples), and the excitation analysis was updated every 5 msec (40 samples). To represent the vocal tract information, 11 low-time cepstrum values (excluding $\hat{h}(0)$ since $\hat{h}(0) = 0$) were used, and in synthesis, a 26 sample normalized impulse response $h(n)$ was derived from the cepstral representation. The cepstrum values were vector quantized using a codebook of 256 entries [3]. In the excitation analysis, a 40-dimensional Gaussian codebook with 128 entries and an excitation signal memory ranging from 32 to 160 samples were used.

The behaviors of $|\beta_1|$ and $|\beta_2|$ with time are distinctly different, but the difference is easily understood. First note that the impulse responses $h(n)$ are all normalized. Then recall that β_2 is the constant multiplier of a portion of the previously computed excitation that contributes to the excitation in the current frame. Notice in Figure 3(b) that $|\beta_2|$ remains fairly constant near unity, except for large spikes and abrupt dips toward zero. This is to be expected, since in steady-state regions, the amplitude in an excitation analysis frame should be about the same as the amplitude in previous frames in that steady-state region. However, in a transition frame from voiced to unvoiced, past excitation amplitudes will be much larger than required, and therefore $|\beta_2|$ will have to be small to compensate. Likewise, in transitions from unvoiced to voiced, the immediate past excitation will be small, while a larger excitation will be required in the current frame. Therefore $|\beta_2|$ must be large to compensate.

In contrast, $|\beta_1|$ tends to track the energy envelope of the speech signal and is somewhat better behaved. Clearly, the amplitude of the residual signal $y_1(n)$ will be proportional to the amplitude of the original speech signal. Therefore, since the codebook sequences all have the same energy, $|\beta_1|$ will track the amplitude of $y_1(n)$ and therefore also the amplitude of the input speech signal.

Figure 3 shows that $|\beta_1|$ and $|\beta_2|$ are not highly correlated, and therefore it would seem that there is little to be gained by jointly quantizing them. However, recall that $|\beta_2|$ is generally close to unity, and $|\beta_1|$ tends to follow the amplitude of the speech signal and the excitation signal. This suggests that if $|\beta_2|$ is normalized by a function of the previous excitation energy, then the correlation with $|\beta_1|$ can be greatly increased. Indeed, Figure 4(a) shows the parameter

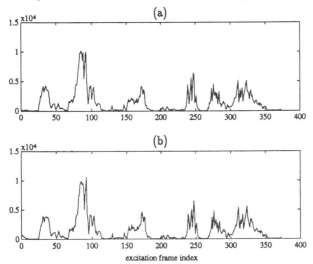

Figure 4: (a) Unquantized normalized self-excitation gain $\alpha|\beta_2|$. (b) Quantized normalized self-excitation gain.

$\alpha|\beta_2|$, where

$$\alpha = \left[\left(\sum_n e^2(n - \gamma_2) \right)^{1/2} \left(\sum_n e^2(n - L) \right)^{1/2} \right]^{1/2} \qquad (8)$$

with L representing the excitation frame length. That is, the gain normalizing factor α is the geometric mean of the energy of the excitation segment beginning at γ_2 and the energy of the just previous excitation frame. This averaging gives a smoothly varying normalizing factor which, as can be seen from Figure 4(a), converts $|\beta_2|$ into a parameter that varies with time in much the same way that $|\beta_1|$ varies.

240

Figure 5 shows the correlation between $|\beta_1|$ and $\alpha|\beta_2|$ more clearly. Indeed, Figure 5 implies that $|\beta_1|$ is proportional to $\alpha|\beta_2|$ to within a constant maximum percentage error. The straight line in Figure 5 is a least squares fit to the

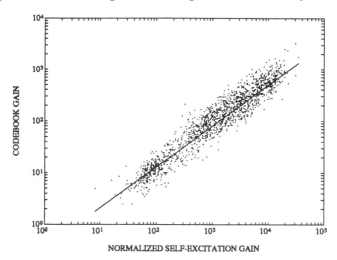

Figure 5: **Illustration of correlation between** $|\beta_1|$ **and** $\alpha|\beta_2|$.

data which include 1536 frames from four different utterances by four different speakers. This linear fit to the log-log data is given by

$$|\beta_1| = 0.3927(\alpha|\beta_2|)^{0.7906}, \tag{9}$$

which serves as an approximate relationship between the codebook gain $|\beta_1|$ and the normalized self-excited gain $\alpha|\beta_2|$.

The 4800 bps homomorphic vocoder that was previously reported [3] used the bit allocation scheme given in the first row of Table 1. In the 4800 bps

Bits/Frame					Samples/Frame		Bit Rate
cepstrum	β_1	β_2	γ_1	γ_2	excitation	cepstrum	(bits/sec)
8	4	4	7	7	40	160	4800
8	1	4	7	7	40	160	4200

Table 1: **Bit allocation for homomorphic vocoders.**

vocoder, each of the two gain parameters was coded using a 3-bit APCM coder to code $|\beta_1|$ and $|\beta_2|$. The results of the previous section suggest that the total bit allocation can be reduced by jointly coding $|\beta_1|$ and $\alpha|\beta_2|$. Clearly, many schemes can be found to take advantage of the correlation illustrated in Figure 5. One approach is simply to code $\alpha|\beta_2|$ using 3-bit APCM. (Figure 4(b) shows an example of this 3-bit quantization.) This information together with one bit each for the signs of β_1 and β_2 completes the representation for a total of five bits instead of eight. With an excitation frame rate of 200 frames/sec, this results in a reduction of 600 bps.

At the receiver, $|\beta_2|$ is derived from the quantized version of $\alpha|\beta_2|$ by dividing by α, which is derivable using (8) from the past excitation. Then $|\beta_1|$ is obtained from $\alpha|\beta_2|$ through (9). As an illustration, Figures 6(a) and 6(b) show the decoded $|\beta_1|$ and $|\beta_2|$ respectively for the corresponding parameters in Figures 3(a) and 3(b).

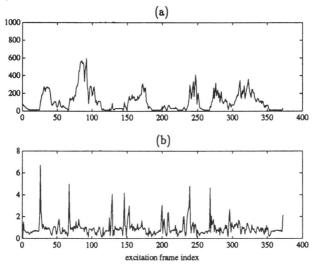

excitation frame index

Figure 6: (a) Decoded codebook gain. (b) Decoded self-excitation gain.

The performance of the 4200 bps homomorphic vocoder is virtually identical to the performance of the 4800 bps version. This is confirmed by careful listening tests and by the fact that over a range of speakers and utterances, the signal-to-noise ratio only decreases from 8.01 dB to 7.62 dB in going from 4800 to 4200 bps.

EFFECTIVE MODELING OF THE EXCITATION SIGNAL

The excitation generator introduced in Section 2 routinely searches through a fixed interval of the past excitation signal, and then searches through the Gaussian codebook without considering the state of the speech segment, i.e., voiced or unvoiced. It is likely that this static excitation modeling scheme is not the best way to describe the nonstationary behavior of the sound sources. In segments of quasi-periodic voiced speech, for example, a sequence selected from a Gaussian codebook is less effective than a sequence selected from the past excitation signal. On the other hand, in segments of random noise-like unvoiced speech, a sequence selected from the past excitation signal is generally less effective than a sequence selected from a Gaussian codebook. More effective excitation modeling is particularly necessary for low bit rate coding because most of the bits in analysis-by-synthesis coding are devoted to the excitation. Therefore, dynamic excitation modeling, i.e., a different excitation modeling for each different mode of the sound sources, may produce higher quality at a given bit rate or allow the bit rate to be lower while maintaining a given level of quality.

Different excitation modeling strategies are used depending on whether a given frame is classified as *voiced, unvoiced,* or *mixed.* For voiced segments, the excitation signal $e(n)$ is comprised of two sequences selected from a time-varying queue of the past excitation history, i.e.,

$$e(n) = \beta_1 e(n - \gamma_1) + \beta_2 e(n - \gamma_2). \tag{10}$$

This emphasizes the pitch dependent periodic nature. For unvoiced frames, the excitation signal $e(n)$ is modeled by a Gaussian codebook sequence, i.e.,

$$e(n) = \beta f_\gamma(n), \tag{11}$$

emphasizing the random nature. Finally, in the case of segments classified as mixed excitation, the excitation $e(n)$ is modeled as the sum of a Gaussian codebook sequence and a sequence selected from the fixed interval of the past excitation history, i.e.,

$$e(n) = \beta_1 f_{\gamma_1}(n) + \beta_2 e(n - \gamma_2). \tag{12}$$

The queue of the past excitation history is updated by whatever model is chosen for a given excitation frame.

Since different voicing states result in different excitation modeling, the classification of a given speech block into a correct voicing state (voiced, unvoiced, or mixed) is important. The cepstrum has been successfully applied as the basis for a voicing decision as well as pitch detection [6,7]. In the homomorphic vocoder, the cepstrum must be computed to obtain the vocal tract impulse response, and therefore the cepstrum is available without any additional computation or delay. Consequently, a voicing classification method primarily based on the cepstrum is adopted in our system.

The state of the speech segment is classified as either periodic (voiced) or non-periodic (mixed and unvoiced) based on the strength of the peak cepstral value. If the value of this peak exceeds a threshold T, the segment is classified as periodic, otherwise as non-periodic. The threshold T is switched between two values T_1 and T_2 $(T_1 > T_2)$ depending on the state of the previous speech segment. More precisely, the value of T_1 is used as a threshold if the previous state is non-periodic, whereas the value of T_2 is used as a threshold if the previous state is periodic. The size of FFT used is 512, and the peak cepstral value is determined over a range from 25 to 100 on the cepstral domain. We have found that values of $T_1 = 0.3$ and $T_2 = 0.2$ yield good performance.

Non-periodic segments are further classified into either mixed or unvoiced (also includes silence) based on the zero-crossing count and the energy of the original speech segment. First, a zero-crossing count of the corresponding speech block is made. If the zero-crossing count either exceeds a threshold Z_1 or does not exceed a threshold Z_2 ($Z_1 > Z_2$), the segment is classified as unvoiced. If the zero-crossing count lies between Z_1 and Z_2, the energy of the block is computed. If the energy of the block exceeds a threshold E, the state of the speech block is classified as mixed. Otherwise it is classified as unvoiced.

Computer simulations were done to evaluate the performance of this dynamic excitation modeling scheme. In the simulations, 11 low-time cepstrum values with a 26 sample normalized impulse response were updated every 20 msec to represent the vocal tract information. A 40-dimensional Gaussian codebook and the past excitation signal ranging from 32 to 160 samples were used in the excitation analysis. The excitation analysis was updated every 5 msec. Using Gaussian codebooks of size 128 or 256, speech signals were synthesized using the dynamic excitation modeling scheme as well as the static excitation modeling scheme. Throughout the experiments, the low-time cepstrum values and the excitation gain parameters were not quantized, since our main interest was to examine the performance improvement of the dynamic excitation modeling scheme against the static excitation modeling scheme.

In our experiments, the dynamic excitation modeling scheme out performed the static excitation modeling scheme both subjectively and objectively. This is a promising result, since a large portion of speech signal is unvoiced, and since unvoiced segments are represented by a single Gaussian codebook sequence in the dynamic excitation modeling scheme. This results in a significant savings in bits for representing unvoiced segments, and those saved bits can be used for other purposes such as more faithful quantization of cepstrum or the excitation gains, channel error correction, etc. For a limited set of speech data, for example, the ratio among the voiced, mixed, and unvoiced was 0.43, 0.20, and 0.37, respectively. Thus, for a 256 size Gaussian codebook, the static excitation modeling would require total 3000 bits per second to represent the index of the selected Gaussian sequence and the position of the chosen past excitation sequence, whereas the dynamic excitation modeling requires total 2396 bits per second on the average.

CONCLUSIONS

This paper has described a homomorphic vocoder with excitation derived by analysis-by-synthesis. It was shown that significant improvements in performance can be achieved by using gain normalization prior to coding the excitation gain parameters and by utilizing the cepstrum as a basis for choosing among three different modeling strategies for each excitation analysis frame.

References

[1] B. S. Atal and J. R. Remde, "A new model of LPC excitation for producing natural-sounding speech at low bit rates," *Intl. Conf. on Acoustics, Speech, and Signal Proc.*, pp. 614-617, 1982.

[2] M. R. Schroeder and B. S. Atal, "Code-excited linear prediction (CELP): high-quality speech at very low bit rates," *Intl. Conf. on Acoustics, Speech, and Signal Proc.*, pp. 937-940, 1985.

[3] J. H. Chung and R. W. Schafer, "A 4.8 Kbps homomorphic vocoder using analysis-by-synthesis excitation analysis," *Intl. Conf. on Acoustics, Speech, and Signal Proc.*, pp. 144-147, 1989.

[4] A. V. Oppenheim, "A speech analysis system based on homomorphic filtering," *J. Acoust. Soc. Am.*, vol 45, pp. 458 - 465, February, 1969.

[5] A. V. Oppenheim and R. W. Schafer, *Discrete-Time Signal Processing*, Prentice-Hall, Englewood Cliffs, N.J., 1988.

[6] A. M. Noll, "Cepstrum Pitch Determination," *J. Acoust. Soc. Amer.*, vol. 41, pp. 293-309, Feb., 1967.

[7] R. W. Schafer and L. R. Rabiner, "System for automatic formant analysis of voiced speech," *J. Acoust. Soc. Am.*, vol. 47, pp. 634-648, Feb., 1970.

23

A 400-BPS SPEECH CODING ALGORITHM
FOR REAL-TIME IMPLEMENTATION

Y. J. Liu

ITT Aerospace/Communications Division
Nutley, New Jersey 07110

INTRODUCTION

An LPC-based speech coder at 2400 bps has been widely used for secure voice communication. An even lower rate at several hundred bps has attracted much interest recently mainly because of its high resistance to channel noise and unfriendly jamming. However, a loss of speech intelligibility, longer throughput delay and other degradations are characteristic of low-bit-rate coding systems. To overcome these difficulties, complicated processing algorithms are often required. But this may make hardware implementation an extremely difficult task. Even though there are many these undesirable consequences associated with low-bit-rate speech coders, ITT Defense Communications Division has developed a high quality speech coder at 400 bps without the above mentioned drawbacks.

The coder starts with a government standard LPC-10 algorithm. The output parameters are then postprocessed to achieve the necessary rate reduction. Only techniques with minimum complexity were considered. These techniques included line-spectrum frequencies, vector quantization and trellis coding. Because of its intended application to successful communication over a high frequency channel, a Reed Solomon coder was also utilized for error-correction. After error-correction, the final transmission rate is increased from 400 bps to 600 bps.

Since the first publication of the coder [1], there were additional efforts to make this coder implementable by a single fixed point digital signal processor. The processor selected is TMS320C25. Because this processor is based upon 16-bit arithmetic, the unlimited accuracy of floating point operation is no longer available. Because both the LPC10-E algorithm and the 400-bps algorithm have to be accommodated on a single TMS320C25 chip, time-consuming operations have to be reduced. The above two requirements have led us to the improvements of solving polynomial roots for line-spectrum frequencies and fast search techniques for vector quantization. The trellis gain coder has also been modified to improve the leading sound match.

The basic algorithm together with these enhancements is discussed. The three-speaker average DRT for the fixed point simulation is 81.3. This score is even better than our earlier floating point algorithm which scored at 80 [1]. Information on error-correction techniques employed can be found in a different paper [2].

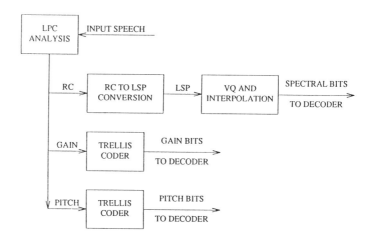

Fig. 1. An overall encoding structure.

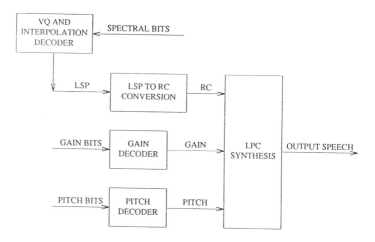

Fig. 2. An overall decoding structure.

SYSTEM CONFIGURATION

An overall encoding structure is shown in Fig. 1. The digitized input speech is first processed through an LPC-10 analysis to generate the required reflection coefficients, gain and pitch. The reflection coefficients are then converted to line-spectrum frequencies (LSFs) [3,4]. For each frame, the LSFs are either encoded by a vector quantizer [5] or interpolated. Both vector quantization index and interpolation information are transmitted. The gain and pitch are encoded by trellis coding.

The decoding follows a reverse process of encoding and is shown in Fig. 2. The decoded line-spectrum frequencies are converted back to reflection coefficients. The restored reflection coefficients together with decoded gain and pitch are processed through LPC-10 synthesis to generate the speech.

A brief description of each block is given in the following sections.

FEATURE VECTOR SELECTION

Spectral quantization in low-bit-rate speech coding is quite important. A good reproduction of spectral information can usually lead to high quality speech. There are two important issues in spectral quantization. One is feature vector selection and the other is encoding technique. LSFs were selected for the feature vector and vector quantization was selected for the encoding. Both are regarded as good coding techniques and have been applied to encode speech at rates such as 800 bps while attaining high intelligibility. The solution of LSFs is discussed in this section while vector quantization is discussed in the next section.

The LSFs are obtained from an LPC analysis filter. Since the government standard LPC-10 algorithm outputs reflection coefficients, they must first be converted to prediction coefficients. In terms of prediction coefficients, the mth order LPC filter is given by the following equation

$$A(z)=1+\sum_{k=1}^{m}a_k z^{(-k)}$$ (1)

where a_k is the kth prediction coefficient. The mth order LPC filter can be extended to the $(m+1)$th order by assuming the $(m+1)$th reflection coefficient to be either 1 or -1. Corresponding to these two different values, a sum filter and a difference filter are subsequently constructed. The sum filter, $P(z)$, and the difference filter, $Q(z)$, are given by the following equations.

$$P(z)=A(z)+z^{-(m+1)}A(z^{-1})$$ (2)

$$Q(z)=A(z)-z^{-(m+1)}A(z^{-1})$$ (3)

$P(z)$ has the property of symmetry while $Q(z)$ has the property of anti-symmetry. After removing the real root, +1, from $P(z)$ and the real root, -1, from $Q(z)$, they can be reduced to two polynomials , $P_1(z)$ and $Q_1(z)$ of the mth order. The roots of both $P_1(z)$ and $Q_1(z)$ are located on the unit circle. Because of this property, a new variable x given below is defined

$$x=cos(w)$$ (4)

where w is a normalized angular frequency. Using this new variable x and the symmetry characteristics, two polynomials $P_2(x)$ and $Q_2(x)$ of the $(m/2)$th order are produced [4]. Even though the roots of $P_1(z)$ and $Q_1(z)$ are in general complex, both $P_2(x)$ and $Q_2(x)$ have real roots only. By taking the arccosine of the real roots of polynomials $P_2(x)$ and $Q_2(x)$, the line-spectrum frequencies are produced.

For synthesis, the LSFs can be converted back to prediction coefficients through a reverse process. From (2) and (3), the restored prediction coefficients can be utilized to generate $A(z)$ through the following equation.

$$A(z) = \frac{P(z) + Q(z)}{2} \qquad (5)$$

Because the government standard LPC-10 algorithm requires reflection coefficients as input, the prediction coefficients are further transformed back to reflection coefficients for synthesis.

LSFs have several important properties. The first is that all roots are located on the unit circle in the z plane. The second is that the roots of symmetric polynomial and anti-symmetric polynomial are interlaced. The third is that the spectral sensitivity of LSFs is localized. This means an error in one LSF causes spectral error only at that particular LSF [3].

For real time implementation, the most difficult task is the solution of polynomial roots. There are several existing algorithms to solve for polynomial roots. For floating point simulation, there are no truncation errors and execution time limitations. Hence, any technique is acceptable as long as reliable roots can be obtained. For fixed point processor implementation, both these two factors and the accuracy requirement must be considered. Our earlier floating point simulation solved each root sequentially. Thus, there is undesirable error propagation from one root to the next. This together with a large dynamic range of certain internal variables made the machine implementation extremely difficult. A new method was subsequently developed without these drawbacks using a two step procedure to solve for the polynomial roots. In the first step, all roots were independently estimated within an accuracy of 15.6 Hz. This is of the same accuracy as the reported FFT technique [6]. In the second step, each estimated root was refined to an average accuracy of less than 0.1 Hz, much better than the refined accuracy of a few Hz claimed by the FFT technique [6].

After obtaining the root, an arccosine is taken to compute the line-spectrum frequency. Series expansion or polynomial approximations are possible ways of machine implementation. However, many terms have to be retained in order to achieve the desired accuracy and the speed requirements prevent their application. Therefore, table look-up techniques are utilized. An arccosine table having a power of two entries is first defined. Three continuous table points closest to the input value are determined. A simple shift operation of the input easily locates the table entry due to the particular table design. Using the three table entries, a quadratic interpolation is applied to compute the arccosine value. The simulation showed the interpolation technique is not only fast but also very accurate. Actually, even a linear interpolation gives pretty accurate values.

VECTOR QUANTIZATION AND INTERPOLATION

After the generation of LSFs, they are encoded using a vector quantization technique. The success of vector quantization depends on the quality of a codebook. The codebook was designed using an algorithm described by the author in a previous paper [7]. A data base consisting of 54 male speakers was utilized. This data base includes all phonemes of English and the normal English accents. There are two steps in the design of a codebook. In the first step, an initial codebook is designed using a procedure that minimizes the guesswork. Instead of choosing all the codewords at random from the start, this procedure selects only one codeword initially from the training data base. The desired number of codewords are subsequently generated through clustering. In the second step, the initial codebook is optimized through an iterative procedure. During the design of a codebook, a unique perceptual distortion measure was adopted. This distortion measure has the advantage of easy implementation and gives greater weights to those LSFs which are considered perceptually important. This same distortion measure was utilized for vector quantization of speech. As a result, better speech intelligibility can be reproduced.

For floating point simulation, a full search technique was utilized to search for a codeword. Under a single TMS320C25 processor restriction, the full search cannot be implemented. For this processor, multiplication is as fast as addition or comparison and there are efficient instructions to perform accumulated sum of products. Hence, there were no attempts to reduce the number of multiplications in exchange for additions or comparisons. Instead, a two level tree codebook was designed. Even though the VQ index selected by the tree search may sometimes be different from the full search, the distortion difference was found to be very small. However, a more strict constraint requiring the same VQ indices was adopted. By adopting a strategy to allow the desired full search codeword to fall into the tree search subspace, the fixed point tree search and fixed point full search generate almost the same VQ indices. Considering voiced frames only, the following results are obtained using the DRT database for each of the three speakers RH, CH and JE.

Table 1. Fixed point VQ accuracy

Speaker	VQ Disagreement Percentage
RH	0.084%
CH	0.061%
JE	0.043%

Table 1 shows the fixed point tree search is virtually identical to the fixed point full search.

To achieve the desired rate at 400 bps, the spectral information cannot be transmitted in every frame. Our earlier coder utilized a variable frame repeat strategy which can cause long throughput delay, less tolerance to bit errors and some design complexities. To overcome these difficulties, a 50% frame interpolation was used to maintain the transmission at a fixed rate. This interpolation information was also encoded and transmitted so the receiver can reconstruct the missing frames.

250

EXCITATION CODING

A fake process trellis coder has previously been employed by ITT to encode pitch and gain. A table look up is an important part of this coder. An M-L search was performed to achieve optimum match between the input data and the decoded data. This table together with certain parameters such as search depth and prediction gain were estimated from a training data base.

An enhanced version of this coder has been developed. Instead of depending upon a fixed table, adaptive tracking techniques were utilized to track closely the input data variations. The performance of this coder is demonstrated in Fig. 3 which is a plot of the input gain and the decoded gain for the word "CHOOSE". The close match is evident in the plot during the whole duration of this word. The signal- to-noise ratio is computed to be about 32 dB.

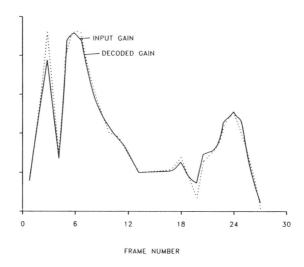

Fig. 3. Typical gain coder performance for the word "CHOOSE".

Even though the gain coder works very well, certain speech frames may still be missed once in a while. Depending upon where the missing frames occur, speech intelligibility may be affected. The situation is classified into three major categories. In the first category, a frame with negligible gain may appear between two frames with very strong gain. This error is easily correctable in the decoder and does not have much effect on speech intelligibility. In the second category, a few frames may be missed near the tailing end of a word. Normally, the missing frames have a gain close to the background noise level and are not of great concern. Also, the speech intelligibility is not affected. In the third category, a few frames may be missed at the beginning of a word. This situation normally happens when the leading sound is a weak fricative or nasal followed by a strong vowel. Due to the tracker's intention to minimize the overall distortion, the good match of weak leading frames is sacrificed. Depending upon the number of missing frames, the word may become totally unintelligible.

Fig. 4 is a plot of the input gain and the decoded gain for the word "POOP".

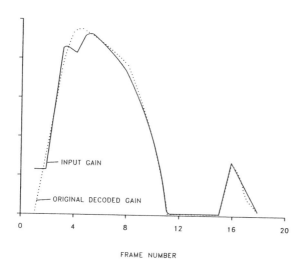

Fig. 4. Input gain and decoded gain using the original algorithm for the word "POOP".

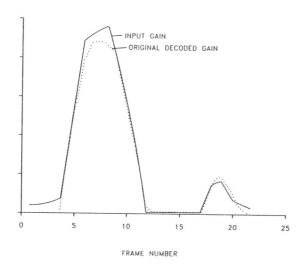

Fig. 5. Input gain and decoded gain using the original algorithm for the word "MIST".

This word has a signal-to-noise ratio of 24.3 dB. For this word, a single leading frame is missed and the DRT test showed no loss of intelligibility. Fig. 5 is a plot of he input gain and the decoded gain for the word "MIST". This word has a signal-to -noise ratio of 14.36 dB. For this word, a few weak frames are missed at the start of

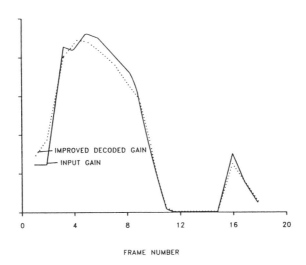

FRAME NUMBER

Fig. 6. Input gain and decoded gain using the modified algorithm for the word "POOP".

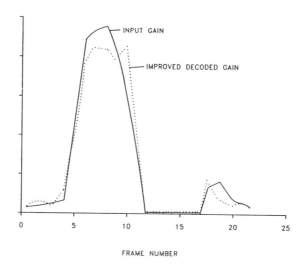

FRAME NUMBER

Fig. 7. Input gain and decoded gain using the modified algorithm for the word "MIST".

the word and the DRT test showed significant loss of intelligibility.

Because the errors caused by the third category have a serious impact on speech intelligibility, a modified algorithm has been developed. This algorithm utilized a sub-optimal approach in terms of overall distortion in exchange for higher intelligibility. Fig. 6 and Fig. 7 are the same plots of the word "POOP" and "MIST" using the new algorithm. No weak frames are missed. However the signal-to-noise ratio drops from 24.35 dB to 19.30 dB for the word "POOP" and drops from 14.3 dB to 11.8 dB for the word "MIST". The DRT test showed both words have perfect intelligibility.

SIMULATION RESULTS

Our basic floating point simulation of the 400-bps speech coder had an average DRT score of 80 for the three speakers RH, CH and JE. For a single TMS320C25 processor implementation, both the LPC10-E algorithm and the 400-bps algorithm have been converted to fixed point using only 16-bit arithmetic. To isolate the effect of the fixed point LPC10-E algorithm, two different DRT tests using the same three speakers were performed. The first test utilized the floating point LPC10-E algorithm and the fixed point 400-bps algorithm while the second test utilized the fixed point LPC10-E algorithm and the fixed point 400-bps algorithm. Both tests were performed by DYNASTAT.

Table 2. DRT test results

LPC10-E	400 bps	Speaker RH	Speaker CH	Speaker JE	All
Floating	Floating	77.3	82.4	80.2	80.0
Floating	Fixed	79.6	82.4	81.9	81.3
Fixed	Fixed	78.1	80.1	84.2	80.8

The test results are shown in Table 2. In the above table, "Floating" represents floating point simulation and "Fixed" represents fixed point simulation. Several points can be concluded from this table. First, the fixed point implementation has little effect on speech intelligibility. Second, a high quality speech coder at 400 bps is implementable using only 16-bit arithmetic. Third, the special tree search has the same performance as the full search. The reasons the fixed point simulation scored higher than the floating point simulation were attributed to two factors. First, a further improved gain coder was utilized in the fixed point simulation but not in the floating point simulation. Second, to save storage space, all speech parameters were stored in fixed point format for both the fixed point simulation and the floating point simulation. Due to the scaling difference, the accuracy of the fixed point simulation was slightly better.

DISCUSSION

Another important issue in low-bit-rate speech coding is associated with speaker independence. Speaker independence is more serious for low-bit-rate speech coding due to the utilization of a codebook. A simple solution is to design an universal codebook that includes all possible accents. In order not to degrade speech intelligibility, the codebook size has to be increased. But the low-bit-rate requirements limit the codebook size to be used. This means some different strategies must be adopted.

A possible strategy is to have a codebook storage module for each possible accent. The only requirement is to let both sides of the communication line insert the appropriate codebook storage module. The information on one speaker's accent can be transmitted to the other speaker during a modem training period. Therefore, there should be no degradation in speech intelligibility due to different accents for low-bit-rate transmission.

Depending upon the specific users of the low-bit-rate coder, two sub-optimal strategies can also be adopted. The first is to design a codebook on a wider selection of training data, including a representative cross section of English accents. The second is to provide for multiple codebooks and a codebook selection algorithm to cover subsets of speaker population.

The DRT results of the fixed point simulation show that it is feasible to use a fixed point digital signal processor for real time implementation without any loss of intelligibility. This fixed point simulation can be quickly transformed to an assembly language program of the TMS320C25 processor. Currently, the real time implementation was complete using a 145 ns cycle time and one wait state for both the program memory and data memory. The execution time was just enough to execute both LPC10-E and 400-bps algorithms within the available 22.5 ms time frame.

Since the work on real time implementation, more improvements have been made on both the spectrum match and the gain match. The most recent DRT test showed the floating point simulation of the 400-bps coder has a DRT score of 82.5. This will be presented in future conferences.

REFERENCES

(1)   Y.J. Liu and Joseph Rothweiler, "A High Quality Speech Coder at 400 bps," ICASSP'89, 1989.

(2)   Joseph Rothweiler and Y.J. Liu, "Low Rate Voice Algorithm for HFECCM Applications," Military Speech Tech., 1988.

(3)   G.S. Kang and L.J. Fransen, "Application of Line-Spectrum Pairs to Low-Bit-Rate Speech Encoders," Proc. ICASSP-85, pp. 244-247, 1985.

(4)   F.K. Soong and B. Juang, "Line-Spectrum Pair and Speech Data Compression," ICASSP-83, 1.10.1-1.10.4, 1983.

(5)   A. Buzo, A.H. Gray, and J.D. Markel, "Speech Coding Based Upon Vector Quantization," IEEE Trans. Acoust., Speech and Sig. Process., ASSP-28 , pp. 562-574, Oct. 1980.

(6)   G.S. Kang and W. M. Jewett, "Error-Resistant Narrowband Voice Encoder," NRL Report 9018, 1986.

(7)   Y.J. Liu, "Improving the Codebook Design for Vector Quantization," Proc. MILCOM'87, pp. 556-559, 1987.

PART VI

SPEECH CODING FOR NOISY CHANNELS AND REAL-TIME IMPLEMENTATIONS

The development of a speech coder is not complete without consideration of both the effects of noisy channels and the requirements for a cost-effective real-time implementation. For relatively high error rate channels, a speech coding algorithm must include techniques to combat the degradations introduced by channel errors. Rather than depend solely on the use of error correcting codes applied to the entire bit stream, it is now recognized that considerable benefit is attainable by selective protection of sensitive parameters, by smoothing parameter values across frames, and by various other techniques that improve robustness to a noisy environment. Real-time implementations force engineers to deal with issues such as algorithm complexity and processor word length and architectural limitations. Often substantial modifications to an algorithm must be made in order to tailor it to the constraints of a hardware environment.

This section reports on some of the issues and considerations given to noisy channels and real-time implementations, primarily for algorithms belonging to the generic family of vector excitation or CELP coders. Techniques for reducing the impact of channel errors to CELP coders are reported in the chapters by Kleijn and by Perkis and Ribbum. A model for joint source and channel coding problems is examined by Soleymani and Khandani. Real-time implementation issues and solutions for two different coders, both based on excitation/filter coding with analysis-by-synthesis, are presented in the final two chapters of this section by McGrath, Barnwell, and Rose and by Chen et al.

24

SOURCE-DEPENDENT CHANNEL CODING AND ITS APPLICATION TO CELP

W. Bastiaan Kleijn

Acoustics Research Department,
AT&T Bell Laboratories,
Murray Hill, NJ 07974, USA

INTRODUCTION

Efforts to minimize the effect of channel errors on low-rate speech coders, such as code-excited linear prediction (CELP), can be divided into methods which change the robustness of the source coder [1], and procedures which add error correction and error detection by means of a separate channel coder. The channel coder often provides several levels of protection, with more error protection being assigned to the parameters or bits which are judged to be more sensitive to channel errors [2-4]. Such protection can be considered to be a form of source-dependent channel coding. This paper focuses on improving source-dependent channel codes through optimization of objective error criteria.

Conventional channel codes, which are designed independently from the source coder, are generally not optimal for speech coders. Channel and source coding design can be separated without affecting performance only if an arbitrarily complex coder-decoder design with arbitrarily long delay is optimized for a particular channel. Then, the source-coder information rate can be matched to the channel capacity. However, this results in suboptimal performance for channels of higher or lower capacity (or different characteristics). Instead, one would like to optimize the mean performance over the ensemble of channels which the speech coder will encounter, while maintaining a delay consistent with its real-time character. To ensure best performance under such conditions, the source distortion has to be considered in the design of the channel-coder.

Thus, the performance of coders employing conventional channel coding or no channel coding can usually be optimized for channels of lower capacity without adding to the redundancy. Gradient-descent techniques to find good channel codes for error protection without redundancy were discussed in [5]. The present paper proposes a design method which globally optimizes channel coding for a given source coder and a given rate of codewords added for the purpose of error protection. This approach results in benefits similar to those of simultaneous optimization of source and channel coding [6] but avoids modification of the complex quantization procedures used in speech coders.

If mainly single-bit errors are assumed to occur, then a good channel code consists of codewords arranged such that a single-bit error results in a codeword associated with a quantization level identical or similar to the originally encoded level. In other words, the codewords have "good" neighbors. Available redundant codewords (which are not transmitted, but may be received) are used for error protection. When used for error correction they are decoded as quantization levels. Alternatively, the redundant codewords can be used for error detection and trigger a recovery scheme which may use information about prior frames, related parameters, and levels associated with neighboring codewords.

The CELP speech-compression algorithm [7] has been shown to provide excellent speech quality at low bit rates [8]. It will be used to illustrate the present method channel coding. As in many other compression algorithms, the issue of channel-error sensitivity of CELP is complicated by its use of feedback. In this paper the optimization of the channel codes is made practical by separating the effect of coding errors into two aspects: the immediate effect of the decoding errors, and the persistence of the resulting distortion of the synthetic speech. The improvement of the channel coder discussed here addresses only the immediate effect; the persistence of the distortion is a function of the source-coder design.

OPTIMIZATION OF SOURCE-DEPENDENT CHANNEL CODES

Definition of the Error Criterion

Let c_m denote codeword m in the binary form in which it is transmitted over the channel. In the following, it is convenient to think of the quantization index i as a mapping of the codeword index m, denoted by $i(m)$. Since each codeword is decoded as a quantization level, this mapping (but not its inverse) is unique. Minimizing the error sensitivity is equivalent to finding the optimal mapping function $i(.)$. The probability that channel code c_m is transmitted is denoted as $P(c_m)$. Let ρ_m be a binary flag indicating whether codeword c_m is redundant ($\rho_m=0$) or not ($\rho_m=1$). Then $P(c_m) = \rho_m P(i(m))$, where $P(i(m))$ is the probability that the quantization level with index $i(m)$ provides the parameter with its best fit. The source-signal value (or vector) associated with the quantization index $i(m)$ is denoted as $r_{i(m)}$. In general, we assume that $r_{i(m)}$ can be the result of multiple quantizations; for example the excitation vector of CELP has a shape quantization as well as a gain quantization.

$D_{i(m)}(r_{i(n)})$ denotes the mean distance between the parameter $r_{i(n)}$ and the unquantized value or vector, given that codeword c_m is transmitted. That is, $D_{i(m)}(r_{i(n)})-D_{i(m)}(r_{i(m)})$ is a penalty function for changing the codeword from m to n under the constraint that $i(m)$ is the quantization index of the value or vector of best fit. Also $f(m,k)$ is the index of the codeword obtained by inverting bit k of codeword c_m and K denotes the number of bits per codeword. Then an appropriate criterion describing the sensitivity of the parameter encoding to single-bit errors is:

$$\varepsilon = \sum_{m=0}^{2^K-1} P(c_m) \sum_{k=0}^{K-1} \left[D_{i(m)}(r_{i(f(m,k))}) - D_{i(m)}(r_{i(m)}) \right]. \tag{1}$$

The reference level $D_{i(m)}(r_{i(m)})$ is constant and can be omitted from criterion (1).

Since speech-coding algorithms have memory, computation of a penalty function $D_{i(m)}(.)$ which is a direct measure of the speech quality is expensive. However, it is sufficient to optimize any criterion which is a monotonically increasing function of a desirable criterion considering the actual speech.

Simulated Annealing for Channel-Code Optimization

Finding the mapping $i(.)$ which minimizes the criterion of (1) is a combinatorial optimization. A variety of nonexhaustive techniques has been used for this type of problem [5, 6, 9]. A particularly powerful technique, which finds good solutions to a variety of combinatorial-optimization problems, is simulated annealing [10, 11]. Here the simulated-annealing algorithm will be used to develop good source-dependent channel codes.

> pick random codeword index m, $\ 0 \le m < 2^K$
> pick random codeword index n, $\ 0 \le n < 2^K$
> $\eta_{orig} = 0$
> for $k = 0$ to $k = K - 1$
> do
> $\qquad \eta_{orig} = \eta_{orig} + P(c_m)D_{i(m)}(r_{i(f(m,k))}) + P(c_{f(m,k)})D_{i(f(m,k))}(r_{i(m)})$
> $\qquad \eta_{orig} = \eta_{orig} + P(c_n)D_{i(n)}(r_{i(f(n,k))}) + P(c_{f(n,k)})D_{i(f(n,k))}(r_{i(n)})$
> end do
> exchange mapping $i(.)$ with trial mapping $j(.)$
> $\qquad$ where $j(q){=}i(q)$, $0 \le q < 2^K$, $q \ne m$, $q \ne n$
> $\qquad \qquad j(m){=}i(n)$
> $\qquad \qquad j(n){=}i(m)$
> $\eta_{trial} = 0$
> for $k = 0$ to $k = K - 1$
> do
> $\qquad \eta_{trial} = \eta_{trial} + P(c_m)D_{j(m)}(r_{j(f(m,k))}) + P(c_{f(m,k)})D_{j(f(m,k))}(r_{j(m)})$
> $\qquad \eta_{trial} = \eta_{trial} + P(c_n)D_{j(n)}(r_{j(f(n,k))}) + P(c_{f(n,k)})D_{j(f(n,k))}(r_{j(n)})$
> end do

Table 1. Perturbation of Channel Code.

During the annealing process an error criterion ε, the "energy" of the system, is minimized. This is achieved by gradually lowering an abstract temperature T while maintaining the system in equilibrium. In equilibrium the system travels through its phase space such that the probability of the system being in a certain state with energy ε_t at time t is proportional to the Boltzmann factor $exp(-\varepsilon_t/T)$. Occurrences of the system states have a Gibbs distribution. An appropriate stochastic motion through phase space is achieved with the Metropolis algorithm [12] which perturbs the system state in a random manner to obtain a trial state of energy ε_{trial}, and then accepts or rejects the trial state as the next system state with probability one if $\varepsilon_{trial} < \varepsilon_t$ and probability $exp((\varepsilon_t - \varepsilon_{trial})/T)$ otherwise. States with low energy (error) are more likely than states with high energy. At high temperature the distribution is less skewed than at low temperature. Because of the fact that the system has memory (it travels through phase space with small, discrete steps), lowering the temperature gradually causes the system to gravitate towards regions of high probability, i.e. deep energy basins.

For the case where no redundant codewords are present, trial channel-code perturbations can be generated by exchanging two randomly selected transmission codewords. The transition probability from the original channel code to the trial channel code depends only on the difference in error criterion (1) before and after this change. To minimize computer run time, the error criterion itself need not be evaluated during each iteration, but only the contributions which are modified by the codeword exchange. If only single-bit errors are considered, only quantization levels with transmission codewords differing by a single bit from the exchanged codewords are involved. The computation of the essential energy contributions for this case, denoted by η_{orig} and η_{trial}, is shown in Table 1.

The result of finding the best neighbors for the ensemble of transmitted codewords can be interpreted as an approximate error correction. If redundant codewords are added, this approximate error correction can be improved by finding better neighbors for the transmitted codewords. Each quantization index $i(m)$ has one or more codewords c_m associated with it; one of these codewords is the transmitted codeword ($\rho_m=1$). If there are a total of 2^K codewords, R of which are transmitted, then the simulated annealing algorithm must be augmented so that any of the $2^K - R$ redundant codewords can point to any of the R valid indices to quantized parameter levels. To perturb the redundant codewords ($\rho_m=0$) one changes the associated index at random to another valid index. An appropriate perturbation is given in Table 2. Thus, for the case where redundant codewords are present, both the perturbations of Table 1 and 2 are performed each iteration, together with the accept/reject logic. The perturbation of Table 1 applies to all codewords, Table 2 to redundant codewords only.

pick random redundant codeword index m: $0 \leq m < 2^K$, $\rho_m = 0$
$\eta_{orig} = 0$
for $k=0$ to $k=K-1$
do
$\qquad \eta_{orig} = \eta_{orig} + P(c_m)D_{i(f(m,k))}(r_{i(m)})$
end do
pick random integer p, $0 < p < R$
exchange mapping $i(.)$ with trial mapping $j(.)$
$\qquad$ where $j(q)=i(q)$, $0 \leq q < 2^K$, $q \neq m$
$\qquad\qquad j(m)=p$
$\eta_{trial} = 0$
for $k=0$ to $k=K-1$
do
$\qquad \eta_{trial} = \eta_{trial} + P(c_m)D_{j(f(m,k))}(r_{j(m)})$
end do

Table 2. Perturbation of Channel Code (Redundant Codewords).

The above procedure for approximate error correction is easily extended to include error detection. An additional, fictitious quantization level is required. Any of the redundant codewords can map into this fictitious level. Receipt of such a redundant codeword would indicate a transmission error, triggering the recovery procedure (for example: repeat the previous-frame value). To evaluate the penalty function values for the fictitious quantization level, the (conditional) performance of the recovery procedure must be evaluated. The design algorithm results in an optimal trade-off between error correction and detection.

The procedures discussed assume that single parameters are to be encoded. However, the procedures readily generalize to include the channel coding of several parameters at once, at the expense of an increase in computational effort. For two parameters x and y, a penalty $D_{i,k}(x_j, y_l)$, describing the performance for all combinations of quantization levels of the two parameters, under the constraint that the combination $x_i y_k$ was obtained by the analyzer, can be used.

AN ERROR CRITERION FOR CHANNEL ERRORS IN CELP

To evaluate the performance of the adaptive codebook of CELP under channel-error conditions, the penalty functions $D_i(.)$ must be defined. A natural method would be to evaluate the quality of the synthetic speech when channel errors occur. However, such an error criterion is unsuitable for the optimization, because of its large computational requirements. Instead, the distortion in the frame where the channel error occurs is used, which usually is a monotonically increasing function of the overall distortion caused by a channel error.

The CELP error criterion can be used as a starting point in the selection of a penalty function which can be evaluated quickly. The codebook search of the CELP algorithm uses a least-squares error measure. The search gives identical results if the least-squares criterion is replaced by a signal-to-noise ratio, or its logarithm. This is important, since the least-squares error criterion is not appropriate for averaging over a large number of frames; it weighs frames with large absolute error unduly heavily. Instead, the segmental signal-to-noise ratio is commonly used for this purpose. Thus, it is reasonable to choose $D_i(r_j)$ to be the mean of the signal-to-noise ratio in dB of the zero-state response of the inverse linear-predictive filter generated with the excitation vector r_j.

The criterion used for the vector quantization of CELP is commonly modified to better model the perceived error [13]. Due to masking, errors in spectral regions with high signal energy are less noticeable than errors in regions with lower signal energy. Here, this type of weighting is included in the evaluation of the penalty function $d_i(r_j)$.

Let H denote the matrix which transforms the excitation vector r_j of CELP into the zero-state response of the inverse linear-predictive filter [14]. Then the matrix $H^T H$ can be interpreted as a spectral-weighting matrix. Further, let t denote the target excitation vector which results in the original speech signal. Assuming that $H^T H$ includes the effect of perceptual weighting, the distance measure $D_i(.)$ for the vector-shape index becomes:

$$D_i(\lambda_k s_j) = E\left[10 \log \left[(\lambda_k s_j - t)^T H^T H (\lambda_k s_j - t) \right] \mid s_i \right], \qquad (2)$$

where s_j is the candidate vector shape associated with index j, which was substituted for the winning candidate vector shape s_i due to a (single-bit) channel error. λ_k is the optimally quantized gain factor for s_i, and $E[. \mid s_i]$ denotes the expectation value under the condition that s_i is the match obtained by the analyzer.

The distance measure for the gain factor λ looks similar:

$$D_i(\lambda_j s_k) = E\left[10 \log\left[(\lambda_j s_k - t)^T H^T H(\lambda_j s_k - t)\right] \mid \lambda_i\right],\qquad(3)$$

where λ_j is the gain quantization level which is substituted for level λ_i due to a channel error. The quantization level λ_i is the analyzer-selected quantizer gain for the winning codebook-vector shape s_k. In the applications in the next section, the expectation values will be approximated by the mean obtained over a large ensemble of frames.

CHANNEL CODING FOR CELP PARAMETERS

In this section source-dependent channel coding is applied to the parameters specifying the adaptive codebook of the CELP algorithm described in [14]. The input signal is sampled at 8000 Hz. The filter coefficients are quantized to 35 bits using absolute line spectral frequencies (this method exhibits low sensitivity to channel errors [3, 15]). The spectral-analysis-window length and the spectral-update intervals are 240 samples while a frame length of 60 samples is used for the quantization of the excitation vector. The stochastic codebook is quantized to 4 bits for gain and 8 bits for shape. The adaptive-codebook gain is quantized to 4 bits. Without error protection, the algorithm used in the following sections requires 4233 or 4366 bits/second for the 7 and 8 bit adaptive codebooks considered.

The results were obtained for 40 seconds of speech, spoken by 19 speakers. To prevent optimization for silence, the penalty functions were optimized for frames with a zero-state-response energy exceeding a threshold. Similarly, the segmental signal-to-noise ratios were obtained for segments where the speech energy per frame exceeded a threshold.

Several reference conditions are provided. The natural binary code (Natural), the Gray code, and the mean of 500 random codes are given. In addition, the performance is reported if a parity bit is employed in combination with a recovery strategy. In this case it was assumed that the parity bit is not affected by the channel, providing an upper bound for the actual performance. If sufficient redundant codewords are provided then the annealing procedure may result in a Hamming code which corrects all single bit errors; its performance is given for the adaptive-codebook gain.

Both the adaptive-codebook gain and its shape index have a nonuniform probability distribution, as is illustrated in Table 3 and Fig. 2. However, their penalty functions are of different character. Fig. 1 illustrates the penalty functions for the indices 6 through 15 of the adaptive-codebook gain, while Fig. 3 provides a penalty function for the shape index. The gain penalty curves are relatively smooth, while the shape penalty curves have distinct, sharp peaks. The latter behavior makes it difficult to find good neighbors for all shape indices simultaneously. Thus, the source-dependent optimization of the channel code will be less effective for the shape index. The performance will improve more if finer resolution is used for the shape index. Recent methods for improved speech quality under clear-channel conditions use such increased resolution [16].

The main results are presented in Tables 4 and 5, which provide both the first-frame penalty obtained after optimization by the annealing procedure, and the segmental signal-to-noise ratio of the speech obtained if in every eighth frame (480 samples) the particular parameter is hit by a bit reversal in a single random bit. The standard deviation of the latter measurement is about 0.1 dB. These actual-speech measurements were performed to verify the monotonic relationship between the mean first-frame distortion used in the penalty function and the total distortion of the synthetic speech, and to provide a more well-known measure of speech quality. The mean penalty given for the adaptive-codebook gain and shape is that obtained prior to adding the stochastic-codebook contribution.

For both the gain and shape indices repeating the previous-frame value was used as an error-detect recovery strategy. In the case of the gain, redundant codewords are more efficiently used to provide better neighbors rather than to provide error detection. However, because of the many sharp peaks of the shape index penalty curve, and the fact that reasonable performance is obtained by

index	level	probability
0	-10.0	0.0026
1	-3.01	0.0090
2	-1.37	0.0163
3	-0.88	0.0333
4	-0.40	0.0251
5	0.00	0.0000
6	0.15	0.0049
7	0.47	0.0518
8	0.69	0.1097
9	0.88	0.1580
10	1.03	0.2566
11	1.32	0.2885
12	2.08	0.0379
13	4.51	0.0048
14	14.9	0.0011
15	20.0	0.0005

Table 3. Gain Probabilities

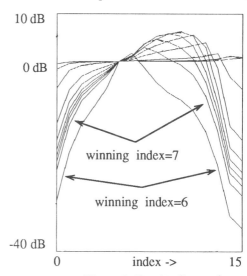

Figure 1: Penalty Curves for Gain Indices 6 through 15.

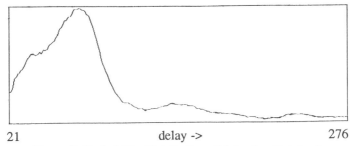

Figure 2: Probability Distribution of Adaptive-Codebook Shape Indices Corresponding to Delays 21 through 276.

264

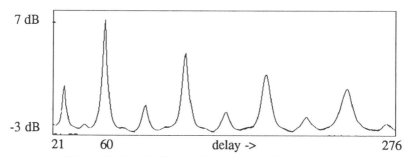

7 dB

-3 dB

21   60   delay ->   276

**Figure 3: Penalty Curves for Adaptive-Codebook Shape
Index Corresponding to Delay of 60 Samples.**

repeating the previous-frame shape index (i.e. pitch), the optimization results in
error detection being the primary method of error protection for the shape index (at
two bit or less redundancy). The advantage of error detection is illustrated for the
case with 128 redundant codewords. If the annealing procedure is not allowed to
use error detection (indicated with * in Table 5), then the performance is more
than a dB lower than that obtained with error detection.

Tables 4 and 5 include examples where noninteger numbers of bits are used
for protection. For the case in Table 4 the number of quantization levels is reduced
from 16 to 12 by eliminating the first four levels. (The sign of the excitation
pulses is usually preserved from one frame to the next.) Not shown in the table is
a minor associated clear-channel performance reduction (5.18 to 5.12 dB for the
penalty function and 9.96 to 9.78 dB for the actual speech). A large performance
improvement with a fractional-bit allotment for error protection is typical of the
channel encoding of parameters with a strongly nonuniform probability distribution
and a large number of potential good neighbors.

Because most of the improvement results from repeating previous-frame
information the effect of optimization of the channel coding of the shape index is
less dramatic than that of the gain index. As a result of the nonuniform probability
distribution, a small number of redundant codewords also provides a relatively large
improvement for the shape index. Minor clear-channel performance losses (less
than .1 dB for both measurements) are associated with the reduction of the range of
delays from 128 to 98.

Method	Bits	Quantizer Levels	Redundant Codewords	Mean Penalty, dB	Speech SNR, dB
Random Code	4	16	0	-5.26	-
Natural Code	4	16	0	-4.12	-1.09
Gray Code	4	16	0	-0.89	2.38
Annealing	4	16	0	0.71	2.69
Annealing	4	12	4	3.59	6.54
Parity	5	16	16	3.91	7.15
Annealing	5	16	16	4.56	8.21
Annealing	6	16	48	4.95	9.37
Hamming	7	16	112	5.18	9.96

Table 4. Performance for Various Encodings of the Adaptive-Codebook Gain Index.

265

Method	Bits	Range of Delays (Samples)	Redundant Codewords	Mean Penalty, dB	Speech SNR, dB
Annealing	7	21-118	30	1.49	3.14
Random Code	7	21-148	0	-1.82	-
Natural Code	7	21-148	0	-0.59	2.08
Gray Code	7	21-148	0	-0.21	2.25
Annealing	7	21-148	0	0.25	2.14
Annealing*	8	21-148	128	1.37	3.08
Annealing	8	21-148	128	2.88	4.22
Parity	8	21-148	128	2.95	4.28
Annealing	9	21-148	384	3.19	4.62
Annealing	8	21-180	76	2.39	3.90
Random Code	8	21-276	0	-1.92	-
Natural Code	8	21-276	0	-0.78	2.01
Gray Code	8	21-276	0	-0.43	2.15
Annealing	8	21-276	0	0.31	2.26

* (no error detection)

Table 5. Performance for Various Encodings of the Adaptive-Codebook Shape Index (Each Index is Associated with a Particular Delay).

CONCLUSION

Using the CELP algorithm as example, it has been shown that source-dependent channel coding can be used to improve the performance of coding algorithms operating in a range of channel-error conditions. The procedure provides a more graceful degradation with increasing channel-error rate.

Optimization of the appropriate error criteria for source-dependent channel codes can be achieved with simulated annealing. In contrast to conventional channel-coding methods, the new method provides optimized error protection at any level of redundancy, including zero redundancy and redundancy of a noninteger number of bits. The optimization results in more probable codewords receiving better protection and an optimal trade-off between error correction and detection.

Although the presentation focused on single-bit errors, the procedure can be generalized to include multiple-bit errors per codeword. Using simultaneous optimization of single-bit and two-bit errors is desirable when higher error rates are expected; in informal tests at random error rates over 1% sensitivity to two-bit errors was observed when the code was optimized for single bit errors only. Also, the applications only dealt with single parameters, but the procedure can be applied for the simultaneous encoding of multiple parameters as well.

The channel-coding optimization methods were applied to the adaptive-codebook quantization of the CELP algorithm. It was found that the improvements were most dramatic for the adaptive-codebook gain, which has a nonuniform distribution and many potential good neighbors for gain values of high probability, a result of the smooth shape of the penalty curves. In contrast, the adaptive-codebook shape has peaky penalty curves. The indices have few potential good neighbors and this, in combination with an effective error-detection strategy, leads to the redundant codewords mostly being used for error detection.

REFERENCES

[1] Cox, R.V., W.B. Kleijn, and P. Kroon, "Robust CELP Coders for Noisy Backgrounds and Noisy Channels", *Proc. Int. Conf. Acoust., Speech and Sign. Process.*, Glasgow, 739-742, 1989.

[2] Tremain, T. E., "The Government Standard Linear Predictive Coding Algorithm", *Speech Technology*, 40-49, April 1982.

[3] J.P. Campbell, V.C. Welch, and T.E. Tremain, "An Expandable Error-Protected 4800 bps CELP Coder", *Proc. Int. Conf. Acoust., Speech and Sign. Process.*, Glasgow, 735-738, 1989.

[4] Cox, R.V., J. Hagenauer, N. Seshadri, C.-E. Sundberg, "A Sub-Band Coder Designed for Combined Source and Channel Coding", *Proc. Int. Conf. Acoust., Speech and Sign. Process.*, New York, 235-238, 1988.

[5] Chen, J-H., G. Davidson, A. Gersho, and K. Zeger, "Speech Coding for the Mobile Satellite Experiment", *Proc. IEEE Int. Conf. Comm.*, 756-763, 1987.

[6] Farvardin, N., and V. Vaishampayan, "Optimal Quantizer Design for Noisy Channels: An Approach to Combined Source-Channel Coding", *IEEE Trans. Information Theory*, Vol. IT-33, No. 6, 827-838, 1987.

[7] Atal B.S. and M.R. Schroeder, "Stochastic Coding of Speech at Very Low Bit Rates", *Proc. Int. Conf. Comm.*, Amsterdam, 1610-1613, 1984.

[8] Kemp, R.D., R.A. Sueda, and T.E. Tremain, "An Evaluation of 4800 bps Voice Coders", *Proc. Int. Conf. Acoust., Speech and Sign. Process.*, Glasgow, 200-203, 1989.

[9] DeMarca, J.R.B., and N.S. Jayant, "An Algorithm for Assigning Binary Indices to the Codevectors of a Multi-Dimensional Quantizer", *Proc. IEEE Int. Conf. Comm.*, 1128-1132, 1987.

[10] Kirkpatrick, S., C.D. Gelatt, M.P. Vecchi, "Optimization by Simulated Annealing", *Science*, Vol. 220, 671-680, 1983.

[11] El Gamal, A.A., L.A. Hemachandra, I. Shperling, and V.K. Wei, "Using Simulated Annealing to Design Good Codes", *IEEE Trans. Information Theory*, Vol. IT-33, No. 1, 116-123, 1987.

[12] Metropolis, M., A.W. Rosenbluth, M.N. Rosenbluth, A.H. Teller, and E. Teller, "Equation of State Calculation by Fast Computing Machines", *J. Chem. Physics*, Vol. 21, 1087-1092, 1953.

[13] Atal, B.S., and M.R. Schroeder, "Predictive Coding of Speech Signals and Subjective Error Criteria", *IEEE Trans. Speech Signal Proc.*, Vol. ASSP-27, No. 3, 247-254, 1979.

[14] Kleijn, W.B., D.J. Krasinski, and R.H. Ketchum, "An Efficient Stochastically Excited Linear Predictive Coding Algorithm for High Quality Low Bit Rate Transmission of Speech", *Speech Communication*, Vol. VII, 305-316, 1988.

[15] Kang, G.S., L.J. Fransen, "Low-Bit Rate Speech Encoders Based on Line-Spectrum Frequencies (LSFs)", *Report Naval Research Laboratory*, January 24, 1985.

[16] Kroon, P, and B.S. Atal, "On Improving the Performance of Pitch Predictors in Speech Coding Systems", *Abstracts IEEE Workshop on Speech Coding for Telecommunications*, Vancouver, 49-50, 1989.

25

VECTOR TRELLIS QUANTIZATION FOR NOISY CHANNELS †

Mohammad Reza Soleymani and Amir Keyvan Khandani

Department of Electrical Engineering
McGill University
McConnell Engineering Building
3480 University Street
Montreal, Quebec, H3A 2A7, Canada

INTRODUCTION

The channel coding theorem of information theory indicates that if the rate of a binary sequence is less than the capacity of the channel over which the binary sequence is to be transmitted, then the source can be reproduced at the channel output with arbitrarily small error probability [1], [2]. Based on this, one can isolate the problem of channel coding from that of source coding. In other words, channel encoder, channel, and channel decoder may be considered as a noiseless link between the output of the source encoder and the input of source decoder, as long as source encoder's output has a rate less than the capacity of the channel [3]. However, this separation is optimal only asymptotically, i.e., in the limit of arbitrarily complex overall encoders and decoders involving arbitrarily long blocklengths. In practice, where we encounter the curse of complexity and are forced to deal with finite blocklengths, such a separation results in a certain degree of sub-optimality.

The above discussion, justifies the recent interest in combined source-channel coding. Recently, Farvardin and Vaishampayan [4] have proposed a procedure for the design of combined source-channel coders for the special case of scalar quantization. They have also reported the extension of their approach to the vector quantization case [5]. Zeger and Gersho have also discussed optimality conditions and presented a design procedure for a vector quantizer for noisy channels [6]. Ayanoglu and Gray have applied the generalized Lloyd algorithm to the design of joint source and channel trellis waveform coders for discrete-time continuous-amplitude stationary and ergodic sources over discrete memoryless noisy channels [7]. Dunham and Gray have established a coding theorem which shows that such a system can perform arbitrarily close to the source distortion-rate function evaluated at the channel capacity [8].

† Research supported by Canada Natural Science and Engineering Council Grant OG PIN 011.

In this work, we present a more general model for the joint source and channel coding problem involving a vector trellis quantizer similar to [9] and [10], however, with a noisy channel. This model can accommodate most of the schemes previously reported as its special cases. The derivation is also general in the sense that we do not make any assumption on the fidelity criterion. However, later we specialize the general procedure for the case of Euclidean distortion measure. The only assumption made on the channel is that present state of the decoder depends only on the present state and present output of the encoder. Hence, the binary symmetric channel assumption made in simulations, does not constitute a necessary condition for the applicability of the model. Finally, simulation results concerning the performance and robustness of channel optimized vector, trellis, and vector trellis quantizers for both a Gauss-Markov source and speech samples over a binary symmetric channel are presented and discussed.

SYSTEM MODEL

The block diagram of the system we wish to analyze is illustrated in Fig. 1.

Fig. 1 Model of the System

The source to be transmitted is modeled as a stationary and ergodic process producing symbols from the alphabet set A with its k-fold cartesian product A^k consisting of all k-tuples $\underline{x} = (x_0, ..., x_{k-1})$.

We assume a discrete channel with input alphabet set B and output alphabet set $\hat{B}$. Let $\hat{A}$ be a finite reproduction alphabet set with $\hat{A}^k$ representing its k-fold cartesian product, and let $\rho : A \times \hat{A} \to [0, \infty)$ be a per letter distortion measure yielding a single-letter fidelity criterion $F_\rho = \{\rho_k, k = 1, 2, ...\}$ defined as:

$$\rho_k(\underline{x}, \underline{\hat{x}}) = k^{-1} \sum_{j=0}^{k-1} \rho(x_j, \hat{x}_j).$$

The objective of the system is to transmit a sequence of L successive source vectors $\underline{x}_l$, $l = 0, 1, ..., L - 1$ under the existing constraints and to represent it at the receiver side by a sequence of L reproduction vectors $\underline{\hat{x}}_l$, $l = 0, 1, ..., L - 1$ in such a way that the average distortion i.e.,

$$D_L = L^{-1} \sum_{l=0}^{L-1} \rho_k(\underline{x}_l, \underline{\hat{x}}_l), \tag{1}$$

is minimized.

The encoder is a sliding block joint source and channel encoder which starting at time zero will view L successive k-dimensional source vectors and based on this observation will produce a sequence of L encoder output symbols or channel input symbols i. We model the encoder as a finite state system by defining its state at time n by $t_{n+1} = \sigma(t_n, i_n)$, $t \in \{0, 1, ..., S_T-1\}$, where S_T is the number of encoder states. In this case the encoder can be represented by a trellis in which the nodes correspond to the states and the branches correspond to the state transitions. Each state at time n can be connected to at most $|B|$ states at time $n + 1$, where $|B|$ is the cardinality of B. If we define an appropriate additive cost $C(t, i, \underline{x})$ as the cost of choosing branch i for source vector $\underline{x}$ when encoder is in state t, the process of encoding a sequence of L successive source vectors can be formulated as finding the length L path through the trellis which, when assigned to the source sequence, will result in the minimum overall cost. Finding the path with the minimum cost through a directed graph is a well known problem which can be optimally handled by Viterbi algorithm [11], or by other suboptimum algorithms [12].

The decoder is also a finite state system with the following state equation

$$r_{n+1} = \delta(r_n, \hat{i}_n), \quad r_n \in \{0, 1, ..., S_R-1\},$$

where S_R is the number of the decoder states. Decoding is achieved by selecting one of the k-dimensional reproduction vectors $\hat{\underline{x}}$ as a function of the decoder state and received symbol. In other words, decoding is a table look-up according to the following formula

$$\hat{\underline{x}}_n = g(r_n, \hat{i}_n).$$

It should be mentioned that in the case of the noisy channel, the channel input symbol at time n, i.e., i_n, is not necessarily equal to the channel output symbols $\hat{i}_n$, and as a result, the encoder and decoder are not able to keep exact track of the state transitions of each other. In other words, the encoder has only statistical information about the reproduction vector that will result from transmitting a special channel symbol and the decoder does not know which symbol has really been sent. Therefore, the encoder trellis should be labelled with the cost averaged over the channel, and the decoder table entries should be selected by taking the channel statistics into account [8]. Another point that should be explained is the effect of the encoder state at time zero. It can be shown [8], that for a large enough search depth L, the effect of the initial state will be negligible. Hence, in the following discussions we do not consider this effect and we assume that the initial state has been chosen in such a way that the system starts at time zero with its steady state conditions.

ENCODER-DECODER STRUCTURE

For large enough search depth L, the average distortion of (1) can be represented as,

$$D = E[\rho_k(\underline{x}, \, \underline{\hat{x}})] = \sum_{\underline{x}} \sum_{\underline{\hat{x}}} P(\underline{x}, \, \underline{\hat{x}}) \, \rho_k(\underline{x}, \, \underline{\hat{x}}).$$

where $P(\underline{x}, \underline{\hat{x}})$ is the probability of receiving $\underline{\hat{x}}$ when $\underline{x}$ is transmitted. We use probability distributions and summations throughout this work merely for notational convenience. However, when dealing with a continuous variable, the corresponding summation and probability distribution can be replaced by integration and probability density respectively. This is true for $\underline{x}$ for which $A^k = \mathrm{R}^k$ in all our examples. At time instant n the value of $P(\underline{x}_n, \, \underline{\hat{x}}_n)$ can be written as:

$$\sum_{t_n} \sum_{r_n} \sum_{i_n} \sum_{\hat{i}_n} P(t_n, \, \underline{x}_n) \, P(i_n \mid t_n, \, \underline{x}_n) P(r_n, \, \hat{i}_n \mid t_n, \, i_n) \, P(\underline{\hat{x}}_n \mid r_n, \, \hat{i}_n)$$

The objective, as already mentioned, is to minimize D by proper choice of the encoder, $P(i_n \mid t_n, \, \underline{x}_n)$, and the decoder $P(\underline{\hat{x}}_n \mid r_n, \, \hat{i}_n)$. In order to derive a recursive optimization algorithm, we find the expression for the optimum decoder for a fixed encoder and the expression for the optimum encoder, when decoder is fixed. For a given encoding rule $P(i_n \mid t_n, \, \underline{x}_n)$, we can minimize D by proper choice of $P(\underline{\hat{x}}_n \mid r_n, \, \hat{i}_n)$. It is easy to show that in order to do so we should assign to each $(r_n, \, \hat{i}_n)$ pair the $\underline{\hat{x}}_n$ which minimizes the following:

$$\sum_{\underline{x}_n} P(\underline{\hat{x}}_n \mid r_n, \, \hat{i}_n) \, \rho_k(\underline{x}_n, \, \underline{\hat{x}}_n) = E[\rho_k(\underline{x}_n, \, \underline{\hat{x}}_n) \mid r_n, \, \hat{i}_n].$$

In other words, the reproduction vector at time n for each receiver state r_n, and received channel symbol $\hat{i}_n$ is

$$\underline{\hat{x}}_n = g(r_n, \, \hat{i}_n) = arg \min \{ E[\rho_k(\underline{x}_n, \, \underline{\hat{x}}_n) \mid r_n, \, \hat{i}_n] \}.$$

Similarly for a given decoder, $P(\underline{\hat{x}}_n \mid r_n, \, i_n)$, we can minimize D by proper choice of the encoding rule $P(i_n \mid t_n, \, \underline{\hat{x}}_n)$. To do this, we should minimize,

$$E[\rho_k(\underline{x}_n, \, \underline{\hat{x}}_n) \mid t_n, \, i_n] = \sum_{\underline{\hat{x}}_n} \sum_{r_n} \sum_{\hat{i}_n} P(r_n, \, \hat{i}_n \mid t_n, \, i_n) \, P(\underline{\hat{x}}_n \mid r_n, \, \hat{i}_n) \rho_k(\underline{x}_n, \, \underline{\hat{x}}_n),$$

for each $(t_n, \, \underline{x}_n, \, i_n)$ triplet. For a deterministic decoder, i.e., when $P(\underline{\hat{x}}_n \mid r_n, \, \hat{i}_n) = 1$ for some $\underline{\hat{x}}_n = g(r_n, \, \hat{i}_n)$, the encoder should pick i_n which minimizes

$$\sum_{r_n} \sum_{\hat{i}_n} P(r_n, \, \hat{i}_n \mid t_n, \, i_n) \rho_k[\underline{x}_n, \, g(r_n, \, \hat{i}_n)].$$

In using Viterbi algorithm, the above expression can be considered as the cost of choosing branch i, when encoder is in state t and $\underline{x}$ is the input vector, i.e.,

$$C(t, \, i, \, \underline{x}) = \sum_{r} \sum_{\hat{i}} P(r, \, \hat{i} \mid t, \, i) \, \rho_k[\underline{x}, \, g(r, \, \hat{i})]. \tag{2}$$

For the special case of the mean square distortion measure, i.e.,

$$\rho_x(\underline{x}, \hat{\underline{x}}) = \| \underline{x} - \hat{\underline{x}} \|^2 = \frac{1}{k} \sum_{j=0}^{k-1} (x_j - \hat{x}_j)^2,$$

it can be easily shown that the encoder cost function (2) reduces to the following form

$$C(t, i, \underline{x}) = -\underline{x}.\underline{Z}(t, i) + C(t, i), \tag{3}$$

where

$$\underline{Z}(t, i) = \sum_r \sum_{\hat{i}} P(r, \hat{i} \mid t, i) \, g(r, \hat{i}) = E[g(r, \hat{i}) \mid t, i], \tag{4}$$

and,

$$C(t, i) = \frac{1}{2} \sum_r \sum_{\hat{i}} P(r, \hat{i} \mid t, i) \, \| g(r, \hat{i}) \|^2 = \frac{1}{2} E[\| g(r, \hat{i}) \|^2 \mid t, i], \tag{5}$$

and in this case the decoding function will be equal to,

$$g(r, \hat{i}) = \sum_{\underline{x}} \underline{x} P(\underline{x} \mid r, \hat{i}) = E(\underline{x} \mid r, \hat{i}). \tag{6}$$

In the special case where the encoder state at time n is defined as,

$$t_n = (i_{n-N_T}, ..., i_n), \tag{7}$$

where N_T is equal to the encoder memory length, the encoders state space is the set B^{N_T+1} and has cardinality $\mid B \mid^{N_T+1}$. Note that, with the inclusion of the present channel input, i_n, in the state, we have chosen to use labelled state notation [10]. This changes the encoder's state space from B^{N_T} to B^{N_T+1}. However, this does not involve any change in the encoding complexity which is still determined by the number of internal states of encoder, i.e., $\mid B \mid^{N_T}$. This is due to the fact that, in this case, the cost associated with each branch only depends on the destination node, and therefore, out of $\mid B \mid$ nodes $t_n = (\alpha, i_{n-N_T+1}, ..., i_n)$, $\alpha \epsilon B$ at stage n, the one with minimum total cost will be the parent node for all nodes $t_{n+1} = (i_n - N_{T+1}, ..., i_n, \beta)$, $\beta \epsilon B$ at stage $n+1$.

Without loss of generality, the decoder state at time n can be defined as $r_n = (\hat{i}_{n-N_R}, ..., \hat{i}_{n-1})$, where N_R is equal to the decoder memory length. In this case the decoder state space will be equal to the set $\hat{B}^{N_R}$ with the cardinality of $\mid \hat{B} \mid^{N_R}$. For such a system the decoding function can be written as,

$$\hat{\underline{x}}_n = g(\hat{i}_{n-N_R}, ..., \hat{i}_{n-1}, \hat{i}_n).$$

This structure can be easily realized by employing a shift register of length $(N_R + 1)\log_2\lceil \hat{B} \rceil$ containing the most recent $N_R + 1$ channel outputs. The contents of this register is used as the table index to select the reproduction vector.

ENCODER AND DECODER DESIGN ALGORITHM

For mean square distortion measure, we can use an iterative algorithm based on (3) and (6) for simultaneously satisfying the optimality conditions for the encoder and decoder structures. Starting from an initial decoder structure, i.e., an initial set of codebooks with memory length N_T, rate r, and dimension k we find the corresponding optimum encoder structure using (4), (5). Then we encode a training sequence of appropriate length according to the new encoding rules and compute $E(\underline{x} \mid t,i)$. In the next step, the new decoder structure will be determined using the following formula,

$$g(r,\hat{i}) = E(\underline{x} \mid r,\hat{i}) = \sum_t \sum_i E(\underline{x} \mid t,i) P(t,i \mid r,\hat{i}).$$

The value of $P(t,\ i \mid r,\ \hat{i})$ in the above formula is calculated according to the Bayes rule, i.e.,

$$P(t,i \mid r,\hat{i}) = \frac{P(t,i)P(r,\hat{i} \mid t,i)}{\sum_t \sum_i P(t,i)P(r,\hat{i} \mid t,i)},$$

where probabilities $P(t,i)$ are calculated in the encoding phase and the probabilities $P(r,\hat{i} \mid t,i)$ are calculated from crossover probabilities of the channel. These steps will be repeated until the algorithm converges. To find an appropriate initial point we first start from the zero rate k-dimensional memoryless codebook, the centroid of the whole training sequence. Then, by applying the splitting technique [13], to the decoder and using a similar iterative procedure, we find the optimum codebook for rate r, dimension k, and zero memory. This case corresponds to the optimum memoryless vector quantizer. In the next step we employ the extension technique [14], with another similar iterative procedure to increase the encoder memory to the desired value of N_T.

ENCODING PROCEDURE

If we assign the numbers 0 to $\mid B \mid -1$ to branch labels, for the case that the encoder state at time n is defined by (7), each node of the trellis can be labelled by the contents of a $\mid B \mid$'ary shift register containing the branch labels of $N_T + 1$ previous branches in the path terminating to that node. In this case to continue the paths from stage n to stage $n + 1$, it is convenient to partition the nodes at stage n into $\mid B \mid^{N_T}$ groups, each of $\mid B \mid$ elements, with the l'th group defined as the set of the states for which N_T least significant $\mid B \mid$'ary digits of the state register contain l. We also partition the nodes at stage $n + 1$ into $\mid B \mid^{N_T}$ blocks, each of $\mid B \mid$ elements, with the m'th block defined as the set of the states for which the most significant digits of the stage register contain m. As already mentioned the elements of m'th group at stage n are competitors for reaching the m'th block at stage $n + 1$, and the node with the minimum total cost will be the parent node of all the elements of the m'th block at stage $n + 1$. The encoding procedure can be formulated as follows:

Receiving a new source vector $\underline{x}$ at time n, we first compare the total cost of the elements of the m'th group at stage n to find the parent node for the m'th block of stage $n + 1$, and update the path of the whole block accordingly. Then the costs of the $|\ B\ |^{N_T+1}$ branches connecting stage n to stage $n + 1$ will be calculated and added to the total cost of the parent nodes at stage n to determine the total cost of each node at stage $n + 1$. The procedure will be repeated for every new source vector.

SPECIAL CASES

As stated earlier, the model considered in this paper is a very general model and can accommodate most cases discussed in the literature as special cases. In this section, we discuss the conditions for specializing our model to those special cases.

Assuming $N_T = N_R = 0$, i.e., assuming that neither the encoder nor the decoder have memory, and $k > 1$, the channel optimized vector quantizer of [5] and [6], is obtained. In this special case, encoding rule consists of minimizing, $-\underline{x}.\underline{Z}(i) + C(i)$, where,

$$\underline{Z}(i) = \sum_{\hat{i}} \hat{\underline{x}}\ P(\hat{i} \mid i) = E[\hat{\underline{x}} \mid i],$$

and

$$C(i) = \frac{1}{2} \sum_{\hat{i}} \|\ \hat{\underline{x}}\ \|^2\ P(\hat{i} \mid i) = \frac{1}{2} E[\|\ \hat{\underline{x}}\ \|^2 \mid i].$$

The decoding rule (centroid calculation), consists of choosing,

$$\hat{\underline{x}} = E[\underline{x} \mid \hat{i}] = \sum_{i} E[x \mid \underline{i}] P(i \mid \hat{i}),$$

for each $\hat{i}$. Assuming that there is no channel noise, i.e., $\hat{i} = i$, results in case of ordinary vector quantization, $N_T = N_R = 0$, $k = 1$ results in channel optimized scalar quantization [4].

For N_T, $N_R > 0$ and $k = 1$, we have the joint source and channel trellis coder of Ayanoglu and Gray [7]. With no channel noise, this case will further be specialized to the ordinary trellis waveform coder of Stewart et al [14].

Finally, assuming N_T, $N_R > 0$ and $k > 1$, and noiseless channel, the vector trellis coder of Juang [9], and Bei and Gray [10], is obtained.

SIMULATION RESULTS

In this section, simulation results concerning the application of the design algorithm to both a Gauss-Markov source and speech samples over a Binary Symmetric Channel will be presented. In all simulations we have assumed $N_T = N_R = N$, i.e., the same memory size, and, therefore, the same number of encoder and decoder states, $S = 2^{kN}$. Table 1 gives the Signal-to-Quantization-Noise ratio (SQNR), for 60,000 samples of a Gauss-Markov source with correlation coefficient $\alpha = 0.9$, for several values of dimension, k, number of states, S, and BSC crossover probability, P_B. The last column, gives the number of multiplications per sample, M, required for encoding as an indication of the complexity involved. The rate is r = 1 bit/sample. The first two rows in Table 1 correspond to the extreme cases of vector quantization (S = 1), and trellis quantization ($k = 1$). The entries of Table 1 indicate that for a given complexity, e.g., M = 64, more can be achieved in terms of SQNR from increasing the number of states than from an increase in dimension. To see this more clearly, we will compare the performance of two extreme cases of vector quantization (V.Q.) and trellis quantization (T.Q.) in more detail. Tables 2 and 3 compare the performance of V.Q.'s with dimensions 2 to 7 with that of equivalent T.Q.'s (T.Q.'s with $K = k$, where $K = N + 1$, is the constraint length of trellis quantizer) for a Gauss-Markov source and speech samples, respectively. In Tables 2 and 3, we have also included the effect of mismatch in order to assess the robustness of each quantizer type. In each case the entry of column 1 is the SQNR resulting from using a quantizer designed for a noiseless channel over a BSC with the given P_B, while the entry of column 2 is the performance of channel optimized quantizer. Columns two to five of table 2 have been compiled based on tables III and V of [7]. The entries of Tables 2 and 3 indicate that for $K \geq 4$ trellis quantizer outperforms the equivalent vector quantizer in terms of SQNR obtained. It can also be seen that T.Q. is more robust than V.Q., i.e., the difference between mismatch (Column 1) and match (Column 2) performance is smaller for trellis quantizer. It is necessary to point out that, in the case of V.Q., the mismatch results for $P_B = 0.01$ were very sensitive to the structure of the codebook used. The entries of column 8 of Tables 2 and 3 are the average of the results for several code books designed with different initial conditions.

k	S	SQNR				M
		$P_B = 0.0$	$P_B = 0.001$	$P_B = 0.005$	$P_B = 0.01$	
6	1	10.90	10.66	9.55	9.09	64
1	32	11.56	11.32	10.50	9.84	64
2	4	10.34	10.11	9.16	8.36	16
2	16	11.70	11.10	9.80	9.60	64
2	64	12.24	11.17	10.05	9.92	256
3	8	10.92	10.77	9.84	9.24	64
4	16	11.26	11.13	9.93	9.69	256

Table 1: Performance of channel optimized vector trellis quantizer for Gauss-Markov source.

K	Trellis Quantizer				Vector Quantizer			
or	$P_B = 0.001$		$P_B = 0.01$		$P_B = 0.001$		$P_B = 0.01$	
k	1	2	1	2	1	2	1	2
2	6.87	6.87	6.26	6.29	7.73	7.78	6.82	6.89
3	8.67	8.68	7.73	7.81	8.84	8.86	7.01	7.57
4	9.98	9.98	8.64	8.82	9.55	8.73	7.49	8.24
5	10.86	10.85	9.01	9.49	10.12	10.31	7.51	8.70
6	11.32	11.32	9.58	9.84	10.55	10.66	7.55	9.09
7	11.63	11.63	9.78	10.12	10.65	10.74	7.35	9.21

Table 2: Comparison of performance and robustness of channel optimized Trellis quantizer with channel optimized vector quantizer for Gauss-Markov Source.

K	Trellis Quantizer				Vector Quantizer			
or	$P_B = 0.001$		$P_B = 0.01$		$P_B = 0.001$		$P_B = 0.01$	
k	1	2	1	2	1	2	1	2
2	5.50	5.75	3.96	4.77	6.32	6.33	4.52	4.86
3	7.13	7.18	5.63	5.66	7.83	7.84	5.32	5.97
4	8.51	8.57	6.45	7.10	8.59	8.73	6.28	7.13
5	9.69	9.76	6.92	7.86	9.17	9.48	6.24	7.79
6	10.47	10.59	7.41	8.69	9.67	10.16	6.51	8.08
7	11.07	11.26	7.80	9.31	10.03	10.69	6.44	8.36

Table 3: Comparison of performance and robustness of channel optimized trellis quantizer with channel optimized vector quantizer for speech samples (120,000 samples, male speaker).

Conclusions

A general model for quantization for noisy channels involving a vector trellis structure was introduced and the corresponding design algorithm was presented. The model and the algorithm provide a general framework for the study of different quantization schemes such as vector and trellis quantization. Simulation results, for a Gauss-Markov source and a binary symmetric channel, indicate that while for the same complexity vector trellis quantizer performs better than vector quantizer, its performance is inferior to a trellis quantizer. However, an advantage of vector trellis quantizer over a trellis quantizer is the possibility of achieving near optimal results using simple sub-optimal encoding methods, such as, M - L algorithm with small M and L. Comparison of trellis quantizers having different constraint lengths, K, with equivalent vector quantizers, i.e., V.Q.'s with dimension $k = K$, indicates that a trellis quantizer outperforms vector quantizer, both in terms of SQNR and robustness.

References:

[1] C.E. Shannon, "A mathematical theory of communication," *Bell Syst. Tech. J.*, vol. 27, pp. 379-423 and 623-656, 1948.

[2] R.G. Galleger, *Information theory and reliable communication*, John Wiley & Sons, N.Y. 1968.

[3] A. J. Viterbi and J. K. Omura, *Principles of Digital Communication and Coding*, McGraw Hill, N.Y. 1979.

[4] N. Farvardin and V. Vaishampayan, "Optimal Quantizer Design for Noisy Channels: An Approach to Combined Source-Channel Coding," *IEEE Trans. Inform. Theory*, pp. 827-888, Nov. 1987.

[5] N. Farvardin and V. Vaishampayan, "Some Issues of Vector Quantizers for Noisy Channels," *IEEE Symp. on Information Theory*, Kobe, Japan, June 19-24, 1988.

[6] K. Zeger, and A. Gersho, "Theory and Design of Optimal Noisy Channel Vector Quantizers," *IEEE Trans. Inform. Theory*.

[7] E. Ayanoğlu and R.M. Gray, "The Design of Joint Source and Channel Trellis Waveform Coders," *IEEE Trans. Inform. Theory*, vol. IT-33, pp. 855-865, Nov. 1987.

[8] J.G. Dunham and R.G. Gray, "Joint Source and Noisy Channel Trellis Encoding," *IEEE Trans. Inform.*, vol. IT-27, pp. 516-519, July 1981.

[9] B. H. Juang, "Design of Trellis Vector Quantizers for Speech Signals," *IEEE Trans. Acousitcs, Speech, and Signal Processing*, Vol. 36, pp. 1423-1431, Sept. 1988.

[10] C.D. Bei and R.M. Gray, "Simulation of Vector Trellis Encoding Systems," *IEEE Trans. Commun.*, vol. COM-34, pp. 214-218, Mar. 1986.

[11] G. D. Forney, JR. "The Viterbi Algorithm," *Proc. IEEE*, vol. 61, pp. 268-178, Mar. 1973.

[12] F. Jelinek and J.B. Anderson, "Instrumentable tree encoding of information sources," *IEEE Trans. Inform. Theory*, vol. IT-17, pp. 118-119, Jan. 1971.

[13] Y. Linde, A. Buzo and R. M. Gray, "An Algorithm for Vector Quantizer Design," *IEEE Trans. Commun.*, vol. COM-28, pp. 84-95, Jan. 1980.

[14] L.C. Stewart, R.M. Gray and Y. Linde "The Design of Trellis Waveform Coders," *IEEE Trans. Commun.* vol. COM-30, pp. 702-710, April 1982.

26

APPLICATION OF STOCHASTIC CODING SCHEMES IN SATELLITE COMMUNICATION

Andrew Perkis, Bernt Ribbum*

Department of Electrical & Computer Engineering
The University of Wollongong
Wollongong, Australia

* Elab-Runt, Trondheim, Norway

INTRODUCTION

As digital communication has become an increasingly important field, a great deal of research is concentrated on developing new applications. Of these, low bit rate speech coding for mobile communication is of special interest. Particularly the Code Excited Linear Predictive (CELP) coders [11] have shown considerable promise in giving high quality speech at the required bit rates. The majority of the proposed coder candidates, however, are considered in the absence of a realistic channel. This paper will describe a possible speech coder candidate for Mobile Satellite Communication, the Self Excited Stochastic Multipulse (SESTMP) coder [2, 7] with a 7.0 kbit/s bit rate, taking into consideration the special features of a satellite channel.

In mobile satellite communication it is necessary to operate within narrow bands at low carrier to noise ratios in order to keep costs down and make efficient use of the available spectrum. For this purpose the behaviour of the SESTMP coder over a noisy channel has been studied [1]. The coder transmits five distinct parameter groups representing the long term spectral information, the fine spectral structure and the residual signal. The bit error sensitivities of the different coder parameters were interpreted both by objective and subjective means. Both measures show that the spectral information, represented by the Linear Predictive Coefficients (LPC) and the stochastic gain factor, are most sensitive. Two different quantization schemes for the LPC coefficients were considered, both based on Line Spectrum Pairs (LSP) [12]. The first one uses both inter- and intra-frame prediction in the quantization, while the second is a non-uniform scalar quantization of the LSP differences.

THE SELF EXCITED STOCHASTIC MULTIPULSE CODER

This study concentrates on one specified speech coder at a given bit rate, 7.0 kbit/s. The speech coder model is based on Linear Prediction, and an analysis-by-synthesis approach for optimizing the innovation sequence for the residual signal.

The coder is shown in Figure 1, and is, in fact, an improved version of the basic Code Excited Linear Predictive Coder (CELP) proposed by Atal and Scroeder, [11].

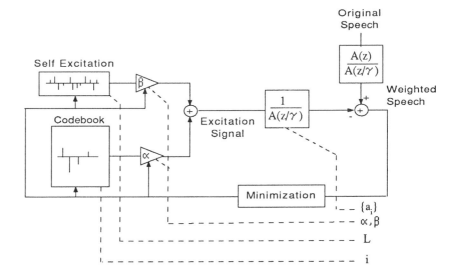

Figure 1: Block Diagram of the Stochastic Multipulse Coder with Self Excitation (SESTMP)

Description of the Speech Coder Parameters

In the SESTMP coder, the input speech is parameterized using five basic parameters which are given in Figure 1 and described in Table 1.

Table 1. Transmitted Parameters for a 7.0 kbit/s SESTMP Codec

Name	Symbol	Values	# Bit/s per value	Frame length	Bit/s
Linear prediction coefficients (LPC)	$\{a_i\}$	10	4	20msec	2000
Self excitation lag	L	1	7	5msec	1400
Self excitation gain factor	β	1	3	5msec	600
Stochastic gain factor	α	1	5	5msec	1000
Codebook index	i	1	10	5msec	2000

The SESTMP is a block-based coder, where the parameters are calculated for each input frame of speech, in our case consisting of 160 samples (20 msec). For each frame a set of LPC coefficients is estimated giving the filter coefficients, $\{a_i\}$, before splitting into four equally sized subframes of 40 samples (5 msec) each of which are matched in an analysis-by-synthesis manner by a self excitation sequence and a codevector.

The parameters are quantized according to Table 1, giving the internal structure of one transmitted frame as shown in Figure 2.

$\{a_i\}$	L	β	α	i	L	β	α	i	L	β	α	i	L	β	α	i
# bits 40	7	3	5	10	7	3	5	10	7	3	5	10	7	3	5	1

Figure 2: Internal Structure of a 20msec Transmitted Frame

Quantization of the LPC Coefficients

The speech spectral information in LPC based coders is perceptually important, making the quantization crucial. In this coder the LSP representation of the speech spectrum is chosen. The LSP frequencies have certain important properties such as limited dynamic range, natural ordering and high correlation between frequencies in adjacent frames making them an attractive alternative to the LPC coefficients for quantization purposes [3]. In addition, the LSP representation provides efficient control with the shape of the spectrum, and allows for a simple check for filter stability.

These properties are utilized in designing two non-uniform scalar quantizers for the LSP frequencies; scalar quantization of ω_k with prediction, both in time and frequency; and scalar quantization of LSP differences.

Scalar Quantization of ω_k with Prediction (PRED).

For scalar quantization of the LSP frequencies, optimal Max quantizers (0,1,2,3,4 and 5 bit) have been designed for each coefficient. Statistical calculations show that there is a high correlation between frequencies in the same frame (correlation in frequency) as well as between frequencies in adjacent frames (correlation in time). These properties are utilized using simple first order prediction schemes in frequency as well as in time.

Scalar Quantization of LSP Differences (DIF).

In [12] it has been observed that the distribution range of LSP frequencies varies significantly with speaker characteristics as well as with recording and transmission conditions. To reduce this variability it was proposed to code the difference between adjacent LSP frequencies, instead of coding the actual LSP frequencies themselves, motivated by a more suitable distribution range. Optimal Max-quantizers (0, 1, 2, 3, 4 and 5 bit) have been designed for each difference.

BIT ERROR SENSITIVITY

The speech codec simulation was written in VAX/VMS/FORTRAN, and run on a VAX 11/750 computer. The simulation model is split into two programs, a coder and a decoder, in order to create a data file containing the quantizer indices for the various parameters only once, while the far less complex decoder can produce the synthesized speech data from any given channel file. The generation of a bit error pattern was produced by a stand alone main program, written in ANSI-C producing a random bit error according to the given probability.

The speech material used for testing the SESTMP consist of four Norwegian utterances (two male, two female speakers) with telephone bandwidth sampled at 8 kHz. Each sentence is 4.6 seconds long (231 frames), including natural silence at the end of each utterance.

The coder was investigated for five typical bit error rates (BER), for a mobile satellite channel, [10]; 10^{-4}, 10^{-3}, $4 \cdot 10^{-3}$, 10^{-2} and 10^{-1}, in order to determine the bit-error sensitivity of the different coder parameters. The speech quality degradation due to the bit errors are measured both by objective and subjective means.

Objective Measures

The simulation was run separately for each parameter, inserting bit errors according to the appropriate rate, leaving the remaining parameters unchanged. In this way the bit error sensitivity was found for the five groups of parameters distinctively. The segmental signal-to-noise ratio (SNR) was calculated for each condition and plotted in a semilog diagram shown in Figure 3. The results show that there is a difference in sensitivity among the parameters. It seems clear that the speech spectral information, mainly reflected by the LPC coefficients is most important.

Subjective Preference

The subjective listening tests have been conducted using the first female utterance, and was run as a standard paired comparison test [8]. Nine different tests have been run, in order to establish the bit error sensitivity of each of the parameters at the bit error rates; 0 (denoted 10^{-5} in the figure), 10^{-4}, $4 \cdot 10^{-3}$, 10^{-3}, 10^{-2}, and also the relative importance among each other at the four different BER. Each test has been run on at least eight listeners, in a silent room. Figure 4 gives the relative sensitivity among the parameters where 0.0 denotes the chosen best quality and -5.0 the worst.

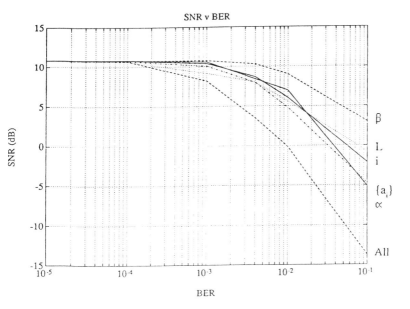

Figure 3: Objective Speech Quality (SNR) as a Function of BER.

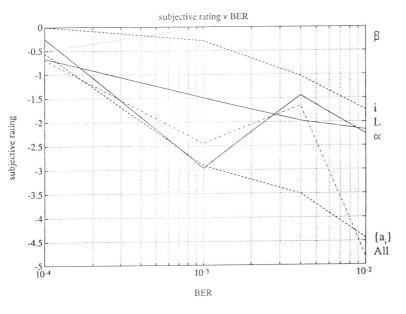

Figure 4: Subjective Rating of the Relative Importance of Bit Errors in the Various Parameters.

The results clearly verify the objective results, in that β is the least sensitive parameter, and no noticeable difference is recorded with BER better than 10^{-2}. The most sensitive parameters are the LPC coefficients and α. Subjectively, the same ranking was obtained as with the objective results, and it is clearly evident that the critical point is somewhere between 10^{-3} and 10^{-2} for all parameters.

A point worth noting is the subjective effect of the bit errors. By inserting errors in β, L and i the speech seems noisier and harder to understand, in a sense of harshness; while errors in the LPC coefficients and α results in a serious degradation of the spectrum and dynamic range. The errors result in gross distortions and large magnitudes, some of which are painful for the ear. As will be shown later, these errors to a large extent, can be minimized by utilizing the ordering characteristics of the LSP frequencies.

SPECTRAL ERROR SENSITIVITY OF LSP FREQUENCIES

To be able to compare the two quantization schemes, PRED and DIF, with respect to bit errors, both SNR and paired comparison tests have been taken into account.

Figure 5 gives the subjective results where, again, 0.0 is the best and -5.0 denotes the worst quality.

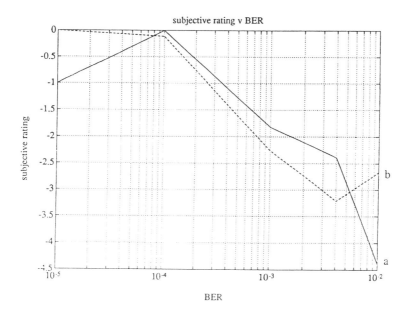

Figure 5: Subjectively Rated Bit Error Sensitivity of the LPC Coefficients.
(a) Using PRED and (b) Using DIF.

From Figure 5 it would appear that DIF is somewhat more robust, and can tolerate BER as high as 10^{-2}. Subjectively the two quantization schemes perform differently. Errors in DIF give a higher noise level and a greater loss of intelligibility in the speech, but far less occurrence of gross distortions. As stated earlier, this degradation of the spectrum causes painful clicks and bangs. These spectral peaks occur when an error moves the LSP frequency ω_i too close to ω_{i+1}, creating a resonance in the spectrum.

It is clear that an error within a frame using PRED is isolated to a small region. Inherent in the prediction schemes is that an error in ω_i will impact on ω_{i+j}, $j=1, \ldots$ (10-i). In addition the predictors for the next frames will suffer greatly. In other words, an error in PRED will result in degradation in an isolated region within a frame, but will, to a great extent, be spread out in time. This again can cause large discontinuities in the spectral estimate from frame to frame partly giving rise to the previously mentioned painful distortion.

Considering DIF, an error will generally spread out in the whole frame. An error in the difference $\Delta\omega_j$ will affect all the differences $\Delta\omega_k$, $k=j+1, j+2 \cdots 10$. Thus an error increasing the LSP difference will shift the spectrum to the right. LSP frequency p+1 is a constant (π), thus a large error in $\Delta\omega_j$ will press the spectrum towards π, and could give a resonant at the high frequencies. Analogously an error decreasing the LSP difference will shift the spectrum to the left.

Therefore it seems clear that the LSP differences $\Delta\omega_1$ -$\Delta\omega_{10}$ will have a decreasing sensitivity with increasing index, i. This has been shown by inserting a 100% BER in the MSB of each LSP frequency and calculating the log rms spectral distortion in each case. In general it can be observed that peaks in log rms spectral distortion occurred in the area one would expect to find the two first formats in human speech (ω_2 -ω_4, ω_5 -ω_7) both for PRED and DIF. In PRED there seems to be no obvious sensitivity difference within the set of LSPs.

In the recommendation for the Pan European mobile telephone system (GSM) [9], the subjective relevance of the speech coder output bits have been classified. By paired comparison listening tests, each of the 260 bits, representing one speech data frame, have been classified to one of three protection classes according to their subjective importance. A small test has been run to investigate the objective importance of each bit in the LSP frequencies, by introducing an error in each bit for all the 10 parameters in PRED and DIF, and calculating the log rms spectral distortion.

As expected, the MSB is most sensitive. From the results for PRED it is noted that the distortion for the LSB is under 1dB, indicating no need for protection. For DIF all the bits would require protection.

Non-redundant Correction Algorithms

Gray Coding.

In instrumentation (especially D/A converters) unit distance codes are commonly used to avoid large errors. The Gray code is a unit distance code that, in addition to a number of instrumentation purposes, has been used for partial protection against channel errors [14]. Gray coding the LSP quantizer indices gives a significant improvement in quality for BER larger than 10^{-2}.

Correction of Unstable Filters.

As mentioned in Section 2, representing the LPC- coefficients by LSP frequencies has certain advantages towards quantization. One of these is the possibility of conserving the filter stability after quantization, by observing that the LSP frequencies (ω_1 -ω_{10}) are interlaced with each other, i.e., $\omega_1 < \omega_2 < .. , \omega_{10}$.

During the quantization process, closely spaced LSP frequencies may change the ordering, resulting in an unstable synthesis filter. A similar problem will occur even if the order is not changed. Frequencies may be moved so close together, resulting in a large resonance in the synthesized speech spectrum. A simple algorithm for avoiding these problems is suggested in [4].

This algorithm is capable of correcting unstable filters only if two consecutive LSP frequencies are interchanged but not, for example, in the case where ω_i and ω_{i+j}, j>1, interchange. Observing the probability density functions (pdf) of the 10 LSP frequencies it can be seen that although the pdfs of ω_i and ω_{i+1} overlap, the probability of these frequencies interchanging will be very small. The large spectral peaks and often painful distortion reported earlier have given rise to suspicion as to whether the filter stability really is conserved. And it is quite true that in the presence of bit errors, the possibility of ω_{i+j}, j>1 and ω_i interchanging may happen, especially at high BER ($7 \cdot 10^{-3}$). A new algorithm for preserving filter stability is therefore proposed, involving a sorting of the LSP frequencies in ascending order before checking for closeness. Additional improvements are proposed in [5,6]. With these algorithms incorporated, the painful distortion is now reduced. At 10^{-1}, however, the coder still breaks down and, while not particularly painful, is definitely unpleasant.

CONCLUSION

The Self Excited Stochastic Multipulse coder (SESTMP) at a fixed bit rate of 7.0 kbit/s, a possible candidate coder for mobile satellite communication, has been simulated under various bit error conditions in order to establish the coders sensitivity to random errors. Results show that in all the curves giving the bit error sensitivities of each individual parameter, there exists a distinct knee in the performance. For the LPC coefficients, the most sensitive parameter, this knee occurs at approximately $5 \cdot 10^{-4}$, while the knee for the self excitation gain factor is found as high as 10^{-2}.

The spectral information, represented mainly by the LPC coefficients, are quantized considering two different schemes: non-uniform Max quantization of LSP frequencies with prediction in both time and frequency (PRED) and non-uniform Max quantization of LSP differences (DIF). The distinctly different behaviour of the two quantization schemes gives rise to an interesting trade-off in selecting a suitable method;

(i)   bit errors in PRED cause large spectral peaks, often painful to ears, but with an overall high intelligibility, while DIF gives a generally higher noise level, lower intelligibility, but smaller spectral peaks; and

(ii)   bit errors in DIF, though smeared across the whole frame, are confined to a single frame, while errors in PRED, affecting only a small region, propagate through several consecutive frames.

REFERENCES

1.   A. Perkis, B. Ribbum and I. J. Fannelop, "A study on the impact of a satellite channel on a 7.0 kbit/s CELP based coder", IEEE Workshop on Speech Coding, Vancouver, September, 1989.

2.   I. J. Fannelop, B. Ribbum and A. Perkis, "Complexity reduction of a 7kbps CELP based speech coder", IEEE Workshop on Speech Coding, Vancouver, September, 1989.

3.   J. P. Campbell, V. C. Welch and T. E. Tremain, "An expandable error-protected 4800 bps CELP coder (U.S. Federal Standard 4800 bps voice coder)", *Proc ICASSP*, 1989.

4.   B. S. Atal, R. V. Cox and P. Kroon, "Spectral quantization and interpolation for CELP coders", *Proc ICASSP*, 1989.

5.   R. Cox, W. Bastiaan Kleijn and P. Kroon, "Robust CELP coders for noisy backgrounds and noisy channels", *Proc ICASSP*, 1989.

6.   A. Perkis, B. Ribbum and T. Svendsen, "A good quality low complexity 4.8 kbit/s stochastic multipulse coder", *Proc. ISCAS*, 1989.

7.   F. K. Soong and B. H. Huang, "Optimal quantization of LSP parameters", *Proc. ICASSP*, 1988.

8.   T. V. Sreenivas, "Modelling LPC residue by components for good quality speech coding", *Proc. ICASSP*, 1988.

9.   GSM Recommendation 06.10., Version 3.1.2: "Gsm full rate speech transcoding", 19th September, 1988.

10.   K. Grythe, personal communication, 1988.

11.   B. S. Atal and M. R. Schroeder, "Stochasitic coding of speech signals at very low bit rates", *Proc ICC*, 1610-1613, 1984.

12.   F. K. Soong and B. H. Juang, "Line spectrum pairs (LSP) and speech data compression", *Proc ICASSP*, 1984.

13.   I. Max, "Quantizing for minimum distortion", *IEEE Transactions on Information Systems*, March, 1960.

14.   CCITT Question U/N (16kbits/s speech coding), "Description of 16 kbits/s low-delay code-excited linear predictive coding (LD-CELP) algorithm", Source AT&T.

27

A REAL TIME IMPLEMENTATION OF A
4800 BPS SELF EXCITED VOCODER.

Stephen J. A. McGrath †, Richard C. Rose ††,
Thomas P. Barnwell III †

† School of Electrical Engineering
Georgia Institute of Technology
Atlanta, Georgia 30332

†† MIT Lincoln Laboratories
244 Wood Street
Lexington, Massachusetts 02173

INTRODUCTION

The purpose of this paper is to describe a fully developed, real-time 4800 bps speech coding system that was designed for use in a nation-wide mobile satellite communications network [1]. The overall goal was to build an inexpensive speech coder capable of producing "near toll quality" coded speech under bursty channel error conditions at bit error rates of approximately 10^{-3}. BER. A form of the self excited vocoder (SEV) was chosen as the speech coding algorithm [2]. Many of the results presented in this paper concern pragmatic issues such as the robust representation of coder parameters, the error protection, and the constraints imposed by a limited hardware environment. A particularly interesting result of this work was that significant subjective speech quality improvements can be achieved under noisy channel conditions by exploiting known properties of the speech coding parameters.

The *self excited vocoder* is a member of a general class of analysis-by-synthesis linear predictive speech coders [3] which represent speech as the output of a slowly time-varying recursive linear filter driven by a block parametric excitation signal. The parameters of this excitation are determined through an analysis-by-synthesis procedure originally attributed to Atal [4]. The individual members of this class of coders differ primarily in the definition of the excitation signal. The entire class contains an infinite variety of coders including many well known coders such as the Code Excited (CELPC) [5], the Multiple Pulse Excited (MPLPC) [4], and the Regular Pulse Excited (RPLPC) [6] linear predictive coders. The SEV was used because of its inherent simplicity and proven performance as compared to the other coders in this class [3].

The goal of the paper is to describe the basic SEV, to describe the techniques used to encode the vocal tract parameters, and to provide a detailed discussion of the pragmatic issues addressed in quantizing the coder parameters.

This work was sponsored by the Jet Propulsion Laboratory.

SELF EXCITATION MODEL

The most common speech synthesis model is composed of three components: a short-term linear predictor, a long-term linear predictor, and an excitation signal. In such systems the function of the short-term predictor is to model the slowly varying spectral envelope of the speech signal, while the long-term predictor is intended to model the pitch redundancy in voiced sounds. Finally, the function of the excitation signal is to excite the system and to model all perceptually important features of the speech signal that are not well modeled by the short-term and long-term predictors.

In order to better understand the relationship of the SEV to other coders in the class, it is more appropriate to use the vocoder synthesizer model illustrated in Figure 1. This model consists of only two (rather than three) components: a short-term predictor, $A(z)$; and a block encoded excitation, $e[n]$, to that filter. The principal difference between this model and the common model is that the function of the long-term predictor has been included explicitly in the excitation model. In general, the excitation signal is composed of a linear combination of K component excitation sequences, $e_k[n]$. The index $k = 1, \ldots, K$ is the component sequence index, and each $e_k[n]$ is an N point sequence chosen from an associated excitation ensemble, $\mathcal{F}_k$.

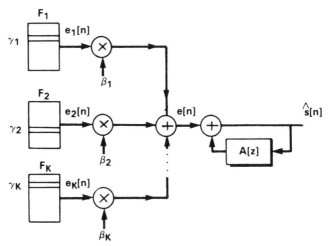

Figure 1: Block diagram of synthesis model used in the analysis-by-synthesis class of predictive speech coders.

The excitation model parameters in this class of coders are the ensemble index γ_k and the ensemble gain β_k corresponding to the excitation ensemble $\mathcal{F}_k$ shown in Figure 1. The optimum excitation parameters for a given component excitation sequence are found by exhaustively searching through the excitation ensemble, $\mathcal{F}_k$, for that ensemble function that minimizes a weighted mean squared error [3]. Each individual type of coder in this class is defined by the component excitation sequences in its excitation function. For example, the excitation function in the CELPC is given by

$$e[n] = \beta_1 e[n - \gamma_1] + \beta_2 v_{\gamma_2}[n] \tag{1}$$

where the first component excitation sequence, $e_1[n] = e[n - \gamma_1]$, corresponds to a sequence from the memory of a first order long-term predictor, and $e_2[n] = v_{\gamma_2}$ is a sequence chosen from a Gaussian ensemble.

A self excited vocoder is simply a LPC vocoder which has one or more long-term predictors and uses no other excitation ensembles. A block diagram of the simplest type of SEV is shown in Figure 2. In terms of the synthesis model of Figure 1, the excitation signal of the SEV is given as $e[n] = \beta_1 e[n - \gamma_1]$. It might seem that such a system is not capable of producing speech because it appears to have no excitation. However, a close examination of the system shows that the SEV is actually deriving the current source excitation from the past history of the excitation signal itself. This is accomplished through proper choice of the predictor delay, γ, and predictor gain, β, using the analysis-by-synthesis procedure discussed above.

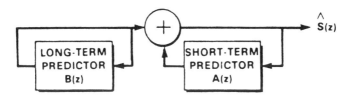

Figure 2: Block diagram of synthesis model for the simplest self excited vocoder.

There are two major advantages of the SEV relative to other coders in its class. The first is that the SEV is an extremely simple, easily implementable coder. Whereas in the CELPC it is necessary to estimate and quantize the excitation model parameters for two excitation sequences, in the SEV there is only a single excitation sequence. The second advantage is that the long-term predictor or "self excitation ensemble" is amenable to a fast ensemble search procedure whose effect is to reduce the computational complexity by over an order of magnitude [3].

A disadvantage of the SEV, which it shares with all coders which use a long-term predictor, is that in non-ideal channel conditions, the SEV is sensitive to corruption of the excitation parameters. Recognizing that there is no fundamental requirement that a particular excitation function ensemble contain signals all of the same type, a more robust excitation ensemble was proposed [7]. A "non-homogeneous" (NH) ensemble was implemented by combining a set of time-varying self excitation sequences with a small number of fixed Gaussian random sequences. The search procedure still chooses a single sequence from the codebook as before, so the determination of the optimum class of sequences is accomplished by choosing the single sequence which provides the least measured distortion. The excitation signal to the short-term predictor is given by $e[n] = \beta z_\gamma[n]$, where

$$z_\gamma[n] = \begin{cases} e[n - \gamma], & 0 \le \gamma \le C \\ v_\gamma[n], & C < \gamma \le F \end{cases} \tag{2}$$

An ensemble of this type was implemented in the real-time coder.

ROBUST VOCAL TRACT MODEL

The vocal tract parameters were quantized using a Line Spectrum Pair (LSP) technique first proposed by Crosmer [8]. This technique can represent the LPC coefficients with less perceptual distortion at the required bit rates (28–30 bits per frame) than the widely used *PARCOR* quantizer. In addition, the LSP representation offers greater robustness against quantization effects. Finally, the LSP representation serves to decouple the effects of the vocal tract parameter quantization from the effects of the residual coding, or source analysis, process [1]. This considerably simplifies the development of the quantization procedures.

The LSP quantization procedure can be broken down into two phases. First, the LSP transformation is used to convert the LPC parameters into a set of Line Spectral Frequencies (LSF's). Second, the LSF's are quantized. Crosmer's LSP quantization procedure is particularly effective because it is based on a perceptual model of the auditory system. Thus, it represents spectral information which is more perceptually important with higher precision than less important information.

The LSP Transformation

Certain of the properties of the Line Spectral Pair representation are functions of the mathematical transformation used to derive the LSP's, and other properties are functions of the interpretation put on the LSP model and it's representation of the LPC spectrum. The interpretation used in this work is sufficiently different from the commonly accepted interpretations, and the properties derived from this interpretation are so significant to the quantization strategy, that a brief explanation of the LSP transformation will be included here. Further details can be obtained in [8].

The LSP transformation of a tenth order LPC analysis will yield ten LSP angular frequencies, lying between 0 and π radians. In the course of the LSP transformation, two polynomials $P(\omega)$ and $Q(\omega)$ are formed. Each of these polynomials has five roots, all of which lie on the unit circle. If we let $\omega_1, \omega_3, \omega_5, \omega_7$ and ω_9 be the roots of $P(\omega)$, and $\omega_2, \omega_4, \omega_6, \omega_8$ and ω_{10} be the roots of $Q(\omega)$, a fundamental property of the LSP transformation is that the LSP frequencies can be labeled so that $\omega_1 < \omega_3 < \omega_5 < \omega_7 < \omega_9$, and $\omega_2 < \omega_4 < \omega_6 < \omega_8 < \omega_{10}$; and further, for stable systems, the two sets of roots are interleaved, so that $\omega_1 < \omega_2 < \omega_3 < \ldots < \omega_{10}$.

The LSP frequencies possess several useful and desirable properties, including a bounded range, a natural ordering, and a simple associated test for synthesis filter stability. It can be shown that a necessary and sufficient condition for synthesis filter stability is the satisfaction of the above ordering property by the LSP frequencies; equivalently, if $\omega_i - \omega_{i-1} > 0$ for $i = 2 \ldots 10$, then the synthesis filter $A(z)$ will be stable. If the difference is identically zero for any i, one of the roots of $A(z)$ will lie on the unit circle. This property provides a method for ensuring filter stability after parameter quantization and coding. Furthermore, although each LSP frequency is bounded such that $0 < \omega_i < \pi$, it has been shown that in speech applications the range of each LSP frequency is actually limited to a fraction of the total range. This property allows for a more efficient quantization strategy.

A further property of the LSP representation of the vocal tract is that there is a strong frame-to-frame correlation of the LSP frequencies, as well as the correlation of frequencies within a frame. This is due to the fact that the vocal tract characteristics change relatively slowly during steady-state speech production, such as for vowels. This allows a differential coding scheme to be effectively used to achieve a reduction in bit rate.

LSP Quantization

It is well known that there is some correspondence between the LSP frequencies and the speech spectral peaks. While the exact nature of this correspondence is difficult to define, it can be shown that the LSP frequencies are related to the poles of the LPC filter. The pole locations of the LPC filter are in turn related to the formant locations of the original speech signal. Generally, each formant yields a complex pole pair in the LPC analysis. It has been found that the LSP frequencies corresponding to a pole pair are close together around the center frequency of the peak for a high-resonance formant, further apart for low-resonance formants, and further still apart where there are no strong peaks in the spectrum. Since the roots of $P(\omega)$ and $Q(\omega)$ are interlaced, such a pair of LSP frequencies will contain one from each of $P(\omega)$ and $Q(\omega)$.

Human listeners have different sensitivities to formant peak locations, formant bandwidths, and overall spectral tilt. To take advantage of these variations, the LSP representation separates the roots of $P(\omega)$ from those of $Q(\omega)$. The roots of $P(\omega)$ are considered to represent formant position information, and are labeled the *Position Coefficients*. The distance from each of the roots of $Q(\omega)$ to the closer of the neighboring roots of $P(\omega)$ contains information about the presence and bandwidth of a formant at that position coefficient. Thus the roots of $Q(\omega)$ are expressed in relation to the position coefficients as

$$|d_i| = \min_{j=-1,1} |\omega_{2i+j} - \omega_{2i}|, i = 1, 2 \ldots 5$$

The sign of d_i is defined as positive if ω_{2i} is closer to the lower position coefficient ω_{2i-1} than to the upper one ω_{2i+1}, and negative otherwise. These d_i are known as the *Difference Coefficients*, and indicate peaks in the speech spectrum; a small positive or negative value indicates a high-Q pole at the lower or upper neighboring position coefficient respectively, while either a large positive or negative value would indicate only a very broad spectral peak or valley. This suggests the use of a three-zone model for coding the difference coefficients, as represented in Figure 3. The first zone is defined by $|d_i|$ less than a minimum value d_{min}, and represents a formant whose bandwidth is too narrow, tending towards instability. The second zone is defined by $|d_i|$ greater than a maximum value d_{max}, and represents the absence of a formant. And the third zone represents the presence of a formant, defined by $d_{min} < |d_i| < d_{max}$. The position and difference coefficients are thus coded according to separate paradigms which are matched to the auditory model.

292

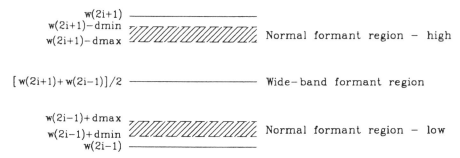

Figure 3: **Classification regions for quantizing the ith difference coefficient** $d_i = \omega_{2i}$.

ROBUST PARAMETER REPRESENTATIONS

The LSP approach makes it possible to implement a coder which is robust to channel errors and whose output will degrade gracefully but remain "speech like" in the presence of increasing channel error rates. This was achieved largely through the parameterization of the variables to be quantized with the addition of some restrictions on parameter behavior. Each parameter was individually evaluated for robustness, and appropriate mechanisms were devised to protect each one. The parameters to be coded belong to two classes: those representing the vocal tract information and those representing the LPC excitation information.

The working parameters of the system were as follows. Speech was band-pass filtered between 300 Hz and 3.3 kHz, and sampled at 8 kHz in a 12-bit linear format. A 10th order LPC analysis was implemented recursively [9]. The full duplex coder was implemented on a system which included three WE-DSP32 processors [1].

Excitation

The excitation analysis is performed by computing a gain value for each candidate sequence from the excitation codebook ensemble, and passing each sequence in turn through the LPC filter. A mean squared error is computed for each candidate sequence by comparison with the original weighted speech, and the gain and index of the sequence which minimizes the mean squared error is selected for quantization and coding.

Frame Rates. Although the LSP representation will tolerate an extraordinarily low LPC frame rate [8], an LPC frame rate of 150 samples (18.75 mS) was found to yield an acceptable trade-off between bit-rate and speech quality. An important benefit of the SEV is that there is only a single component excitation sequence in each excitation analysis frame, as opposed to most other implementations of analysis-by-synthesis predictive speech coders which have at least two component excitation sequences per frame. This allows the SEV to use a much shorter excitation frame, which yields increased time resolution and correspondingly better subjective quality for a given size codebook. Simulations and early experiments indicated that an excitation analysis frame of 20 samples (2.5mS) yielded acceptable speech quality, with noticeable improvements but also unacceptable increases in complexity as frame size was reduced. Increases in the length of the excitation frame produced perceptible degradation, which accelerated noticeably beyond 30 samples. An excitation analysis frame rate of 30 samples was eventually selected as a compromise between bit allocation and speech quality.

Gain Quantization. Quantization of the excitation gain parameter was a challenging task due to its erratic behavior during speech transitions. A modified APCM quantizer, based on the Jayant APCM quantizer, was designed based on the characteristic behavior of the gain parameter [1]. This novel quantizer tracks the gain parameter more closely with four bits per sample than the Jayant quantizer using five bits. The quantizer works by normalizing the gain by multiplying it by the energy of the selected excitation sequence to produce a much more predictable gain parameter. With reference to Figure 4(a), the original gain parameter is well behaved during steady-state speech but exhibits large isolated peaks, up to several orders of magnitude larger than the average, during transitions. Figure 4(b) shows the effect of the gain normalization. The new parameter has an offset proportional to the excitation signal energy, and is well-behaved around this offset. The quantizer thus concentrates its levels about this offset by means of a table of stepsize multipliers. Rather than using integer multiples of the stepsize as the quantizer levels as in the Jayant quantizer, giving levels at $\delta, 2\delta, 3\delta \ldots$, the table of stepsize multipliers $m_1, m_2, m_3 \ldots$ is used to shift the quantizer levels to an offset corresponding to the energy in the signal, giving levels at $m_1\delta, m_2\delta, m_3\delta \ldots$. Further, the multipliers $m_1, m_2, m_3 \ldots$ are selected to spread the available levels exponentially around the offset rather than linearly, providing high resolution for low levels around the offset. The stepsize parameter is modified in the same manner as in the Jayant APCM quantizer, with the result that both the offset and the range of the quantizer levels are proportional to the energy in the signal.

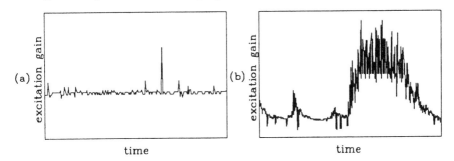

Figure 4: Typical behavior of (a) original gain parameter and (b) normalized gain parameter (not to scale).

Error protection of the gain parameter was accomplished by imposing limits on the stepsize adjustment. The effect of these limits was to periodically synchronize the stepsize at the receiver and at the transmitter by making it saturate to one level or the other. This strategy was carefully optimized to minimize the perceived distortion added by the stepsize limits, while synchronizing the gain parameter often enough during normal speech to recover from any loss of synchronization quickly.

Size and Structure of Excitation Ensemble. The size of the excitation ensemble which could be searched was fundamentally limited by the constraints of real-time operation on the WE-DSP32. Several complexity-reduction methods were employed to reduce the number of computations for each candidate sequence [3], but the 16MHz WE-DSP32 on which the excitation ensemble search was realized could search a maximum of 140 sequences per frame. Thus an ensemble size of 128 was used, requiring seven bits per excitation frame. As defined in (2), this ensemble comprised 120 self-excitation sequences and eight fixed Gaussian random sequences. Experiments have shown that the use of such a small ensemble results in a performance penalty, albeit a minor one [3], and so the constraints of real time operation on the selected processor architecture were fundamental limitations to the quality of the final 4800bps realization.

As well as providing a means for coder initialization, the Gaussian random codebook provides an error protection mechanism for the excitation signal. Corruption of the data stream could result in mismatches between the excitation signal at the transmitter and that at the receiver through two mechanisms: corruption of the gain parameter, and corruption of the excitation index. The gain parameter is protected by the imposition of limits on the stepsize as described in the previous section. Since the excitation signal is time varying, the only way to correct for errors is to periodically purge the memory of the long-term predictor and set it to a known state. This is achieved by periodically (every five seconds) forcing the transmitter to select its excitation sequence from the Gaussian codebook for enough (six) consecutive excitation frames to totally repopulate the memory of the long-term predictor. This update interval was chosen to be short enough to recover from errors within a few spoken words, but long enough to have a minimal impact on the perceptual quality of the speech. In the event that the 4800bps bitstream gets sufficiently corrupted to cause a total loss of synchronization between the transmitter and the receiver, the channel controller signals the receiver to silence its output, and waits until the receiver recognizes a sequence of six consecutive excitation indices from the Gaussian codebook.

Vocal Tract Parameters

The quantization strategy used for the Line Spectral Pairs was developed to match the characteristic behavior of these parameters. As described above, the position coefficients are related to the peaks in the frequency response of the LPC filter. These coefficients need to be tracked quite closely, and small shifts need to be tracked more closely than large ones. Errors in the upper frequency ranges are less perceptible than those at low frequencies, so more bits need to be allocated to tracking lower frequency position coefficients. The most efficient way of achieving these requirements is to code the position coefficients differentially, coding the shifts from each position coefficient to the corresponding coefficient in the next frame. This difference is coded using an exponential quantizer to provide fine resolution for smaller frequency shifts from frame to frame and increasingly coarser resolution for larger shifts.

The difference coefficients, on the other hand, are related to the formant bandwidth information of the speech signal, and as such may be coded with less accuracy than the position coefficients. It has been found [8] that high quality speech can be produced by discarding most of the bandwidth information and coding the difference coefficients as either the presence or absence of a formant. In this case, $d_{min} = d_{max} = d_0$ and a three-level quantizer can be used to code the levels d_0, d_m and $-d_0$, where d_m is halfway between the neighboring position coefficients ω_{2i-1} and ω_{2i+1}. A higher quality representation can be obtained by coding the difference frequencies falling between d_{min} and d_{max} with higher resolution.

Error protection of the LSPs is achieved using one bit per frame. There is some error protection inherent in the LSP model used. The position coefficients are checked for adherence to the ordering property rule described previously. Any coefficient not in order is replaced by its value in the previous frame. Since the difference coefficients are coded relative to the position coefficients, this check does not apply to them. However, there are several rules which they must obey. If d_i is too close to a position coefficient, it is moved out to d_{min}; if a negative difference coefficient falls closer to the lower neighboring position coefficient than to the upper one (or vice versa), there has been an error in transmission and the offending difference coefficient is moved halfway between the two; and if d_i is negative and d_{i+1} is positive, placing both close to the same position coefficient p_{i+1} and representing an unnaturally high-Q formant, d_{i+1} is moved halfway between p_{i+1} and p_{i+2}. The loophole in this protection strategy is that any errors in transmission of the position coefficients will be cumulative. The mechanism which protects against this accumulation of errors is based on the fact that during steady state speech events, the difference coefficients will change very little or not at all from frame to frame as the formant characteristics change very slowly. This is especially true of the upper formants. A single bit per frame is used to indicate when the upper three difference coefficients have not changed from those of the previous frame, and the absolute (as opposed to differential) value of a single position coefficient, along with its index, is sent in the place of those three difference coefficients. Each of the position coefficients is sent in turn using this strategy; during normal speech, the complete set of position coefficients is sent several times a second.

Table 1 shows the bit allocation in the final version of the real-time implementation. This implementation achieves good communication quality speech at 4800 bits per second.

CONCLUSION

It is currently an issue of debate, as applications increasingly require implementations of low bit rate speech coders in real channels, whether error correction and protection should be provided within the bit rate assigned to the coder or in addition

296

Parameter	Bits/Parameter	Bits/Frame
LSP Position Coeffs	18	18
LSP Difference Coeffs	14	14
Excitation Gain	4	20
Excitation Index	7	35
Syncronization/ECC	3	3
Total: 90 bits/frame at 53.3 frames/sec		

Table 1: Bit allocation for parameters of 4800 bps. SEV.

to that bit rate. The implementation of channel protection schemes independent of the coding scheme can typically add a substantial amount of transmitted information to the coder bitstream; 50–70% is not uncommon, particularly when the transmission path may include a satellite link. One particularly interesting result of the work described here has been to show that certain classes of coders and coding strategies can provide a significant amount of the required error protection at a very low cost in either bit rate or subjective speech quality. The cost of providing this protection using the properties of the speech parameters is substantially less than providing the same level of protection by other means.

A second result of this work has been to demonstrate that state-of-the-art medium-rate speech coders are sufficiently complex to require more computational capacity than was provided by DSP microprocessors which were available. This implementation required three AT&T WE-DSP32's, and was still limited by both computational and memory constraints of the processors. With the generation of DSP microprocessors which has emerged since this work was completed, the implementation described above should fit comfortably on a single floating-point processor. The next generation of floating-point devices should provide enough computational capacity to search larger codebooks, and provide the flexibilty to evaluate and implement other features and trade-offs.

REFERENCES

[1] S. J. A. McGrath and T. P. Barnwell III. Development, design, fabrication, and evaluation of a breadboard speech compression system at 4800 bps – phase four final report. Technical report, School of Electrical Engineering, Georgia Institute of Technology, 1988.

[2] R. C. Rose and T. P. Barnwell III. The self excited vocoder – an alternate approach to toll quality at 4800 bps. *Proc. Inter. Conf. on Acoustics, Speech, and Signal Proc.*, pages 453–456, April 1986.

[3] R. C. Rose. *The Design and Performance of an Analysis-by-Synthesis Class of Predictive Speech Coders*. PhD thesis, Georgia Institute of Technology, Atlanta, Ga., 1988.

[4] B. S. Atal and J. R. Remde. A new model of LPC excitation for producing natural sounding speech at low bit rates. *Proc. Inter. Conf. on Acoustics, Speech, and Signal Proc.*, pages 614–617, April 1982.

[5] M. R. Schroeder and B. S. Atal. Code excited linear prediction: High quality speech at very low bit rates. *Proc. Inter. Conf. on Acoustics, Speech, and Signal Proc.*, pages 937–940, April 1985.

[6] P. Kroon, E. F. Deprettere, and R. J. Sluyter. Regular–pulse excitation: A novel approach to effective and efficient multipulse coding of speech. *IEEE Trans. Acoust., Speech, and Sig. Processing*, ASSP-34(5):1054–1063, Oct. 1986.

[7] R. C. Rose, T. P. Barnwell III, and S. McGrath. The design and performance of a real–time self excited vocoder. *Proc. Mobile Satellite Communications Conference*, May 1988.

[8] Joel R. Crosmer. *Very Low Bit Rate Speech Coding Using the Line Spectrum Pair Transformation of the LPC Coefficients*. PhD thesis, Georgia Institute of Technology, June 1985.

[9] T. P. Barnwell III. Recursive windowing for generating autocorrelation coefficients for LPC analysis. *IEEE Trans. Acoust., Speech, and Sig. Processing*, ASSP-29(5):1062–1066, Oct. 1981.

28

A REAL-TIME FULL DUPLEX 16/8 KBPS CVSELP CODER WITH INTEGRAL ECHO CANCELLER IMPLEMENTED ON A SINGLE DSP56001

Juin-Hwey Chen, Ronald G. Danisewicz,
Richard B. Kline, Dennis Ng, Reinaldo A. Valenzuela, and
Bruce R. Villella

CODEX Corp.,
20 Cabot Boulevard, Mansfield MA 02048

INTRODUCTION

The trend in today's private digital networks is to integrate data with voice traffic. This offers considerable cost savings and convenience by allowing data and digitized voice to share the same transmission media, and much of the hardware and software. Code excited linear prediction (CELP) [1] is a speech coding method that has tremendous potential for providing the high quality that network voice applications demand for rates at and below 16 Kb/s. One difficulty with CELP is that it requires tremendous computational resources and memory for its codebook search. Just as there is a cost associated with the maximum bandwidth that a given network link can accommodate, there is a cost associated with the hardware necessary to support speech compression. With new telecommunications technologies promising to significantly reduce the cost of network bandwidth, the cost of the speech processing hardware may someday become the single most important consideration. Since digital signal processing chips (DSPs) and the fast memory chips supporting them consume a relatively major portion of a speech coding board's real estate and power consumption, coding schemes that reduce the number of required DSPs and memory will significantly reduce the overall cost of a network.

Several speech coding algorithms in existence today retain the basic CELP configuration without requiring nearly as many computations as a true CELP. Most of these coding algorithms achieve these economies by forcing constraints on the codebook which facilitate the codebook search. Vector Sum Excited Linear Prediction (VSELP) [2] is one such coding algorithm. Codevectors in the VSELP codebook are linear combinations of a small set of basis vectors. This arrangement elicits a very fast codebook search requiring minimal memory. However, a 16 Kb/s VSELP still demands more computations than deemed economical for implementation in network environments. Codex's version of VSELP, CVSELP, reduces complexity even further, particularly in the long-term predictor, and permits the implementation of a full duplex encoder/decoder with integrated echo canceller on a single fixed-point DSP. The echo canceller has 128 taps (16 msec span) and meets or exceeds all performance requirements of CCITT recommendation G.165.

CVSELP

CVSELP was developed from VSELP in order to achieve implementation of a 16 Kb/s speech encoder/decoder with 128-tap echo canceller on a single fixed-point DSP. In many ways CVSELP is merely a pared-down version of VSELP in that Codex has removed several enhancements from VSELP to significantly reduce loading without critically reducing the quality. CVSELP does, however, have a long-term predictor (LTP) search strategy that is significantly different from that of VSELP. The text that follows will describe CVSELP while highlighting the pronounced differences between it and its more powerful and more computationally demanding predecessor.

Figure 1 is a block diagram of the CVSELP encoder. With reference to the encoder, a frame of speech consists of the concatenation of N_V vectors ($N_V=4$ for the 8 Kb/s version, $N_V=8$ for the 16 Kb/s), each vector having N samples. Let N_F be the frame length (160 samples), $N_F=N_VN$. The block diagram can be partitioned into a frame processing section (top) and a vector processing section (bottom). The routines in the frame processing section operate on a frame of input speech samples from s(n), these routines generate the following open-loop parameters: pitch, frame gain, and linear predictive coding (LPC) coefficients. The frame processing section also passes one frame of s(n) through the perceptual weighting filter, whose transform is denoted by W(z), generating the output s'(n).

The LPC block extracts one set of LPC coefficients for every frame of input speech samples and parameterizes the weighting filter and LPC synthesis filter with these coefficients. In the encoder the LPC synthesis filter always appears in cascade with the weighting filter[1]; in the diagram, the transform of this cascade is denoted by H(z). Unlike VSELP, CVSELP does not interpolate the LPC coefficients across the vectors, so this parameterization is only done once per frame.

The pitch analysis block extracts one pitch value for every frame of the input speech and passes this pitch to the pitch mapping block. The pitch mapping block takes the pitch and produces a set of candidate lag values as its output. When processing each of the N_V vectors in the current frame, the closed-loop LTP only searches through the candidates in this set. Neither of these modules exist in VSELP.

The frame gain extraction block uses the LPC coefficients to parameterize an inverse LPC filter whose output is the residual of the speech signal after linear

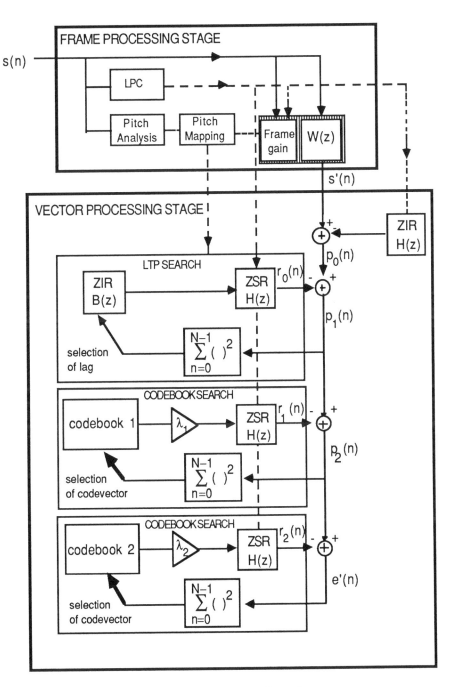

Figure 1: CVSELP Encoder

prediction. The residual after linear prediction is then input to the inverse pitch filter. The root-mean-square value of the residual after both linear prediction and pitch prediction is calculated and used as the frame gain λ_f. Instead of quantizing the excitation gains λ_1 and λ_2, CVSELP quantizes λ_1/λ_f and λ_2/λ_f. This is different than in VSELP[2]. Since the numerator polynomial of the weighting filter's z-transform is the z-transform of the inverse LPC filter[1], computing the residual after linear prediction is the first step in weighting the speech signal. Because of this, in our implementation the weighting filter and the frame gain extractor are implemented in the same module.

Once per vector the zero-input response (ZIR) subtraction block subtracts N samples of the zero-input response of the filter H(z) from the kth vector of s'(n), yielding $p_0(n)$. The closed-loop LTP search block takes the vector $p_0(n)$ as its input and attempts to find a vector in the LTP delay line which, after being filtered by H(z), is closer to $p_0(n)$ (least-squares sense) than the filtered version of any other vector in the delay line. Refer to Figure 2. which depicts the LTP delay line. When the search is complete, the LTP gain, β, is fixed to the quantized version of the optimal gain value corresponding to the chosen L. In VSELP the entire range of closed-loop lags are tested, CVSELP's range is reduced by using the open-loop pitch estimate to derive a more narrow search range.

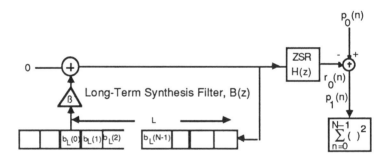

Figure 2: LTP Search

The codebook search is implemented in two stages. The first codebook search takes $p_1(n)$, the residual after long-term prediction, as its input and finds the excitation vector, $v_1(n)$, that after being filtered by H(z), is closer to $p_1(n)$ than the filtered version of any other excitation vector in codebook 1. The second codebook search functions in a similar manner with $p_2(n)$ and codebook 2. The excitation gains λ_1 and λ_2 take on quantized versions of the optimal values given the code vectors selected.

Rather than computing residuals at each stage as in CVSELP, VSELP adopts a "decorrelation" procedure [2] which improves performance but requires computing the norm squares of the filtered codevectors once per vector. Since CVSELP neither decorrelates nor interpolates LPC coefficients, the filtered basis vectors and the norm squares of the code vectors require computation only once per frame. The VSELP decorrelation procedure is performed in conjunction with vector quantization of the excitation gains. Without this decorrelation procedure vector quantization of the excitation gains is not as profitable, so CVSELP performs scalar quantization of these gains.

OPEN-LOOP PITCH EXTRACTION

In the open-loop pitch extractor, the pitch is determined directly from the input signal (or residual signal after linear prediction). One method of extracting the pitch is to assign the pitch to the delay, k, that maximizes the value of Eq. 1, in which s(n) may be either the original signal or the LPC residual.

$$f_{s,s}(k) = \frac{\left(\sum_{n=0}^{N_F - 1} s(n)\, s(n-k) \right)^2}{\sum_{n=0}^{N_F - 1} s(n-k)^2} \qquad \text{(Eq. 1)}$$

For voiced speech, Eq. 1 usually exhibits multiple peaks. Significant peaks (let us call peaks with a function value above 75% of the maximum function value significant peaks) usually occur around integer multiples of the smallest value of k for which a significant peak appears. Since the closed-loop lag search is designed to search around multiples of the pitch, the CVSELP pitch extractor strategically selects the smallest value of k where a significant peak appeared as the pitch estimate, provided that the lag that achieved the largest peak is an integer multiple of that pitch estimate. If it is not, it tries the next value of k where a significant peak occurred until the criterion is met.

To further reduce the loading, the CVSELP pitch extractor makes use of a decimation scheme similar to the one that Chen and Gersho have adopted [3]. In the reference, Chen and Gersho describe a pitch extraction algorithm in which they use the autocorrelation function of the decimated signal to find a *single* autocorrelation peak. They then use the undecimated signal to calculate the autocorrelation function in a neighborhood centered about that *single* peak location (after that location has been properly scaled by the decimation factor). In CVSELP, an initial pitch search uses the decimated signal to pass *multiple* peaks to a refinement procedure. CVSELP's refinement uses the autocorrelation function of the undecimated signal to gain better resolution around each of these peak locations. The CVSELP strategy costs slightly more but drastically reduces the likelihood of missing the highest peak due to the the reduced resolution associated with the autocorrelation function of the decimated signal.

CLOSED-LOOP LAG EXTRACTION

In the VSELP closed-loop LTP, the lag, L, and the LTP gain, ß, are chosen to minimize the mean square of the weighted error between the speech signal and the output of the cascade of the long-term synthesis filter and short-term (LPC) synthesis filter. To find the lag that achieves the minimum, the LTP search tests all lags in the range from 20 to 147. The main computational bottlenecks in the implementation of this closed-loop LTP search result from the presence of H(z): for every lag tested, the zero-state response of H(z) to the vector from the delay line corresponding to that lag must be computed; the norm square of the *filtered* vector corresponding to lag k cannot be updated from the norm square of the *filtered* vector corresponding to lag k-1 as it can be in the pitch search. These factors in addition to more complicated fixed-point scaling issues make a closed-loop search significantly more involved computationally than an open-loop pitch search.

Observations indicate that in the case of voiced speech, the closed-loop lag very often occurs around some multiple of the pitch. So during voiced speech, it is plausible that the required bit-rate and computational loading can be reduced by transmitting the pitch at a low bit-rate and by performing the closed-loop search for lags only in a restricted range about multiples of that pitch. The bits freed up by doing so can be donated to the excitation codebook. CVSELP adopts this strategy for both voiced and unvoiced speech so that no voiced/unvoiced decision is required. For the case of unvoiced speech, this does not seem to result in any significant degradation, probably because random vectors can provide the same benefits as previous samples in the delay line of the long-term synthesis filter. For the case of voiced speech, in many instances this strategy selects the same lag that would be obtained through a full closed-loop search, yet avoids wasting computations required for testing lags that have slim chances of being selected.

CVSELP maps every pitch in the range from 20 to 147 to a set of candidate lags. Presently the mapping is ad hoc and is constructed so that the set of lags includes entries around multiples of the pitch and so that no lag exceeds 147. For example, the 8 Kb/s coder uses 7 bits per frame to transmit the pitch. Once per frame each pitch value is mapped to a unique set of 8 lags. The closed-loop LTP search which will use this set as a search range each of the 4 times it is called in that frame. Thus CVSELP uses 7+4*3=19 bits per frame for transmission of the lag parameters as opposed to the 4*7=28 that would have been spent if it performed the full closed-loop search. If the frame's pitch were 40, throughout that frame the closed-loop LTP would only search the range {39,40,79,80,81,119,120,121}. Codex is currently considering an optimization scheme to construct the mapping.

ECHO CANCELLER

Figure 3 illustrates the flow of data from the point of view of the echo canceller and the operating system. In the implementation depicted by this diagram, a μ-law codec converts the analog signal to μ-law PCM samples. These samples are transmitted serially to the coder by means of the analog interface. An interrupt driven routine puts these μ-law PCM samples in the buffer ATD_BUFF. Once per frame, a μ-law conversion routine and an echo cancellation routine convert N_F of these μ-law samples to N_F linear, echo cancelled samples which are placed in NEW_BUFF. In NEW_BUFF, these samples await processing by the coder.

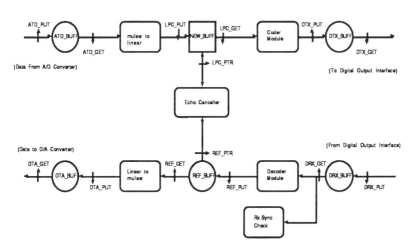

Figure 3: Echo Canceller Placement

Every frame, the operating system calls the coder as a subroutine to generate N_F/N_C bytes of coded data, where N_C is the compression ratio. This outgoing digital data is passed to the DTX_BUFF where it awaits servicing by the routines that manage the digital interface. Service routines in the operating system place data coming into the coder from the digital interface into DRX_BUFF. The decoder generates linear speech samples, one vector's worth at a time, from the coded data in DRX_BUFF. The decoder puts these decoded linear speech samples in REF_BUFF. The operating system invokes routines that take samples from REF_BUF, convert them to μ-law samples and place them in DTA_BUFF. In DTA_BUFF, outgoing μ-law samples await shipment on the analog interface. The echo canceller uses the linear samples residing in REF_BUFF as the reference signal.

The echo canceller uses a delayed least-mean-square (DLMS) algorithm [6] to update its taps. Let r(n) be the echo estimate at time n, let y(n) be the reference signal at time n, let c(n,k) be the kth coefficient in the finite impulse response filter (FIR) used to derive r(n) from samples in the delay line where the reference signal is stored, and let σ denote the step size. Eqs. 2a, 2b, and 2c specify the iteration used for the DLMS. On the DSP56001, the DLMS permits the computation of these equations using 2 instructions per tap whereas a standard LMS [7] implementation would require 3 instructions per tap. The remainder of the echo canceller's loading is for implementation of the near-end speech detector, which turns off the coefficient update when the near-end speaker is talking, and residual-echo suppressor, which opens the echo path when only the remote speaker is talking [7].

$$r(n) = \sum_{k=0}^{N_{tap}-1} c(n,k)y(n-k) \qquad (Eq.\ 2a)$$

$$c(n+1,k) = c(n,k) + \sigma e(n-1)y(n-k-1) \qquad (Eq.\ 2b)$$

$$e(n) = s(n) - r(n) \qquad (Eq.\ 2c)$$

PARAMETERIZATIONS AND LOADING

Bit allocations for the frame parameters at both 8 Kbps and 16 Kbps bit rates are: 36, 7, and 5 bits for the LPC coefficients, pitch, and frame gain respectively. For the parameters updated at the vector rate, the bit allocations are 4, 3, and 3 bits for gains β, λ_1, and λ_2, while the lag is coded with 3 bits and the codewords are coded with 7 and 6 (8 Kb/s) or 9 and 10 (16 Kb/s). Table 1 gives the loading for each of the major subdivisions in the CVSELP 8 Kb/s and 16 Kb/s coders. The loading for the full LTP is also given to demonstrate the amount of loading that is saved by using the CVSELP constrained search. The other major savings are incurred by filtering the basis vectors and performing the norm-square calculations N_v times less often than in VSELP. Table 2 gives the SNRs for both VSELP (as it stood in Nov. 1987) and CVSELP for a data base containing 20 seconds of speech from 6 males and 6 females.

CONCLUSIONS

At 16 kb/s, CVSELP attains speech quality that is below, but very near the quality attained at the same rate by an algorithm that supports many of the features of its more complex parent, VSELP. This is supported both through informal listening tests and SNRs. At lower rates, the difference in quality is much more pronounced, with VSELP clearly superior in quality to CVSELP. However, at 16 Kb/s, VSELP requires nearly three times the computational power of CVSELP. Because of its reduced computational loading, the CVSELP coding algorithm seems particularly well suited to network applications that require compression at 16 Kb/s with the lower rate, 8 Kb/s, offered as a fall back in times of network congestion.

TABLE 1: DSP LOADING FOR CVSELP

COMPUTATION	LOADING 8 Kb/s (INSTR./SAMPLE)	LOADING 16 Kb/s (INSTR./SAMPLE)
LPC analysis and quantization	19.52	19.52
Pitch analysis	84.33	84.33
Parameterize filters	4.70	4.70
Frame gain and weighting	38.64	38.64
Filter basis vectors	46.20	33.18
Energy of filtered basis vectors	31.08	106.51
Receiver	76.01	76.50
ZIR subtraction	17.30	17.30
LTP search	140.10	175.02
Codebook search	80.71	550.56
Filter update	39.11	50.45
Bit packer	8.01	12.52
Echo canceller	300.00	300.00
TOTAL	885.71	1469.23
Full-complexity LTP	410.00	710.00

TABLE 2: SNR MEASUREMENTS

SNR MEASUREMENTS	CVSELP 8 Kb/s (db)	CVSELP 16 Kb/s (db)	VSELP 8Kb/s (db)	VSELP 16Kb/s (db)
Ordinary SNR	11.02	18.90	13.55	19.50
Segmental SNR	9.20	16.00	12.23	17.10

308

References

[1] M.R. Schroeder and B.S. Atal, "Code-excited linear prediction (CELP): high quality speech at very low bit rates," *Proceedings of the IEEE International Conference on Acoustics, Speech, and Signal Processing*, pp. 937-940, Tampa, March 1985.

[2] I. Gerson and M. Jasiuk, "Vector sum excited linear prediction (VSELP) speech coding at 8 Kb/s," *Proceedings of the IEEE International Conference on Acoustics, Speech, and Signal Processing*, Albuquerque, New Mexico, April 1990.

[3] J. H. Chen and A. Gersho, "Real-time vector APC speech coding at 4800 bps with adaptive postfiltering," *Proc. IEEE International Conf. on Acoustics, Speech, and Signal Processing*, vol. 4, pp 2185-88, April 1987.

[5] A. Chrysafis and S. Landsdowne, "Fractional and integer arithmetic using the DSP56000 family of general-purpose digital signal processors," Motorola Applications Report, 1988.

[6] "The DLMS algorithm suitable for the pipelined realization of adaptive filters," *Proc. IEEE Acoustics, Speech, and Signal Processing Workshop*, Academia Sinica, Beijing, 1986.

[7] D.G. Messerschmitt, "Echo Cancellation in Speech and Data Transmission," *IEEE Journal on Selected Topics in Communications*, SAC-2, No. 2, 283-303, March 1984.

PART VII

TOPICS IN SPEECH CODING

This section is dedicated to new techniques that improve the performance of some of the existing speech coding systems. Kang and Fischer show how the excitation quantization can be improved in Regular Pulse Excitation (RPE) by using pyramid vector quantization, trellis coded quantization, and a modified distortion measure. The next two chapters refer to improvements in pitch prediction in CELP environment. Kroon and Atal present pitch predictors with non-integer delays that can achieve similar or better performance than higher-order integer-delay predictors. Yong and Gersho introduce a restricted deviation pitch coding technique which takes advantage of the limited variation of the pitch lag within a 20 ms speech frame. Shoham presents a technique which improve the speech quality of the basic CELP codec by adaptively reducing the amount of stochastic excitation during voiced sounds. Ramabadran and Sinha present a system which uses Kalman estimation and replaces the usual excitation signal by sparsely transmitted measurements of the prediction error. Chang and Wang present a technique for enhancing the unvoiced sound quality in Sine Transform Coding. Finally, Yuan et al. present a system based on time domain harmonic scaling and an improved adaptive delta modulation scheme using high-order linear prediction.

29

IMPROVED EXCITATION SEQUENCE QUANTIZATION IN REGULAR PULSE EXCITATION SPEECH CODING

Sangwon Kang † and Thomas R. Fischer ‡

† Department of Electrical Engineering
Texas A&M University
College Station, Texas 77843

‡ Department of Electrical and Computer Engineering
Washington State University
Pullman, Washington 99164

INTRODUCTION

Good quality speech coding at bit rates under 16 kbps has a growing number of applications for efficient digital transmission and storage. Traditional waveform coders can provide good quality encoded speech at bit rates above 16 kbps, but their performance drops rapidly for lower rates. On the other hand, traditional vocoder techniques enable one to encode speech at very low rates, but the perceptual quality is limited, even for relatively large bit rates (4 to 9.6 kbps).

At bit rates in the range of about 8 to 16 kbps, good quality encoders have been developed as a hybrid of methods, and include code-excited linear prediction (CELP) [3], multi-pulse excitation (MPE) [2], and regular-pulse excitation (RPE) [1]. CELP provides good quality speech at low bit rates (below 8 kb/s), while MPE and RPE are very efficient at medium bit rates. The CELP encoding quality should generally be superior to MPE and RPE at medium bit rates, but the encoder is also substantially more complex. CELP, MPE, and RPE replace the traditional pitch pulse and white noise excitation used in vocoders by, respectively, code vectors from a given codebook, a set of pulses located at non-uniformly spaced intervals, or a set of uniformly spaced pulses. At an encoding rate of 16 kbps, the performance of the RPE coder is roughly comparable to the MPE coder, but with a lower implementation complexity.

In this work, we will concentrate on the RPE encoding structure. A significant limitation to RPE encoding performance is the quantization of the excitation sequence amplitudes. We examine the effect of the excitation sequence quantization, and show that the typically large gain of the synthesis filter causes significant amplification of the quantization noise. To reduce the quantization distortion, two alternative approaches to encoding the RPE sequence ampli-

311

312

tudes are studied. The first approach simply uses multidimensional encoding methods (vector quantization and trellis coding) to better encode the excitation sequence. The second approach uses a modified distortion measure to encode the RPE sequence amplitudes. The effect of these schemes on the speech quality is described by segmental signal-to-noise ratio (SEGSNR) comparisons, and by informal listening tests.

BASIC RPE SYSTEM

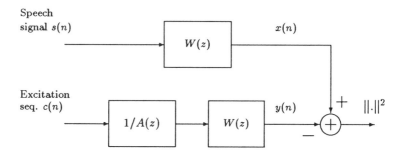

Figure 1. Basic structure of RPE coder.

The basic structure of the RPE speech coder [1] is shown in figure 1. The Pth order LPC parameters, $\{\tilde{a}_k, k = 1, 2, .., P\}$, are first found from the input speech, and then encoded as $a_k, k = 1, \ldots, P$, and transmitted. The synthesis filter used to model the short-term correlation in the speech is $1/A(z)$, where $A(z) = 1 - \sum_{k=1}^{P} a_k z^{-k}$. As suggested in [4], a weighting filter $W(z)$ is used to reduce the loudness of the quantization noise (by taking advantage of the frequency masking effect of the human ear). This filter is chosen to be

$$
\begin{aligned}
W(z) &= \frac{A(z)}{A(z/\gamma)} \\
&= \frac{1 - \sum_{k=1}^{p} a_k z^{-k}}{1 - \sum_{k=1}^{p} a_k \gamma^k z^{-k}} ,
\end{aligned}
\tag{1}
$$

where γ is a perceptual weighting factor between 0 and 1 (and selected to be 0.8 in the present study).

As developed in [1], the RPE system blocks the speech into frames of L (typically $L = 40$) samples, and for each frame a regular pulse pattern is selected. The allowable pulse patterns are sparse, with NS $- 1$ zeros between each pulse amplitude. Since the first nonzero pulse can occur in any of the first NS positions, there are NS possible regular pulse patterns for each excitation frame. The encoding procedure determines the position of the first pulse within a frame and the amplitude of each pulse in the excitation sequence. If the (noise shaping) filtered original speech is denoted as $x(n)$, and the weighted synthesized speech is denoted as $y(n)$, then the squared difference between $x(n)$ and $y(n)$ is used as the distortion measure for finding the excitation sequence. The excitation selection algorithm computes, for each of the NS regular pulse patterns, the optimal amplitude vector to minimize the distortion. The particular pulse pattern that provides the smallest distortion is then selected (and denoted as **b**), and the encoded offset and quantized pulse amplitudes are then transmitted or stored. The decoding procedure involves filtering the quantized excitation sequences with the synthesis filter.

AMPLIFICATION OF QUANTIZATION NOISE

For the encoding structure in figure 1, the spectrally weighted synthesized speech signal is

$$
\begin{aligned}
y(n) &= \sum_{k=1}^{p} a_k \gamma^k y(n-k) + c(n) \\
&= \hat{y}(n|n-1) + c(n),
\end{aligned} \tag{2}
$$

where $\hat{y}(n|n-1)$ may be thought of as a "prediction" of the filter output. It is desired to minimize the MSE over a frame of L samples,

$$
\begin{aligned}
\sum_{n=n_0}^{n_0+L-1} (x(n) - y(n))^2 \\
&= \sum_{n=n_0}^{n_0+L-1} (x(n) - \hat{y}(n|n-1) - c(n))^2 \\
&= \sum_{n=n_0}^{n_0+L-1} (\hat{x}(n|n-1) - c(n))^2, \tag{3}
\end{aligned}
$$

where $\hat{x}(n|n-1) \equiv x(n) - \hat{y}(n|n-1)$ is thought of as a prediction error.

If the excitation sequence, $c(n)$, is constrained to be in the RPE format (and denoted, for a single frame, as the vector **c**), then **c** = **b** is the optimum sequence to minimize (3). The problem with this approach is that the quantization of **b** (or **c**) introduces a quantization noise which builds up over the duration of the L sample frame, due to the synthesis filtering at the decoder. Although the RPE sequence amplitudes are selected to minimize the frame squared error distortion, the subsequent scalar quantization of these amplitudes results in an overall suboptimum encoding. That is, scalar quantization of the (real-valued) optimized pulse amplitudes is not equivalent to minimizing the distortion subject to a quantization constraint on the pulse amplitudes. Since the prediction gain of the synthesis filter is often quite large (especially for voiced segments) even though the RPE encoding procedure prevents accumulation of the quantization noise from frame to frame, within a frame this noise can be significant.

Table 1 lists the average filter gain for the excitation sequence quantization noise within a frame. In computing the quantization noise gain, a scalar quantizer designed using the generalized Lloyd algorithm [5] was used to quantize the excitation sequence amplitudes. The roughly 12 dB of quantization noise amplification by the synthesis filter causes a significant degradation in the quality of the encoded speech.

The synthesis filter parameters are specified by the linear prediction analysis of the input speech, so that the gain of the synthesis filter is a characteristic of the speech, and hence cannot readily be reduced. To reduce the quantization noise in the reconstructed speech, it is necessary to improve the encoding of the excitation sequence amplitudes. Two methods of improved excitation sequence encoding are next described: the first uses vector quantization or trellis encoding to reduce the encoding distortion; the second uses predictive coding and a modified distortion measure to also reduce the encoding noise.

TABLE 1. Average synthesis filter gain
for the quantization noise.

sentence	Average Gain (dB)
1	13.97
2	13.3
3	11.85
4	11.65
5	12.64

VECTOR AND TRELLIS QUANTIZATION

The quantization noise in RPE coding can be reduced by using vector quantization or trellis coding to better encode the nonzero amplitudes of the excitation sequence. Since the excitation pulse amplitudes are typically quantized with 3 or 4 bit scalar quantizers, it is not feasible to use generalized Lloyd algorithm vector quantizers, because of complexity limitations. Instead, we consider the use of the pyramid vector quantization (PVQ) [6] and trellis coded quantization (TCQ) [7] as low-complexity multidimensional encoders that offer significant improvement in encoding performance.

The PVQ is an efficient vector quantization scheme, motivated by the asymptotic equipartition principle of information theory. It is a type of lattice quantizer, with the codewords selected as the cubic lattice points which lie on the surface of a pyramid. The encoding and decoding algorithms are simple to implement, with an encoding complexity (per dimension) that grows only linearly with the vector dimension.

For the memoryless Laplacian source, the PVQ encoding performance is, for large dimension, equivalent to that of entropy-constrained scalar quantization [8]. However, the PVQ requires no variable-length (noiseless) coding, as is typical in entropy-coded scalar quantization. The PVQ is useful as a quantization scheme for the nonzero RPE excitation sequence amplitudes which are typically modeled as Laplacian. Since the PVQ implementation complexity is very modest, vector dimensions as large as 40 are easily handled, even at large encoding rates.

Trellis coded quantization is a promising source coding technique based on Ungerboeck's signal set expansion and set partitioning ideas in trellis coded modulation [9]. TCQ is a type of trellis encoding that labels the trellis branches with subsets of reproduction symbols. The actual TCQ encoding is accomplished by using the Viterbi algorithm [10], or other algorithm, to search the trellis, and the resulting computational burden is quite low. TCQ can also be used for predictive encoding of speech [11].

For memoryless sources, the PVQ and TCQ multidimensional encoding schemes offer significant improvement over scalar quantization. Table 2 lists the mean-square error encoding performance of several encoding techniques for a memoryless Laplacian source. Since the RPE excitation sequence amplitudes have been found to be well-modeled as Laplacian [1], both the PVQ and TCQ are appropriate for encoding the RPE pulse amplitudes, and will be effective at reducing the quantization noise.

TABLE 2. The encoding performance of several encoding techniques
for a memoryless Laplacian source.

Rate (bits)	Lloyd-Max SQ (dB)	PVQ (dB) (dim.= 20)	TCQ (dB) 8-state	TCQ (dB) 16-state
1	3.01	5.14	4.47	4.92
2	7.54	10.58	9.56	10.47
3	12.64	16.43	15.00	16.20

MODIFIED DISTORTION MEASURE

The excitation sequence used in RPE encoding is selected by first computing the optimum (real-valued) pulse amplitudes to minimize the frame distortion, and then scalar quantizing these amplitudes. As shown previously, the quantization noise is amplified significantly by the synthesis filter. An alternative to the RPE sequence encoding procedure is the differential encoding process in which the excitation sequence amplitudes, $c(n)$, are selected on a pulse-by-pulse basis to minimize $(\hat{x}(n|n-1) - c(n))^2$, where $\hat{x}(n|n-1)$ is a prediction of $c(n)$. Generally, this encoding does not provide the global minimization desired for the criterion in (3). However, we note that in differential encoding there is no amplification of the quantization noise by the synthesis filter.

In an attempt to reduce the synthesis filter amplification of the quantization noise in RPE encoding, consider a distortion measure of the form

$$J = \lambda(b(n) - c(n))^2 + (1 - \lambda)(\hat{x}(n|n-1) - c(n))^2, \qquad (4)$$

where λ is a parameter between 0 and 1, and $b(n)$ is the sequence of ideal (unquantized) pulse amplitudes found using the RPE encoding method. The modified distortion measure, J, is the weighted sum of the typical (implicit) distortion criterion used in RPE encoding to scalar quantize the pulse amplitudes, and the predictive coding distortion measure. Clearly, if $\lambda = 1$, then we just have the RPE distortion measure for minimum scalar quantization of the pulse amplitudes, while if $\lambda = 0$, we have the mean-square error distortion criterion for differential encoding. The modified distortion measure is admittedly *ad hoc*, but allows, at least intuitively, a tradeoff between the RPE pulse amplitude quantization distortion measure, for which there is large synthesis filter amplification of the excitation sequence encoding noise, and the predictive encoding criterion, which does not optimize the excitation sequence amplitudes on a frame basis, but has no amplification of the quantization noise. An optimal value of λ can be determined empirically.

Selecting $c(n)$ to minimize J yields the ideal pulse amplitudes

$$c(n) = \lambda b(n) + (1 - \lambda)\hat{x}(n|n-1). \qquad (5)$$

The encoded excitation sequence amplitudes are computed as follows. For each frame, $b(n)$ and the position of the first pulse are computed in the standard RPE fashion. Then, $c(n)$ is selected according to (5) for all n in the frame. However, $c(n)$ is constrained to be zero whenever the RPE pulse pattern (i.e., b) is zero. Finally, $c(n)$ is optimally scalar quantized to minimize (4). The encoding with the modified distortion measure is illustrated in figure 2.

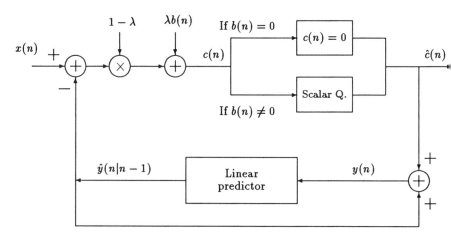

Figure 2. The RPE encoding using the modified distortion measure.

EXPERIMENTAL RESULTS

The effectiveness of the pyramid vector quantization, the trellis coded quantization, and the modified distortion measure in RPE encoding was evaluated by simulation. We considered the five English sentences in Table 3, pronounced by 5 different speakers (two females and three males). The speech sentences were bandlimited to 3.2 kHz and sampled at 8 kHz with 12 bit resolution. A 10th order LPC analysis was performed using the autocorrelation method with a 20 msec Hamming window, and updated every 15 msec. The gain parameter of the excitation sequence was determined by use of the **b** vector computed in the RPE encoding method. The computation of **b** and the adjustment of the position of the first pulse were done every 5 msec for NS= 2, and every 3.75 msec for NS= 3. The quantization levels for the scalar quantization of the pulses (either **b** for the RPE or **c** for the modified distortion measure) were determined by use of the generalized Lloyd algorithm [5] for the case of $\lambda = 1$. This designed scalar quantizer was then also used for every other value of λ. The overall SEGSNR was used as an objective performance measure and informal listening tests were also performed.

TABLE 3. Sentences used to test encoding performance.

1. The pipe began to rust while new.	(Female)
2. Add the sum to the product of these three.	(Female)
3. Oak is strong and also gives shade.	(Male)
4. Thieves who rob friends deserve jail.	(Male)
5. Cats and dogs each hate the other.	(Male)

Tables 4 and 5 give the value of the overall SEGSNR for RPE with the modified distortion measure. A value of $\lambda = 1$ corresponds to the basic RPE system. For each pulse spacing factor NS, the optimal λ value decreases as the bit rate used to encode the excitation pulse amplitudes decreases (presumably, because the decrease of the bit rate improves the value of the predictive criterion, and hence reduces the amplification of the quantization noise). Also, the optimal λ value decreases as the pulse-spacing-factor NS decreases. In other words, as the number of pulses in a frame increases, the value of the predictive criterion increases. At NS= 2 and a bit rate of 3 bits/amplitude, the modified distortion measure yields a coder providing 0.7 to 1 dB improvement in SEGSNR.

Table 6 lists the results of 16 kbps RPE encoding with the modified distortion measure. In our informal listening tests, the speech quality of the RPE coder with the modified distortion measure was judged to be perceptually smoother and clearer than that of the typical RPE system, thus indicating the value of the modified distortion measure.

TABLE 4. SEGSNR improvement with the new distortion measure.

pitch pred. order	bit rates per amplitude	NS= 2		NS= 3	
		optimal λ	SEGSNR (dB) improvement	optimal λ	SEGSNR (dB) improvement
0	2	0.6	2.23	0.8	0.67
tap	3	0.7	0.97	0.9	0.13
1	2	0.7	2.26	0.9	0.53
tap	3	0.9	0.68	NA	NA

TABLE 5. SEGSNR (dB) for various λ at NS= 2 and
a bit rate of 3 bits/amplitude. ($\gamma= 0.8$)

pitch pred. order	λ					
	1	0.9	0.8	0.7	0.6	0.5
0 tap	14.13	14.63	14.9	15.1	14.83	14.69
1 tap	16.33	17.01	16.72	16.41	NA	NA

TABLE 6. The performance of RPE coder with the modified
distortion measure at 16 kbps. (SNR/SEGSNR (dB), $\gamma= 0.8$)

Sent.	$\lambda= 1$	$\lambda= 0.9$
1	18.59/17.37	19.28/18.11
2	19.76/17.17	20.61/17.97
3	17.14/14.92	17.56/15.51
4	16.17/16.03	16.66/16.67
5	17.10/16.16	17.63/16.81
Ave.	17.75/16.33	18.35/17.01

The results of encoding the excitation sequence with the PVQ and TCQ appear in Table 7. For comparison, the performance of encoding with scalar quantization of the excitation sequence is also listed. The PVQ and TCQ were based on a Laplacian model for the excitation sequence amplitudes. The scalar quantizer was designed using the generalized Lloyd algorithm with speech training data. The performance without quantizing any of the parameter values or excitation sequence amplitudes is also listed as the upper bound to the RPE encoding performance. Each entry is of the form SNR/SEGSNR.

The PVQ excitation sequence encoding offers improvement of about 1 dB in overall segmental signal-to-noise ratio, while the trellis coded quantization encoding of the excitation sequence provides about 1.3 dB improvement in overall

SEGSNR. This roughly 1 dB improvement in SEGSNR is perceptually significant, and the improved quality of the PVQ and TCQ encoder systems is evident
in informal listening tests.

TABLE 7. The performance of encoding excitation amplitude
with PVQ and TCQ at 16 kbps. (SNR/SEGSNR (dB), $\gamma= 0.8$)

Sent.	LBG SQ	PVQ	TCQ	Unquantized
1	18.59/17.37	19.32/17.89	19.68/18.64	21.65/24.61
2	19.76/17.17	20.57/17.54	20.90/18.46	23.92/23.15
3	17.14/14.92	18.54/16.03	18.34/16.31	21.51/20.62
4	16.17/16.03	17.40/17.28	16.95/17.14	18.79/20.74
5	17.10/16.16	18.71/17.72	18.07/17.55	20.68/21.90
Ave.	17.75/16.33	18.79/17.62	18.91/17.29	21.31/22.20

CONCLUSION

The RPE system introduced in [1] achieves good speech coding results with a
lower complexity than MPE. However, the distortion introduced by quantizing
the excitation sequence is amplified by the synthesis filter. Two techniques are
considered to reduce the quantization noise. First, pyramid vector quantization
and trellis coded quantization are used to directly quantize the RPE excitation
sequence amplitudes. This provides average improvements in overall SEGSNR
of about 1 (dB) for the PVQ and about 1.3 (dB) for the TCQ. This improvement is noticeable perceptually in the improved quality of the reconstructed
speech. Second, a modified distortion measure is used for efficient 16 kbps RPE
speech coding, and shown, by simulation, to provide increases in SEGSNR of
about 0.7 to 1 dB. This objective performance improvement is noticeable in the
quality of the encoded speech.

It is possible to combine the TCQ with the modified distortion measure
to encode the RPE excitation sequence amplitudes. Unfortunately, the coding
gains of the individual methods are not directly additive, and an improvement
in SEGSNR of only about 0.2 dB, over the TCQ encoding alone, is achieved.

ACKNOWLEDGEMENT

This work was supported, in part, by the National Science Foundation under Grant No. MIP-8619888.

REFERENCES

[1] P. Kroon, E. F. Deprettere, and R. J. Sluyter, "Regular-pulse excitation: A novel approach to effective and efficient multipulse coding of speech," *IEEE Trans. Acoust., Speech, Signal Processing*, vol. ASSP-34, pp. 1054-1063, Oct. 1986.

[2] B. S. Atal and J. R. Remde, "A new model of LPC excitation for producing natural-sounding speech at low bit rates," in *Proc. IEEE Int. Conf. Acoust., Speech, Signal Processing*, pp. 614-617, Apr. 1982.

[3] M. R. Schroeder and B. S. Atal, "Code-excited linear prediction (CELP): high-quality speech at very low bit rates," in *Proc. IEEE Int. Conf. Acoust., Speech, Signal Processing*, vol.3, pp. 937-940, March, 1985.

[4] B. S. Atal, "Predictive coding of speech at low bit rates," *IEEE Trans. Commun.* vol. COM-30, pp. 600-614, April 1982.

[5] Y. L. Linde, A. Buzo, and R. M. Gray, "An algorithm for vector quantizer design," *IEEE Trans. Commun.*, vol. COM-28, pp. 84-95, Jan. 1980.

[6] T. R. Fischer, "A pyramid vector quantizer," *IEEE Trans. Inform. Theory*, vol. IT-32, pp. 568-583, July 1986.

[7] M. W. Marcellin and T. R. Fischer, "Trellis coded quantization of memoryless and Gauss-Markov sources," *IEEE Trans. Commun.*, vol. COM-38, pp. 82-93, Jan. 1990.

[8] N. Farvardin and J. W. Modestino, "Optimum quantizer performance for a class of non-Gaussian memoryless sources," *IEEE Trans. Inform. Theory*, vol. IT-30, pp. 485-497, May 1984.

[9] G. Ungerboeck, "Channel coding with multilevel/phase signals," *IEEE Trans. Inform. Theory*, vol. IT-28, pp.55-67, Jan. 1982.

[10] G. D. Forney, Jr., "The Viterbi algorithm," *Proc. IEEE (Invited Paper)*, vol. 61, pp. 268-278, Mar. 1973.

[11] M. W. Marcellin and T. R. Fischer, "A trellis-searched 16 kbps speech coder with low delay," *Advances in Speech Coding*, May 1990.

30

ON IMPROVING THE PERFORMANCE OF PITCH PREDICTORS IN SPEECH CODING SYSTEMS

Peter Kroon and Bishnu S. Atal

Acoustics Research Department
AT&T Bell Laboratories
Murray Hill, NJ 07974, USA

INTRODUCTION

Pitch prediction [1] plays an important role in many speech coding systems such as multipulse [2] and vector or code-excited linear predictive coders [3]. The pitch predictor removes the redundancy in a periodic speech signal by predicting the current signal from a linear combination of past versions of this signal. The general form of an odd-order pitch predictor with delay M and predictor coefficients $b(k)$ is given by:

$$P(z) = 1 - \sum_{k=-(p-1)/2}^{(p-1)/2} b(k)z^{-(M+k)}, \quad p = 1,3...$$ (1)

The value of M is the equivalent in number of samples of a delay in the range from 2 to 20 ms. For periodic signals, this delay would correspond to a pitch period (or possibly an integral number of pitch periods). The delay would be random for nonperiodic signals.

The prediction gain is a good measure of the effectiveness of a linear predictor [4]. Its value depends on many factors, such as how frequently the predictor parameters are updated, the predictor order, and the amount of periodicity in the input signal. For sampled signals the prediction gain also depends on the sampling frequency f_s. Increasing the sampling frequency f_s, increases the average prediction gain [5, 6]. Typical values for the prediction gain for periodic signals are in the range from 6 to 15 dB.

Higher-order predictors result in a higher prediction gain, but more bits are needed to encode the additional coefficients (2 to 3 bits/coefficient [1]). Multiple coefficients provide interpolation between the samples, if the pitch delay does not correspond to an integer number of samples. Moreover, multiple coefficients allow representation of a frequency-dependent gain factor which is useful because most speech signals exhibit less periodicity at high frequencies than at low frequencies. However, from examining the coefficients obtained for higher-order predictors, we find that for periodic speech their major role is to provide interpolation.

For a first-order predictor, the input signal $x(n)$ is approximated by the predictor signal $y(n)$, which is a function of both the delay M and the predictor coefficient b.

322

The squared prediction error for a frame of N samples is then given by:

$$E(M, b) = \sum_{n=0}^{N-1} [x(n) - y(n)]^2 = \sum_{n=0}^{N-1} [x(n) - bx(n - M)]^2 \qquad (2)$$

For a given delay value M the optimal value of b is found by setting the derivative of E with respect to b to zero, which leads to:

$$b = \frac{\sum_{n=0}^{N-1} x(n)x(n - M)}{\sum_{n=0}^{N-1} x^2(n - M)} . \qquad (3)$$

Substituting this optimum value of b into (2), leads to the error function:

$$E(M) = \sum_{n=0}^{N-1} x^2(n) - E'(M) \qquad (4)$$

$$with \; E'(M) = \frac{\left[\sum_{n=0}^{N-1} x(n)x(n - M) \right]^2}{\sum_{n=0}^{N-1} x^2(n - M)}$$

This function $E'(M)$ is computed for all possible values of M, and its maximum indicates the best choice for the delay M. For a periodic signal, this function has local maxima at delays corresponding to the pitch period and its multiples. For nonperiodic signals, the function $E'(M)$ has a more erratic behavior and shows no distinctive peaks.

Based on the observation that higher-order predictors provide mainly interpolation, we propose to use a first-order predictor having a delay with arbitrary temporal resolution [5, 6]. Such a predictor is conceptually simpler than a higher-order predictor, and can provide a more efficient representation of the predictor coefficients.

In the next sections, we discuss the realization of such a predictor and compare its performance with conventional higher-order predictors.

INCREASING THE TEMPORAL RESOLUTION

The delay M is expressed as an integer number of samples at sampling frequency f_s. A higher temporal resolution can be obtained by specifying the delay as an integer number of samples plus a fraction of a sample l/D, where $l = 0, 1, .., D - 1$, and l and D are integers. A noninteger delay l/D at the original sampling frequency f_s

corresponds to an integer delay l at a frequency Df_s. In other words, to implement a delay of l/D samples, the sampling frequency is increased by a factor D (by inserting $D-1$ zero-valued samples between each sample of the input signal). The resulting signal is then low-pass filtered at $f_s/2$ to obtain an interpolated version of the input signal. This interpolated signal is then delayed by l samples at the high sampling frequency and the delayed output is downsampled to the original sampling frequency f_s. The resulting signal is the original signal delayed by the noninteger delay l/D. In addition, a constant integer delay is introduced due to the delay of the low-pass interpolation filter. This interpolation filter $h(n)$, $n = 0, 1, .., N - 1$ is chosen to be an FIR filter with exactly linear phase and an impulse response of N samples. Its delay at the high sampling frequency Df_s is $(N - 1)/2$ samples. Since we want to compensate for this overall delay at the lower sampling frequency, N must be chosen such that $(N - 1)/2$ is an integer multiple of D, or equivalently:

$$N = 2ID + 1, \tag{5}$$

where I is the delay at the lower sampling frequency.

The fractional delays can be efficiently implemented with a polyphase structure as described in [7]. The polyphase filters $p_l(k)$ are obtained from the coefficients of the low-pass filter $h(n)$ according to

$$p_l(k) = h(kD - l) \quad 0 \le l \le D - 1 \ , k = 0, 1, .., q - 1, \tag{6}$$

where $h(n) = 0$ for $n < 0$, such that $p_l(0) = 0$ for $l > 0$. The number of coefficients of the polyphase filter p_l is given by

$$q = 2I + 1. \tag{7}$$

For each value of the delay l/D the corresponding l-th polyphase filter branch is used and the output is given by

$$y(n) = \sum_{k=0}^{q-1} p_l(k)x(n - k). \tag{8}$$

Taking into account the delay I of the low-pass filter, the expression for the pitch predictor with an effective noninteger delay $M + l/D$ becomes:

$$P(z) = 1 - \beta \sum_{k=0}^{q-1} p_l(k)z^{-(M-I+k)}, \tag{9}$$

where β represents the pitch predictor coefficient. For values of $I \le M$, we see that the filter is causal and no additional overall delay is required.

The design of the low-pass filter should be such that the aliasing components due to the downsampling process are sufficiently attenuated. This means that it should have its stop-band cut-off frequency at $1/2D$ (normalized to the high sampling frequency Df_s) and a stop-band ripple that is sufficiently small.

PERFORMANCE EVALUATION

The prediction gain was used to compare the performance of integer delay pitch predictors versus the performance of noninteger delay predictors. In computing the overall prediction gain, segments whose prediction gain was below 1.2 dB were not included since those segments usually represent silence or nonperiodic speech. The performance of the various pitch predictors was evaluated using 90 seconds of speech from a wide variety of speakers (11 male and 11 female). The speech signals were bandlimited to 100-3600 Hz and sampled at 8 kHz. The pitch prediction was done on the LPC residual signal, obtained from a 10-th order autocorrelation analysis performed every 20 ms. The pitch predictor parameters were computed every 5 ms and the delay was limited to the range from 2.5 ms to 18.375 ms (20 to 147 samples). Different choices of interpolation filters are possible. For example, one could use a low-pass filter that matches the bandwidth of the input signal or design an interpolation filter that minimizes the mean-squared interpolation error for a given spectral distribution of the input signal [8, 9]. The interpolation filter we used was a $\sin(x)/x$ function weighted with a Hamming window, which approximates an ideal low-pass filter with a pass-band between 0-4000 Hz. The filter length is chosen according to (5), and depends on the value of I. It was found that values of I in the range from 4 to 16 produce acceptable results and in our experiments we used I equal to 16.

The resulting prediction gains are listed in Table 1. The averaged prediction gain increases for larger interpolation factors D, but does not significantly improve for

Table 1: Average Prediction Gains for the Various Pitch Predictors Obtained From 90 S of Speech. The D-1 Fractions Are Distributed Uniformly Between Each Integer Delay Value.

Predictor order (p)	Interpolation factor (D)	Prediction Gain (dB)	
		female	male
1	1	6.6	5.6
1	2	7.9	7.2
1	4	8.7	7.9
1	8	8.9	8.2
1	16	9.0	8.2
2	1	7.9	7.2
3	1	8.6	8.0

values of $D > 8$. The first-order predictors with interpolation factors 2 and 4 produce a similar prediction gain as 2nd and 3rd order predictors without interpolation. This leads to a more efficient coding since at least 2 bits/coefficient are needed to encode the coefficients of a higher-order predictor [1], while only $\log_2 D$ bits are needed for encoding the fractions of a first-order noninteger-delay predictor.

Fast Search Procedures

The search for noninteger delays increases the complexity of the pitch predictor. For each fractional delay value the predicted signal has to be generated by convolving delayed samples with the interpolation function. The effect of the fractional delays is that we obtain a finer sampling of the error of (4) as function of the delay. By appropriate sampling of this function we can determine possible positions of its extremes, and do a subsequent fine search of those areas to find the best delay value. The prediction gain obtained with such a fast search is almost identical to that obtained with the exhaustive search procedure.

Smoothing of Pitch Delay Contours

For periodic signals the current period is not only similar to the previous period but also to periods that occurred multiple periods ago. A mismatch between pitch period and pitch predictor delay produces a prediction error whose values will depend on the difference between period and delay [5, 6]. When such a mismatch occurs, the exhaustive search procedure may find the best delay value to be equal to a multiple of the pitch period. Although this locking to a multiple of the pitch period does not affect the operation of the predictor, it has disadvantages for coding since a smooth pitch contour can be encoded more efficiently. Pitch predictors with high temporal resolution suffer less from this pitch multiplication effect than low resolution predictors. To make the curves even smoother we can use the observation that the local maxima that correspond to the pitch period or multiples have error values that are close to each other. By searching local maxima whose peak values are within a specified tolerance from the global maximum, and selecting the one corresponding to the smallest delay value, we obtained smooth curves with only a small loss (less than 0.5 dB) in prediction gain.

APPLICATION TO SPEECH CODING

Pitch predictors play an important role in many analysis-by-synthesis adaptive predictive coders [10] such as multipulse and various versions of vector or code-exited coders. Since the performance of these systems is directly related to the prediction gain of the predictors used, any of these coders will benefit from the use of a noninteger predictor. However, there exists a trade-off between the improved performance and the extra bits required for transmitting the fractions.

To study this trade-off we incorporated the noninteger delay predictor in a CELP coder [3]. Instead of using extra bits for encoding the fractions, one could use these bits to increase the size of the stochastic codebook or use a multiple-coefficient integer-delay predictor. To examine the difference between these approaches we used a CELP coder whose 10-th order LPC filter is updated every 20 ms and whose excitation parameters (including the pitch predictor parameters) are updated every 5 ms. The

Table 2: Different CELP Systems Used for Evaluating Pitch Predictors with Noninteger Delays. Only the Features That Are Different Are Listed. The Pitch Predictor Coefficients Were Not Quantized.

Coder	Pitch Predictor		Codebook	SNRSEG (dB)	
	Order	Delay (bits)	Size (bits)	female	male
A	1	integer (7)	8	12.7	10.9
B	1	noninteger (8)	8	13.6	11.5
C	1	integer (7)	9	13.3	11.6
D	2	integer (7)	8	13.5	11.5

LPC parameters, stochastic-codebook gain and pitch predictor coefficients were not quantized. The systems that were compared are listed in Table 2. System A serves as a reference system, while the other systems are identical except for the following. System B uses noninteger delays whose fractions are nonuniformly distributed according to the distribution of average pitch period durations for a large speaker population. The resulting table of delay values contains 256 entries (8 bits), with no fractions for the delays from 147 to 100 and nonuniformly-spaced fractions for the shorter delays. Compared to system A, only one more bit is required to encode the fractional delays. To investigate the net effect of this extra bit, we included system C, which is similar to system A but has a codebook that is twice as big. System D is similar to system A but has a 2nd order integer-delay pitch predictor. Table 2 shows the resulting segmental SNR values measured between the original and reconstructed speech. In addition, we performed an informal listening test using a subset of the test data base (4 female and 4 male speakers). Although all three systems B,C and D improved the quality it was concluded from the listening tests that the use of a noninteger delay predictor (system B) improved the speech quality more than the other systems. We also observed that the use of noninteger delays increased the significance of the pitch predictor in the error matching procedure, and decreased the role of the stochastic codebook excitation.

CONCLUSION

Pitch predictors with noninteger delays can achieve a similar or better performance than higher-order integer-delay pitch predictors. In addition, the pitch predictor parameters can be encoded more efficiently.

The performance of analysis-by-synthesis adaptive predictive speech coders such as multipulse and vector or code-excited coders is strongly related to the prediction gain of the pitch predictor, which means that the use of noninteger-delay predictors can improve the coder performance.

As an example we incorporated the noninteger-delay predictor in a CELP coder and it was found that the resulting improvement in performance was more noticeable than the improvement obtained by increasing the size of the stochastic codebook.

ACKNOWLEDGEMENT

The authors thank J.S. Marques and J.M. Tribolet of INESC, Lisbon, Portugal, for the fruitful discussions during the early stages of this work.

REFERENCES

[1] B. Atal, "Predictive coding of speech at low bit rates," *IEEE Trans. Communications*, vol. COM-30, no. 4, pp. 600–614, April 1982.

[2] B. Atal and J. Remde, "A new model of LPC excitation for producing natural-sounding speech at low bit rates," *Proc. IEEE Int. Conf. Acoust., Speech, Signal Processing*, pp. 614–617, 1982.

[3] M. Schroeder and B. Atal, "Code-excited linear prediction (CELP): high quality speech at very low bit rates," *Proc. IEEE Int. Conf. Acoust., Speech, Signal Processing*, pp. 937–940, 1985.

[4] N. Jayant and P. Noll, *Digital Coding of Waveforms*. Englewood Cliffs, NJ: Prentice Hall, 1984.

[5] P. Kroon and B. Atal, "On improving the performance of pitch predictors in speech coding systems," *Proc. IEEE Workshop on Speech Coding for Telecommunications*, pp. 49–50, 1989.

[6] P. Kroon and B. Atal, "Pitch predictors with high temporal resolution," *Proc. IEEE Int. Conf. Acoust., Speech, Signal Processing*, 1990.

[7] R. Crochiere and L. Rabiner, *Multirate Digital Signal Processing*. Englewood Cliffs, NJ: Prentice Hall, 1983.

[8] G. Oetken, T. Parks, and H. Schüssler, "New results in the design of digital interpolators," *IEEE Trans. Acoust., Speech, Signal Processing*, vol. ASSP-23, no. 3, pp. 301–309, 1975.

[9] T. Parks and D. Kolba, "Interpolation minimizing maximum normalized error for band-limited signals," *IEEE Trans. Acoust., Speech, Signal Processing*, vol. ASSP-26, no. 4, pp. 381–384, 1978.

[10] P. Kroon and E. Deprettere, "A class of analysis-by-synthesis predictive coders for high quality speech coding at rates between 4.8 and 16 kb/s," *IEEE J. on Selected Areas in Communications*, vol. SAC-6, no. 2, pp. 353–363, February 1988.

31

EFFICIENT ENCODING
OF THE LONG-TERM PREDICTOR
IN VECTOR EXCITATION CODERS

Mei Yong† and Allen Gersho

Center for Information Processing Research
Department of Electrical and Computer Engineering
University of California
Santa Barbara, CA 93106

INTRODUCTION

Speech coding algorithms based on Code-Excited Linear Prediction (CELP) [1], have been widely studied in the past few years for low bit rate speech coding and many improvements and variations of the basic algorithm have since emerged. We use Vector Excitation Coding (VXC) as a generic name to represent the class of coders using vector quantization (VQ) coding of filtered excitation signals with "closed-loop" codebook search [2,3].

A VXC coder generally uses a synthesis filter consisting of the cascade of a short-term (ST) and a long-term (LT) synthesis filter. These two filters are based respectively on a short-term predictor (STP) and long-term predictor (LTP) which remove the redundancies in a speech waveform. Since 1970, considerable effort has been devoted to efficient coding of the STP. On the other hand, relatively little attention has been given to the LTP which is essential to the performance of a VXC coder,

This chapter introduces a new method, *Restrictive Pitch Deviation Coding*, for LTP representation and gives a comparative analysis of the performance of different LTP encoding techniques. We show that the new method offers the possibility of reducing both bit rate and complexity in coding of the LTP.

LONG-TERM PREDICTOR DESIGN IN VXC

In a VXC decoder excitation vectors are successively fed into the time-varying LT and ST synthesis filters generating the reconstructed speech. For each given input speech vector, the encoder selects the optimal excitation codevector from a codebook by minimizing the perceptually weighted distortion between the input and reconstructed speech vectors. The LT and ST synthesis filters have the transfer function $[1 - P_j(z)]^{-1}$ where the label j is either l for long or s for short,

---

Mei Yong is currently with Codex Corporation, Mansfield, MA 02048.

$P_l(z) = \sum_{i=-I}^{J} \beta_i z^{-T+i}$, $P_s(z) = \sum_{i=1}^{M} a_i z^{-i}$ are respectively the transfer functions of the LTP and the STP, and T is the estimated pitch period for a given segment of speech.

In the *open-loop* approach, the LTP parameters are computed by directly analyzing a block of input speech samples, minimizing the sum of the squared prediction errors. In the *closed-loop* approach, the parameters are computed by minimizing the energy of the overall reconstruction error sequence between the input and reconstructed speech. The closed-loop LTP design requires much more computation than in the open-loop case; however, it usually outperforms the open-loop predictor. In addition, when the closed-loop LTP is used, the size of the excitation codebook can be reduced so that the overall computation is reduced.

Computing Open-Loop Predictor Parameters

An LTP is specified by a pitch period T (also called pitch lag) and a set of prediction coefficients or taps for time lags ranging from $T-I$ to $T+J$ where I and J are nonnegative integers. Applying linear prediction analysis, the set of prediction coefficients can be chosen by minimizing the energy of the prediction error sequence in a frame. Let $\{d(1), d(2), \cdots, d(L)\}$ be a frame of input speech samples to an LTP, and assume for the moment that the pitch lag is known. The minimization of the prediction error energy leads to the equation $\Phi\beta = c$, where Φ is an $I+J+1 \times I+J+1$ square matrix whose ijth element is $\phi(i,j) = \sum_{n=1}^{L} d(n-i)d(n-j)$ and the indices i and j range from $T-I$ to $T+J$, $\beta = [\beta_{-I} \beta_{-I+1} \ldots \beta_J]^T$ is the vector of LTP coefficients, and

$$c = [\phi(0,T-I) \ \phi(0,T-I+1) \ \ldots \ \phi(0,T+J)]^T .$$

There are several "open-loop" methods to estimate the pitch period directly from a block of speech samples. An effective approach [4] is to find the *optimal* pitch lag that minimizes the energy of the one-tap prediction error. A simplified method called the *correlation-peak-picking* method, is to find the pitch lag that maximizes the correlation value $|\phi(0,T)|$.

Computing Close-Loop Predictor Parameters

The closed-loop LTP method was first proposed for the multi-pulse excitation coder [5] and later applied to vector excitation coders [6] [7] [8]. The pitch lag and predictor coefficients of a closed-loop LTP are chosen in such way that the mean square of the perceptually weighted reconstruction error vector is minimized.

For a one-tap LTP ($I = J = 0$), the predictor parameters can be determined in two steps: a) find the pitch lag T (from a predefined range) such that A^2/B is maximized where $A = \langle \mathbf{x}, \mathbf{F}\bar{\mathbf{d}}_T \rangle$ and $B = \|\mathbf{F}\bar{\mathbf{d}}_T\|^2$ b) compute the prediction coefficient using the equation: $\beta = A/B$, where $\mathbf{x}$ is the weighted input speech vector after subtracting out the zero-input response of the weighted ST synthesis filter $H_s(z/\alpha)$, $\mathbf{F}$ is a Toeplitz triangular matrix composed of the samples of the impulse response of the filter $H_s(z/\alpha)$ as given in [9], and the vector $\tilde{\mathbf{d}}_T$ contains the previous outputs of the LT synthesis filter, i.e.,

$$\tilde{\mathbf{d}}_T = [\tilde{d}(1-T)\tilde{d}(2-T) \ . \ . \ . \ \tilde{d}(k-T)]^T \ ,$$

where k is the dimension of a speech vector.

In the closed-loop LTP method, the pitch lag ordinarily has to be greater or equal to the speech vector dimension in order to obtain the previous LTP output vector $\tilde{\mathbf{d}}_T$. Hence, the vector dimension, which is also the adaptation interval of the LTP, needs to be reasonably small to handle short pitch periods. Decreasing the adaptation interval increases the bit rate needed to code the LTP parameters.

ALTERNATIVES FOR LONG-TERM PREDICTOR ADAPTATION

Frequently, in VXC coders, the same update rate is used for adjusting both predictors. However, since the LTP models the effect of the glottis which usually changes faster than the vocal tract shape, it is sometimes important to update the pitch parameters more frequently than the formant parameters. In addition, for low overall bit rates, careful consideration of the tradeoff in allocating bits for coding the LTP versus bits for coding the innovation sequence is needed, taking into account the complexity implications. A reduced excitation codebook size can generally be achieved by more effective LTP tracking.

To evaluate the performance of different LTP methods we use the open-loop LT prediction gain. Although not equivalent to the closed-loop LT prediction gain, it gives an indication of the asymptotic performance of the closed-loop LTP since both measures will approach the same value as the overall coding rate increases. In our experiments, adaptive ST prediction was performed first, where the predictor is updated every 20 ms frame, and the ST prediction residual is used as the input to the LT prediction error filter. For LT prediction, a frame was further divided into subframes (or vectors) so that the LTP can be adapted at subframe intervals. Several cases of adaptation are discussed below.

Restrictive Pitch Deviation Coding

A straightforward implementation of an adaptive LT predictor is to update the pitch lag and predictor coefficients every subframe. As the subframe length decreases, the prediction gain will generally increase. However, the bit rate for encoding the LTP with a fixed number of bits will also increase as the frame rate increases. Thus, we need a way of avoiding a high bit rate while maintaining accurate LTP tracking

We have found that within a duration of about 20 ms the pitch lag usually makes only small fluctuations around a constant average pitch period. Based on this observation, we propose a new method to represent the time-varying pitch lag, called *restrictive pitch deviation coding* (RPDC). We first compute an average pitch lag of a frame using, for example, the correlation-peak-picking method. Then for each individual subframe, a new optimal pitch lag (optimized for the given subframe) is found within some predefined offset limits of the average pitch value. The encoder then codes the set of offset values and the average pitch lag. With sufficiently small offset limits, the bit-rate for pitch encoding can be considerably reduced.

Figure 1 shows some simulation results for the LT prediction gain versus subframe length for several different offset limits (in samples), where the same value is used for both positive and negative offsets. In the figure, zero offset (or offset = 0) means that the pitch lag for each subframe is set to the average pitch lag. The "unlimited offset" (or offset = unlimited) means that there is no limitation on the pitch variation in each subframe.

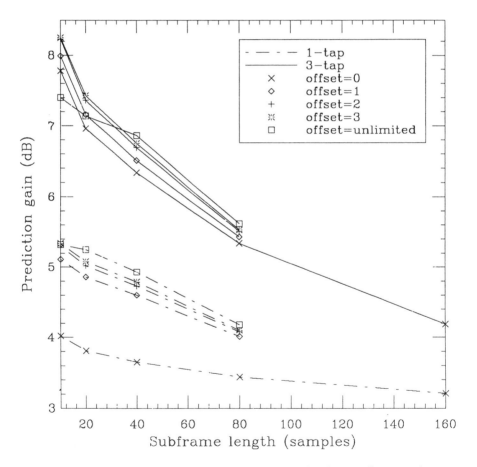

Fig. 1   Long-Term Prediction Gain vs. Subframe Length for Different Pitch Offset Limits

Note that a significant increase in prediction gain is achieved when the offset limit is increased from zero to one sample for a one-tap predictor. When the limit is increased further, even to "unlimited", only marginal improvement is obtained. (In the 3-tap unlimited case, the performance drop for small subframe lengths is due to the suboptimality of the correlation-peak-picking method.) This result shows that RPDC can significantly reduce the bit rate and computation for coding the pitch lag with little loss in the LT prediction gain.

Updating LTP Parameters With Different Rates

In this section, we consider reducing the bit rate for the LTP encoding by restricting some parameters' update rate. Our intention is to decide which parameters in a time-varying LTP really need to be adapted more frequently, and which can be adapted less frequently, so we can efficiently make use of available bits.

The first choice we examine is to use a constant pitch lag in an entire frame, but update the predictor coefficients every subframe. This is just a special case of using RPDC, with the zero pitch offset limit. The predictor designed in this way has a slowly varying pitch lag but more rapidly varying coefficients.

Alternatively, a predictor can also be designed in such way that the pitch lag is updated every subframe but the same set of predictor coefficients is used through an entire frame. To obtain a set of optimal predictor coefficients for a time-varying pitch lag, the equation $\Phi\beta = c$, is still valid, but the correlation values $\phi(i, j)$ of the matrix Φ need to be computed based on the time-varying pitch lag.

The performance comparison of the two types of adaptive LTP is given in Fig. 2. The curves are labeled so that the first character indicates the update rate of the pitch lag and the second character indicates the update rate of the LTP coefficients, where "F" is frame-based and "S" is subframe-based adaptation; the third character, a number, indicates the order of the LTP (one-tap or three-tap). For example, FS3 indicates that the pitch lag in a three-tap LTP is updated once per frame and the coefficients are updated once per subframe. For the cases SF1 and SF3, the pitch lag is estimated using RPDC with the pitch offset limit equal to one. We also include the performance of the fully adaptive predictor (symbolized as SS), in which both the pitch lag and predictor taps are updated every subframe without restriction, for reference.

From the figure, we see that FS3 is significantly better than SF3 but the opposite is true for FS1 and SF1. Thus, for a three-tap predictor the information needed for effective prediction is largely contained in the coefficients, while for a one-tap predictor the needed information is contained in the pitch lag. In particular, for a one-tap LTP, the pitch lag needs to be updated more frequently than the coefficient. It should not be surprising that the higher the predictor order, the less critical the value of pitch lag.

Similar experiments have also been performed separately on voiced and unvoiced speech. For unvoiced signals which do not have a clear periodic structure, neither FSn nor SFn is comparable to the fully adaptive predictor (SS),

334

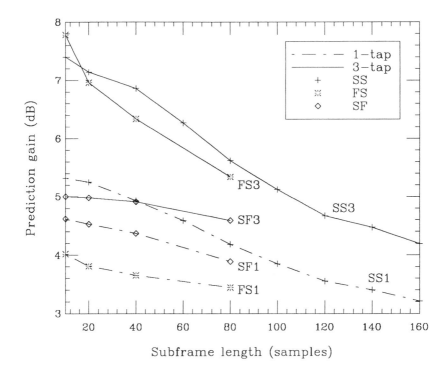

Fig. 2 Performance Comparison of Three Types of Adaptive LTP

especially when the subframe length becomes small. Hence, it is almost equally important to update both pitch lag and predictor taps for this kind of signals, which has relatively little periodicity. Furthermore, the prediction gain increases approximately linearly as the adaptation interval decreases for voiced speech, but exponentially for unvoiced speech.

CLOSED-LOOP PREDICTOR WITH RESTRICTIVE PITCH DEVIATION CODING

To achieve high performance with a closed-loop LTP, it is necessary to keep the subframe short so that the reconstructed speech samples of the previous pitch period can be exploited. Reducing the subframe length implies that the bit rate for encoding the LTP parameters needs to be increased while the bit rate for other parameters such as the STP parameters needs to be decreased. One compromise is to use the RPDC method where an average pitch lag over a frame is first found using an open-loop method and then a closed-loop search for the subframe pitch lag offset value is performed. The use of RPDC not only cuts down the bit rate for pitch encoding, but also greatly reduces the computational cost associated with coding of the closed-loop LTP. The pitch lag is determined by maximizing $\dfrac{A^2}{B}$ (as defined previously) over all candidate lag values. If the pitch deviation is not restricted, the number of pitch values, T, to be tested is 128 for a 7 bit representation of T. On the other hand, if the pitch deviation is restricted, T is allowed to vary from $T_0 - \Delta T$ to $T_0 + \Delta T$, where T_0 represents the average pitch lag and ΔT the offset limit. For example, if ΔT equals one (sample), only three tests need to be conducted to find a pitch lag, offering a significant saving in computation. The closed-loop LTP with RPDC is in fact a combination of the open-loop and closed-loop LTP, where an approximate pitch lag is first found using the open-loop approach, and the predictor is finally determined using the closed-loop approach.

A simulation of a conventional VXC/CELP coder using the RPDC method has been conducted and the result is presented in Fig. 3, where SNR and SNRSEG of the coded speech versus pitch offset limit are plotted. The results were based on coding 20 seconds of speech including both male and female speakers. In this coder, the STP parameters were updated every 20 ms, the LTP parameters every 5 ms, and the duration of the excitation vector was also 5 ms. The excitation codebook contained 32 40-dimensional Gaussian random vectors. The LPC parameters, excitation gains, and LTP coefficient were left unquantized. The average pitch period was searched in the range from 21 to 148 samples. The pitch offset limit varied from zero to "unlimited".

From the figure, it can be observed again that a significant improvement is achieved when the pitch offset limit is increased from zero to one (as was the case in the open-loop predictor simulation); but the improvement is very small as the offset limit is further increased up to 6. If the subframe pitch deviation is allowed to be "unlimited", an additional increase in SNR and SNRSEG occurs. A relatively high increase in SNRSEG, but not in SNR, at the offset limit of "unlimited" indicates that the improvement with non-restrictive pitch deviation coding is mainly achieved for the lower energy frames which mostly contain unvoiced speech. Informal listening tests showed that the degradation of the synthetic speech quality by using RPDC (offset = 1) compared to that using non-restrictive pitch coding (offset = "unlimited") is only slightly noticeable. Therefore, the considerable reduction achieved in both bit rate and complexity makes this new method highly attractive.

SIMULATION RESULTS

We have extensively tested the VXC coder and compared its performance using different LTP coding schemes but under the same overall coding rate. The results are summarized in this section.

With the open-loop LTP, we found that the three-tap LTP substantially outperforms the one-tap LTP. Furthermore, noticeable improvement can be achieved by increasing the update rate of the LTP coefficients (keeping the pitch lag fixed) in a frame. For example, it was judged that by updating the LTP coefficients twice per frame, the coded speech quality at the excitation bit rate of 7 bits per

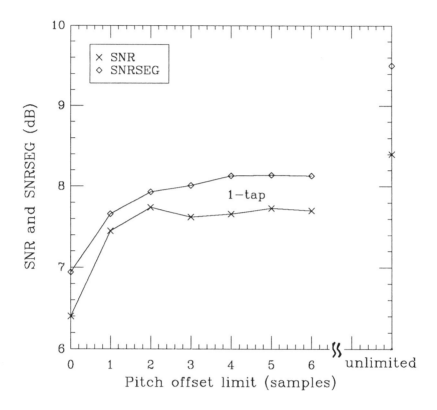

Fig. 3 VXC SNR vs. Pitch Offset Limit for One-Tap LTP

vector (with vector dimension of 40 samples) was slightly better than that at 9 bits per vector when the LTP was updated once per frame, although the latter strategy gave a higher SNR. Generally, the more bits are allocated to the excitation vectors, the higher is the SNR attained. However, the SNR performance is of very limited value in assessing perceptual speech quality in low bit-rate speech coding. In the closed-loop implementation, we choose to use the closed-loop LTP with RPDC (with pitch offset limit of unity), because of its efficiency in reducing both bit rate and complexity. With this closed-loop strategy, we found that using 6 bits per vector to code the excitation sequence was comparable to using 8 bits per vector when the open-loop LTP is used (where the LTP coefficients were updated twice per frame).

In quantization of the LTP parameters, an average pitch lag is represented by an integer in the range of 21 to 148 (for the 8 kHz sampled speech signal).

The set of pitch offsets is vector quantized. The single coefficient of a one-tap LTP can be adequately quantized using a three-bit Lloyd-Max scalar quantizer. For a three-tap open-loop LTP, we found that at least six bits or more are needed to vector-quantize the three coefficients. In the case where the coefficients of an LTP are updated twice per frame, a group of six coefficients are treated as one vector and quantized using an eight or nine-bit VQ. With the above bit alloca-tion, the drop in both SNR and SNRSEG caused by quantization of the LTP is less than 0.2 dB, according to our experiments. The difference in perceptual qual-ity is almost unnoticeable.

As a final comparison, two fully implemented 4.8 kbps VXC coders, one using the open-loop LTP (denoted by VXC-OL), the other using the closed-loop LTP (denoted by VXC-CL), were simulated and their objective and subjective performances were compared. The open-loop coder used an excitation codebook size of 256, a frame size and STP update interval of 20 ms, and an LTP having 3 tap coefficients with update interval of 10 ms and pitch lag with update interval of 20 ms. The closed-loop coder used a codebook size of 128, a 25 ms frame size and STP update interval, and a one-tap LTP with update interval of 5 ms. and pitch offset limit of one sample. Both coders used 40 sample excitation vectors, a tenth order STP computed with the stabilized covariance method, and the correla-tion peak-picking method for finding the average pitch lag. The simulation was conducted by coding a 20-second speech file containing several sentences spoken by one male and one female speaker. Our informal listening tests indicated that VXC-CL produces higher perceptual speech quality than that produced by VXC-OL, although the latter gives slightly higher SNR. Furthermore, the overall encoding complexity of VXC-CL is significantly lower than that of VXC-OL, since a much smaller excitation codebook is used in the closed-loop version.

SUMMARY

In this chapter, we have presented a study of LTP encoding in vector exci-tation (or CELP) coders. Both open-loop and closed-loop predictor design algo-rithms were described. The performance of several different forms of adaptive LTP was carefully examined. This investigation provides some insight on how to efficiently represent the effect of the glottal excitation signal on the speech waveform. According to this study, we have found that allowing a small variation of the pitch lag in a one-tap LTP within a 20 ms frame period is very helpful in improving the LTP performance, yet this method (RPDC) demands less bit rate and complexity in coding the LTP than conventional non-restrictive pitch coding. When a three-tap LTP is used, we do not find much advantage in updating the pitch lag more than once in a frame, but frequent updating of the predictor coefficients is of significant help in improving the overall coder performance. On the other hand, for a one-tap LTP, an accurate estimate of the pitch lag is more important for effective prediction.

VXC coders using both the open-loop LTP and the closed-loop LTP with RPDC have been simulated at the bit rate of 4.8 kbps. Their performance and

338

complexity were compared. Our simulations indicated that VXC with the closed-loop LTP produces higher perceptual speech quality compared to the speech produced by VXC with the open-loop LTP, although the latter gives a higher SNR. Furthermore, the closed-loop version requires significantly lower encoding complexity than the open-loop version.

References

1. M. R. Schroeder and B. S. Atal, "Code-Excited Linear Prediction (CELP): High-Quality Speech at Very Low Bit Rates," *Proceedings of IEEE International Conference on Acoustics, Speech, and Signal Processing*, pp. 937-940, Tampa, March 1985.

2. G. Davidson and A. Gersho, "Complexity Reduction Methods for Vector Excitation Coding," *Proceedings of IEEE International Conference on Acoustics, Speech, and Signal Processing*, pp. 3055-3058, Tokyo, Japan, April 1986.

3. G. Davidson, M. Yong, and A. Gersho, "Real-Time Vector Excitation Coding of Speech At 4800 bps," *Proceedings of IEEE International Conference on Acoustics, Speech, and Signal Processing*, vol. 4, pp. 2189-2192, Dallas, April 1987.

4. B. S. Atal and M. R. Schroeder, "Adaptive Predictive Coding of Speech Signals," *Bell System Technical Journal*, vol. 49, pp. 1973-1986, October 1970.

5. S. Singhal and B. S. Atal, "Improving Performance of Multi-Pulse LPC Coders at Low Rates," *Proceedings of IEEE International Conference on Acoustics, Speech, and Signal Processing*, vol. 1, pp. 1.3.1-1.3.4, San Diego, March 1984.

6. R. C. Ross and T. P. Barnwell, "The Self-Excited Vocoder," *Proceedings of IEEE International Conference on Acoustics, Speech, and Signal Processing*, vol. 1, pp. 453-456, Japan, April, 1986.

7. P. Kabal, J. L. Moncet, and C. C. Chu, "Synthesis filter Optimization and Coding: Applications to CELP," *Proceedings of IEEE International Conference on Acoustics, Speech, and Signal Processing*, vol. 1, pp. 147-150, New York, April, 1988.

8. W. B. Kleijn, D. J. Krasinski, R. H. Ketchum, and Improved Speech Quality and Efficient Vector Quantization in SELP, *Proceedings of IEEE International Conference on Acoustics, Speech, and Signal Processing*, vol. 1, pp. 155-158, New York, April, 1988.

9. B. S. Atal, "High-Quality Speech at Low Bit Rates: Multi-Pulse and Stochastically Excited Linear Predictive Coders," *Proceedings of IEEE International Conference on Acoustics, Speech, and Signal Processing*, vol. 3, pp. 1681-1684, Japan, April 1986.

32

CONSTRAINED-STOCHASTIC EXCITATION CODING OF SPEECH AT 4.8 KB/S

Yair Shoham

AT&T Bell Laboratories
600 Mountain Ave.
Murray Hill, NJ 07974

INTRODUCTION

In the last few years, Code-Excited Linear Predictive (CELP) coding has emerged as the most prominent technique for digital speech communication at rates of 8 Kb/s and below, and it is now considered the best candidate coder for digital mobile telephony and secure speech communication. While the CELP coder is able to provide fairly good-quality speech at 8 Kb/s, its performance at 4.8 Kb/s is yet unsatisfactory for many applications. The novelty in the CELP coding concept, namely, the *stochastic* excitation of a linear filter, also constitutes a weakness of this method: the excitation contains a noisy component which does not contribute to the speech synthesis process and can not be completely removed by the filter. It is a common opinion among speech communication researchers that new forms of excitations need to be studied in order to improve the CELP performance at low bit rates.

This paper reports on one study in this direction. It is proposed in this study to adaptively constrain the amount of stochastic excitation by linking its level to a performance index of the long-term (pitch-loop) sub-system. This operation reduces the noisy effects of the excitation, enhances the synthesized speech periodicity and hence, the perceptual quality of the coder.

The next section briefly reviews the basic CELP coder. Then, the concept of constrained excitation is introduced and the algorithm is discussed. Finally, listening test results are presented to demonstrate the subjective improvement of this coder over the basic CELP.

THE BASIC CODING SYSTEM

The coding system is based on the standard Codebook-Excited Linear Predictive (CELP) coder which employs the traditional excitation-filter model. A brief description of the system follows as a necessary introduction to the

Constrained-Stochastic-Excitation Coding (CSEC) concept. More details on the CELP system can be found in numerous previous papers, e.g., [1]-[9].

The speech signal $s(n)$ is processed frame by frame and the frames are contiguous and equal in size. Throughout this paper, we use the convention that the current frame corresponds to the time window $[n = 0,..,N-1]$, N being the frame size.

$s(n)$ is filtered by a pole-zero, noise-weighing linear filter to obtain $X(z) = S(z)A(z)/A'(z)$ where $x(n)$ is the *target signal* used in the coding process. $A(z)$ is the standard LPC polynomial corresponding to the current frame, with coefficients a_i , $i=0,..,M$. $(a_0=1.0)$. $A'(z)$ is a modified polynomial, obtained from $A(z)$ by shifting the zeroes towards the origin in the z-plane, that is, by using the coefficients $a'_i = a_i \gamma^i$ with $0. < \gamma < 1$. (typical value: $\gamma=0.8$). This pre-filtering operation reduces the quantization noise in the coded speech spectral valleys and enhances the perceptual performance of the coder [6].

The LPC filter $A(z)$ is assumed to be a quantized version of an all-pole filter obtained by the standard autocorrelation-method LPC analysis. The LPC analysis and quantization processes are independent of the other parts of the CELP algorithm and will not be discussed here.

The coder attempts to synthesize a signal $y(n)$ which is as close to the target signal $x(n)$ as possible, usually, in a mean-square-error (MSE) sense. The synthesis algorithm is based on the following simple equations

$$\sum_{i=0}^{M} a'_i y(n-i) = r(n) \tag{1}$$

$$r(n) = \beta r'(n,P) + g c(n) \tag{2}$$

$$r'(n,P) = \begin{cases} r(n-P) & , \quad n < P \\ r'(n-P,P) & , \quad n \geq P \end{cases} \tag{3}$$

β and P are the so-called pitch tap and pitch lag respectively. g is the excitation gain and $c(n)$ is an excitation signal. Each of the entities β, P, g, $c(n)$ takes values from a predetermined finite table. In particular, the table for the excitation sequence $c(n)$ (the excitation codebook) holds a set of N-dimensional codevectors.

The task of the coder is to find a good (if not the best) selection of entries from these tables so as to minimize the distance between the target and the synthesized signals. The sizes of the tables determine the number of bits available to the system for synthesizing the coded signal $y(n)$.

Notice that Eq. (2) and (3) represent a 1st-order pitch-loop (with periodic extension [7]). Higher-order pitch loops could also be used. However, spreading the limited number of bits for transmitting parameters of more than one pitch loop has not been found to yield higher performance.

The actual output signal, denoted by $z(n)$ ($Z(z)$ in the z-domain), is obtained by using the inverse of the noise-weighting filter. This is accomplished simply by computing $Z(z) = R(z)(1/A(z))$ where $R(z)$ is the z-domain counterpart of $r(n)$. Note that, in general, minimizing the MSE distance between $x(n)$ and $y(n)$ *does not* imply the minimization of the MSE between the input $s(n)$ and the output $z(n)$. Nevertheless, the noise-weighting filtering has been found to significantly enhance the perceptual performance the CELP coder.

A key issue in CELP coding is the strategy of selecting a good set of parameters from the various codebooks. A global exhaustive search, although possible in principle, is prohibitively complex. Therefore, sub-optimal procedures are used. A common and sensible strategy is to separate the pitch parameters P and β from the excitation parameters g and $c(n)$ and to select the two groups independently. P and β are found first and then, for a fixed such selection, the best g and $c(n)$ are found. $y(n)$ can be expressed in the form

$$y(n) = y_0(n) + \beta r'(n,P)*h(n) + g c(n)*h(n) \qquad (4)$$

where $y_0(n)$ is the response to the filter initial state without any input and $h(n)$ is the impulse response of $1/A'(z)$ in the range $[0,..,N-1]$. The notation * denotes the convolution operation. The best P and β are given by

$$P^*, \hat{\beta} = \underset{P,\beta}{\operatorname{argmin}} \| \ x(n) - y_0(n) - \beta r'(n,P)*h(n) \ \| \qquad (5)$$

where the search is done over all the entries in the tables for β and P. The notation $\| \ . \ \|$ indicates the Euclidean norm of the corresponding time-sequence. The values for P are typically in the integer range $[20,..,147]$ (7 bits). The table for β typically contains 8 discrete values (3 bits) in the approximate range $[0.4,..,1.5]$. Numerous low-complexity methods have been used for minimizing (5). A common sub-optimal method is to first minimize (5) for P with an *unquantized* β and, then, to quantize β that corresponds to the best P [3].

Once $\hat{\beta}$ and P^* are found, the coder attempts to find a best match to the resulting error signal $d(n) = x(n) - y_0(n) - \hat{\beta} r'(n,P^*)*h(n)$ by finding

$$\hat{g}, \hat{c}(n) = \underset{g,c(n)}{\operatorname{arg\,min}} \| \ d(n) - g c(n)*h(n) \ \| \qquad (6)$$

where the search is performed over all entries of the gain table and the excitation codebook. As for the pitch loop, the search for g, $c(n)$ can be performed suboptimally by first searching for the best excitation with an unconstrained (unquantized) gain and, then, quantizing that gain.

The CSEC system departs from the basic CELP described above at the stage of selecting g and $c(n)$. In the CSEC system, these parameters are selected in such a way as to constrain the level of the excitation and make it adaptive to the performance of the long-term subsystem. The concept behind this approach is discussed next.

THE CONCEPT OF CONSTRAINED EXCITATION

The CELP coding approach is based on the observation that the LP residual is essentially white and can be characterized as a Gaussian process. The CELP coder attempts to replace the ideal LP residual by external pseudo-random Gaussian excitation sequences, held in a finite-size codebook, with the hope of obtaining a reasonable match to the source spectrum. Since these artificial excitation signals have nothing to do with the source, they poorly represent the *perceptually relevant* component of the residual and contain a significant amount of irrelevant noise.

There is no known explicit way of identifying or reducing the noisy components in the excitation codebook, based on the local source characteristics. Therefore, our philosophy is to treat the excitation as mainly a noise signal and to restrict its use in order to reduce the amount of noise injected into the system.

The two components of $y(n)$ in Eq. (4) which carry new information about the source are the "pitch" signal $p(n) = \beta r'(n,P)*h(n)$ and the filtered excitation $e(n) = gc(n)*h(n)$. $p(n)$ is the result of attempting to utilize the periodicity of the source. There is no additive noisy component in it and the new information is introduced by modifying the delay P and the scale factor β. It is therefore expected to be perceptually more appealing than the excitation noisy component $e(n)$. Fortunately, in voiced (periodic) regions, $p(n)$ is the dominant component. Figure 1 shows the RMS ratio (dB) of $p(n)$ to $e(n)$ as a function of time for a typical speech segment (shown at the top of the figure) composed of voiced and unvoiced regions. The RMS values were calculated over frames of 50 samples. In voiced regions, the RMS of $p(n)$ is about 15 dB higher than that of $e(n)$. Our aim will be to make $p(n)$ even more dominant in voiced regions by appropriately suppressing the signal $e(n)$.

We propose to reduce the level of the noisy excitation and to impose a heavier reconstruction burden on the pitch signal $p(n)$. However, since $p(n)$ is not always efficient in reconstructing the output, particularly in unvoiced and transitional regions, the amount of excitation reduction should depend on the efficiency of $p(n)$. The efficiency of $p(n)$ should reflect its closeness to $x(n)$ and may be defined in various ways. In this work we use the (signal to noise) ratio

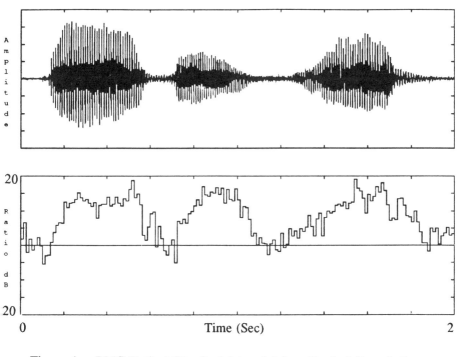

Figure 1. RMS Ratio (dB) of $p(n)$ to $e(n)$ in a Typical Speech Segment.

$$S_p = \frac{\| x(n) \|}{\| x(n) - y_0(n) - p(n) \|} \qquad (7)$$

The quantity S_p is used in controlling the level of the excitation. Recalling that the excitation is perceived as essentially a noisy component, we define the signal-to-noisy-excitation ratio

$$S_e = \frac{\| x(n) \|}{\| e(n) \|} \qquad (8)$$

The basic requirement now is that S_e be higher than some monotone-nondecreasing threshold function $T(S_p)$:

$$S_e \geq T(S_p) \qquad (9)$$

Figure 2 shows the empirical function $T(S_p)$ on a dB scale, used in this work. It consists of a linear slope followed by a flat region. When S_p is high namely, $p(n)$ is capable of efficiently reconstructing the output, S_e is forced to be high and $e(n)$ contributes very little to the output. As S_p goes down, the constraint on $e(n)$ is relaxed and it gradually takes over, since $p(n)$ becomes inefficient. $T(S_p)$ is shaped by a slope factor α and a saturation level f. Based on limited listening to coded speech, we use the preliminary parameters $\alpha = 6.0$ and $f = 24.0$ dB. However, these parameters should be a subject of more careful optimization by intensive listening to coded speech.

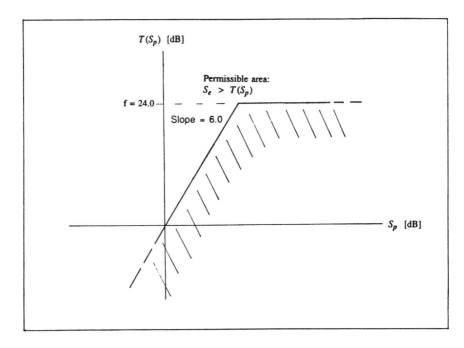

Figure 2.   The Threshold Function $T(S_p)$

The procedure for constraining the excitation, whose details are discussed next, is quite simple: the system measures S_p for the current frame, determines the threshold using $T(.)$ and selects the best excitation $\hat{c}(n)$ and the best gain $\hat{g}$ subject to the constraint of Eq. (9).

CONSTRAINED-EXCITATION SEARCH ALGORITHM

The objective is to find the best gain and excitation vector from the corresponding codebooks, under the constraint of Eq. (9). Defining the unscaled excitation response $c_h(n) = c(n)*h(n)$, the minimization problem is, therefore, stated as:

$$\hat{g} , \hat{c}(n) = \underset{g , c(n)}{\operatorname{argmin}}\{-2 g < d(n) , c_h(n) > + g^2 \| c_h(n) \|^2\} \qquad (10)$$

subject to:

$$| g | \ \| c_h(n) \| \ \leq \ \frac{\| x(n) \|}{T(S_p)} \qquad (11)$$

where $< .,. >$ denotes the inner product of the arguments. The minimization range is the set of all the entries of the gain and excitation codebooks. It is clear from the quadratic form of the problem that for a fixed excitation $c(n)$ the best quantized gain is obtained by quantizing the unconstrained optimal gain, given by

$$g^* = \frac{< d(n) , c_h(n) >}{\| c_h(n) \|^2} \qquad (12)$$

Thus, for a given $c(n)$ the best quantized gain is:

$$\hat{g} = \underset{g}{\operatorname{argmin}} \| g - g^* \| \qquad (13)$$

subject to Eq. (11).

The search procedure is to obtain the best gain for each excitation vector as in (13), record the resulting distortion and to select the pair $\hat{g}$, $\hat{c}(n)$ corresponding to the lowest distortion.

Notice that the constraint (11) is "soft" in the sense that if it is satisfied for the truly optimal gain no restriction takes place. In other words, the standard best excitation is used. This happens when the quantized version of the optimal gain g^* falls within the permissible range for the gain. This situation occurs mainly when $T(S_p)$ is low (unvoiced and transitional segments) which essentially defaults the system to a regular CELP. It may also happen in (voiced) regions of a very high LPC prediction gain, that is, very low excitation power. Note, however, that a pair $\hat{g}$, $\hat{c}(n)$ for which the constraint does not apply is not necessarily the best choice. There may be another pair with a constrained gain that actually yields a lower distortion.

346

There may arise a situation in which the constraint (11) can not be satisfied for any pair of gain and excitation vector. This could be easily remedied by including a zero-valued gain in the table. However, such a solution would be inefficient from a coding efficiency standpoint since such a gain would rarely be used. Another practical solution to this problem is to somewhat relax the constraint by applying it to the optimal (unquantized) gain rather than to the quantized one. In this approach a modified optimal gain is defined as

$$g^{**} = \min\{ \; |g^*| \; , \; \frac{\| x(n) \|}{\| c_h(n) \| \; T(S_p)} \; \} sign(g^*) \qquad (14)$$

This gain is computed and quantized for each excitation codevector and the pair $\hat{g}^{**}$, $\hat{c}(n)$ minimizing the distortion, is selected.

Figure 3 shows the ratio in dB of a regular gain (basic CELP) to a constrained gain (CSEC) as a function of time, for the same speech segment. As shown, the gain reduction is high in voiced regions (up to 15 dB) and it is around zero in unvoiced regions.

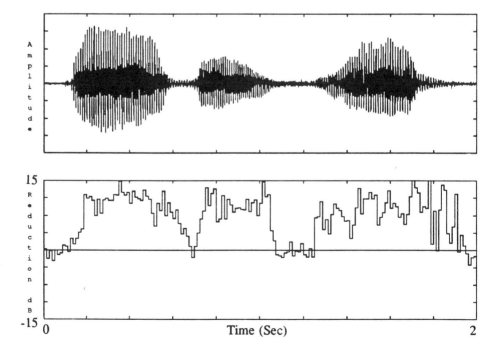

Figure 3. Gain Reduction (dB) vs. Time, as Determined by the CSEC System, for a Typical Speech Segment.

Since the gain of the excitation is constrained, the highest possible SNR cannot, in general, be achieved. Therefore, it is sensible to use some gain-independent performance criterion. The correlation between the target $x(n)$ and the reconstructed signal $y(n)$ is a natural measure to think of. This measure turns out to be particularly simple if the inequality in the constraint (11) is replaced by an equality. In this case maximization of the correlation amounts to

$$\hat{c}(n) = \underset{c(n)}{\operatorname{argmax}} \, \hat{g} < x(n) \,, \, c_h(n) > \qquad (15)$$

with

$$g = \frac{\| x(n) \|}{\| c_h(n) \| \, T(S_p)} \, sign(< x(n) \,, \, c_h(n) >) \qquad (16)$$

and $\hat{g}$, the quantized version of g. Based on a limited comparative listening test we obtained the impression that the correlation method performed slightly better than the MSE method.

PERFORMANCE

The subjective performance of the proposed coder was measured by a so-called informal A-B comparison listening test and by a formal Mean Opinion Score (MOS) test. In the A-B test, 10 speech sentences were processed by CELP and CSEC coders. 10 listeners took part in this test and voted for the better coder in their judgement. The CSEC was similar to the CELP in all respects, except for the excitation search. The intent was to show the improvement obtained by the constrained-excitation principle with the rest of the system parameters unchanged. Both coders ran at 4.8 Kb/s. The speech sentences were taken from the "Phoneme-Specific" database [8]. The test scores, defined as an average percent votes in favor of a given system, are given in Table 1. As shown, the total average scores are 85% in favor of the CSEC system in contrast to 15% in favor of the basic CELP.

Listener	1	2	3	4	5	6	7	8	9	10	Average
CELP	0	20	20	20	30	10	10	30	10	0	15
CSEC	100	80	80	80	70	90	90	70	90	100	85

Table 1.   Listening Test Scores (%) of a Comparison Between CELP and CSEC at 4.8 Kb/s.

In a formal MOS test, a 6.6 Kb/s CSEC coder was compared to two other 6.6 Kb/s CELP systems. The first one was the AT&T version of CELP implemented in hardware (real-time), which scored 3.45. The second was the software version of the first coder, with a mu-law input. It scored 3.44. The CSEC scored in this test 3.73, noticeably higher than the two other coders.

The above results shows that the CSEC coder performs distinctly better the CELP coder. The complexity of the CSEC coder is essentially the same as that of the CELP since the same type and amount codebook-search arithmetic is needed in both coders. Also, most of the complexity-reducing "tricks" that have been proposed for the CELP algorithm can be combined with the CSEC method. Therefore, the CSEC method is essentially a no-cost improvement of the CELP algorithm.

REFERENCES

[1]   B.S. Atal, M.R. Schroeder , "Stochastic Coding of Speech Signals at Very Low Bit rates", Proc. IEEE Int. Conf. Comm., May 1984, P. 48.1

[2]   M.R. Schroeder, B.S. Atal, "Code-Excited Linear Predictive (CELP): High Quality Speech at Very Low Bit Rates", Proc. IEEE Int. Conf. ASSP., 1985, pp. 937-940.

[3]   P. Kroon, E.F. Deprettere "A Class of Analysis-by-Synthesis Predictive Coders for High-Quality Speech Coding at Rate Between 4.8 and 16 Kb/s.", IEEE J. on Sel. Area in Comm. SAC-6(2), Feb. 1988, pp. 353-363.

[4]   P. Kroon, B.S. Atal, "Quantization Procedures for 4.8 Kb/s CELP Coders", Proc. IEEE Int. Conf. ASSP 1987 pp. 1650-1654.

[6]   B.S. Atal, M.R. Schroeder, "Predictive Coding of Speech Signals and Subjective Error Criteria", IEEE Tr. ASSP, Vol. ASSP-27, No. 3, June 1979, pp. 247-254.

[7]   W.B. Kleijn, D.J. Krasinski, R.H. Ketchum, "Improved Speech Quality and Efficient Vector Quantization in SELP", Proc. IEEE Int. Conf. ASSP, 1988, pp. 155-159.

[8]   A.W.F. Huggins, R.S. Nickerson, "Speech Quality Evaluation Using Phoneme-Specific Sentences", J.Acoust. Soc.Am. 77(5) pp. 1896-1906, May 1985.

[9]   G. Davidson, A. Gersho, "Complexity Reduction Methods for Vector Excitation Coding", Proc. Int. Conf. ASSP 1986 pp. 3055-58

33

SPEECH CODING USING
LEAST-SQUARES ESTIMATION

Tenkasi V. Ramabadran   and   Deepen Sinha

Department of Electrical Engineering and Computer Engineering
Iowa State University, Ames, Iowa 50011

INTRODUCTION

An important goal in current speech coding research is providing high-quality speech at low bit rates (4.8 - 16 Kbps). Several methods [1]-[3] have been proposed recently to achieve this end. Compared to the conventional linear predictive (LP) vocoder [4], these methods employ an enhanced speech production model to synthesize speech. For example, instead of a single stage, the modulation filter now typically consists of two stages: i) a short-delay filter modeling the spectral envelope of speech, and ii) a long-delay filter modeling the spectral fine structure. Both are time-varying, all-pole filters and are derived from the original speech through LP analysis. Also, some information is provided about the excitation signal, which is selected by means of an analysis-by-synthesis procedure whereby a perceptually weighted error criterion is minimized In the multi-pulse linear predictive coder (MPLPC) [1], the excitation signal is a sequence of appropriately located and scaled impulses. In the code excited linear predictive coder (CELPC) [2], it is an entry from a codebook of white, gaussian noise sequences. In the self excited vocoder (SEV) [3], it is selected from the past history of the source excitation. As a result of these improvements, the above coders are able to synthesize high-quality speech at low bit rates.

In this paper, we describe a somewhat different approach to achieving the above goal of providing high-quality speech at low bit rates. We regard speech as a piecewise-stationary random signal, and treat the synthesis operation as an estimation procedure for this signal. Accordingly, we employ the Kalman estimator [5][6] for this purpose in the decoder which is known to provide optimal linear estimates, i.e., linear estimates with minimum mean squared error. The Kalman estimator requires for its operation a signal model and a sequence of measurements of the signal. As in the other coding methods, a two-stage, time-varying, all-pole filter excited by white noise is used as the speech signal model. Linear combinations of speech samples taken at sparse, but regular (periodic), intervals serve as measurements. The job of the encoder is now seen as providing the necessary information, viz., the parameters of the signal model and the signal measurements, to the decoder which then uses the information to estimate (or synthesize) speech.

In the following, we first discuss the Kalman estimation algorithm briefly. Next, we describe a new speech coder based on Kalman's algorithm. Finally, some experimental results which are indicative of the performance of the new coder are presented.

THE KALMAN ESTIMATOR

A classical problem in communications and control is the estimation of a random signal based on the measurements of a correlated signal. In Kalman's solution to this problem [5], the random signal to be estimated is modeled as the output of a linear dynamic system excited by an uncorrelated random input ("white noise"). The dynamic system is described using a state-space formulation, and the measurements are expressed as linear combinations of states. The solution consists of a set of recursive equations which can be used to estimate the states of the dynamic system and hence the system output. The estimates are linear, and optimal in the sense that the sum of the mean squared values of the state estimation errors is minimized.

The discrete-time version of the Kalman state estimation algorithm [6] is presented below. We begin by assuming that the random process to be estimated can be modeled as

$$x_{k+1} = \Phi_k x_k + w_k \qquad (1)$$

and the measurement of the process as

$$z_k = H_k x_k + v_k \qquad (2)$$

where,

$x_k - (n \times 1)$ process state vector at time t_k,
$\Phi_k - (n \times n)$ state transition matrix at time t_k relating x_k to x_{k+1},
$w_k - (n \times 1)$ process noise vector at time t_k,
$z_k - (1 \times 1)$ measurement vector at time t_k,
$H_k - (1 \times n)$ measurement matrix at time t_k relating x_k to z_k, and
$v_k - (1 \times 1)$ measurement noise vector at time t_k.

The process and measurement noise vectors w_k and v_k are assumed to be zero mean, white (time uncorrelated) noise sequences with known covariance structures given by

$$E[\mathbf{w}_k \mathbf{w}_i{}^T] = \mathbf{Q}_k \, \delta_{ik} \quad \text{and} \quad E[\mathbf{v}_k \mathbf{v}_i{}^T] = \mathbf{R}_k \, \delta_{ik} \tag{3}$$

where $E[\cdot]$ denotes the expectation operator, the superscript T indicates a vector transpose, and δ_{ik} is the Kronecker delta function. It is also assumed that $\mathbf{w}_k$ and $\mathbf{v}_k$ are mutually uncorrelated, and both $\mathbf{w}_k$ and $\mathbf{v}_k$ are uncorrelated with the initial state vector $\mathbf{x}_0$.

The goal of the Kalman estimator is to estimate the state vector $\mathbf{x}_k$, $k \geqslant 0$, based on the measurements $\mathbf{z}_k$, $k \geqslant 0$. We will use the "hat" to denote an estimate. Depending on whether the measurements $\mathbf{z}_0$ through $\mathbf{z}_{k-1}$ or $\mathbf{z}_0$ through $\mathbf{z}_k$ are used in the estimation of the state vector $\mathbf{x}_k$, we have two estimates, viz., the *a priori* estimate $\hat{\mathbf{x}}_k{}^-$ and the *a posteriori* estimate $\hat{\mathbf{x}}_k$. The Kalman estimator uses a recursive procedure to estimate $\mathbf{x}_k$. Assuming that $\hat{\mathbf{x}}_k{}^-$ is available, it uses the current measurement $\mathbf{z}_k$ to form the estimate $\hat{\mathbf{x}}_k$, which is then used to compute $\hat{\mathbf{x}}_{k+1}{}^-$, and so on. To initiate the procedure, we need $\hat{\mathbf{x}}_0{}^-$ which is usually chosen as $E[\mathbf{x}_0]$, if known. In this case, it turns out that both $\hat{\mathbf{x}}_k{}^-$ and $\hat{\mathbf{x}}_k$ are *unbiased*, which we will assume henceforth. The errors corresponding to the two estimates can now be defined as

$$\mathbf{e}_k{}^- = \mathbf{x}_k - \hat{\mathbf{x}}_k{}^- \quad \text{and} \quad \mathbf{e}_k = \mathbf{x}_k - \hat{\mathbf{x}}_k. \tag{4}$$

Since the estimates are assumed to be unbiased, the estimation errors have zero mean and we can define the corresponding error covariance matrices as

$$\mathbf{P}_k{}^- = E[\mathbf{e}_k{}^- \mathbf{e}_k{}^{-T}] \quad \text{and} \quad \mathbf{P}_k = E[\mathbf{e}_k \mathbf{e}_k{}^T]. \tag{5}$$

The error covariance matrices play an important role in the recursive estimation procedure. Once again, to initiate the procedure, we need $\mathbf{P}_0{}^-$ which is usually taken to be $E[(\mathbf{x}_0 - E[\mathbf{x}_0]) (\mathbf{x}_0 - E[\mathbf{x}_0])^T]$, if known. The recursive equations defining the Kalman estimator are given by

$$\mathbf{K}_k = \mathbf{P}_k{}^- \mathbf{H}_k{}^T (\mathbf{H}_k \mathbf{P}_k{}^- \mathbf{H}_k{}^T + \mathbf{R}_k)^{-1}, \tag{6}$$

$$\hat{\mathbf{x}}_k = \hat{\mathbf{x}}_k{}^- + \mathbf{K}_k (\mathbf{z}_k - \mathbf{H}_k \hat{\mathbf{x}}_k{}^-), \tag{7}$$

$$\mathbf{P}_k = (\mathbf{I} - \mathbf{K}_k \mathbf{H}_k) \mathbf{P}_k{}^-, \tag{8}$$

$$\hat{\mathbf{x}}_{k+1}{}^- = \Phi_k \hat{\mathbf{x}}_k, \quad \text{and} \tag{9}$$

$$\mathbf{P}_{k+1}{}^- = \Phi_k \mathbf{P}_k \Phi_k{}^T + \mathbf{Q}_k. \tag{10}$$

It is clear that the above equations can be used recursively to estimate the state vector $\mathbf{x}_k$, $k \geqslant 0$, based on the measurements $\mathbf{z}_k$, $k \geqslant 0$, provided the system matrices Φ_k and $\mathbf{H}_k$ as well as the noise covariance matrices $\mathbf{Q}_k$ and $\mathbf{R}_k$ are known and suitable values are assumed for $\hat{\mathbf{x}}_0{}^-$ and $\mathbf{P}_0{}^-$. The Kalman estimator is optimal in the sense that the trace of the *a posteriori* error covariance matrix $\mathbf{P}_k$ is minimized.

352

APPLICATION TO SPEECH CODING

In applying the Kalman state estimation algorithm to speech coding, the speech signal has to be first modeled as the output of a linear dynamic system excited by white noise. This is accomplished easily if we identify the two-stage (short-delay and long-delay), time-varying, all-pole filter that can be obtained from the original speech through LP analysis with the linear dynamic system. Such a model is shown in Figure 1, in which the quantities $r(k)$, $d(k)$, and $s(k)$ denote respectively the white noise input, the output of the long-delay filter, and the speech signal output at time t_k. In Z-transform notation, the system function of the short-delay predictor can be expressed as

$$P_s(z) = \sum_{i=1}^{p} \alpha_i \, z^{-i} \tag{11}$$

where, α_i, $i = 1,2, \cdots ,p$, denote the short-delay predictor coefficients. Similarly, the system function of the long-delay predictor can be expressed as

$$P_d(z) = \sum_{j=M}^{M+q-1} \beta_{j+1-M} \, z^{-j} \tag{12}$$

where, β_j, $j = 1,2, \cdots ,q$, denote the long-delay predictor coefficients and M is the approximate pitch period.

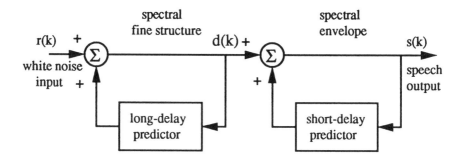

Figure 1. **Speech signal model with long-delay and short-delay predictors**

Letting $R(z)$ and $S(z)$ denote respectively the Z-transforms of $r(k)$ and $s(k)$, and by combining the two filter stages of Figure 1, the overall transfer function of the speech signal model can be expressed as

$$H(z) = \frac{S(z)}{R(z)} = \frac{1}{(1 - \sum_{j=M}^{M+q-1} \beta_{j+1-M} z^{-j})(1 - \sum_{i=1}^{p} \alpha_i z^{-i})}$$

$$= \frac{1}{1 - \sum_{l=1}^{M+p+q-1} \gamma_l z^{-l}} = \frac{1}{1 - P_c(z)} \qquad (13)$$

where $P_c(z)$ denotes the system function of the combined predictor and γ_l, $l = 1,2, \cdots ,M+p+q-1$, denote the corresponding coefficients. In computing the combined predictor coefficients γ_l in terms of β_j and α_i, it is important to take the time-varying nature of the latter coefficients into consideration.

A state-space representation of the speech signal model shown in Figure 1 can be obtained by realizing the all-pole filter specified by (13) in direct form, and then defining the outputs of the different delay elements as states. In this case, we have

$$\mathbf{x}_k = [\, s_{k-1} \, s_{k-2} \, s_{k-3} \, s_{k-4} \, \cdots \, s_{k-(M+p+q-1)} \,]^T,$$

$$\Phi_k = \begin{vmatrix} \gamma_1 & \gamma_2 & \gamma_3 & \cdots & \gamma_{M+p+q-2} & \gamma_{M+p+q-1} \\ 1 & 0 & 0 & \ldots & 0 & 0 \\ 0 & 1 & 0 & \ldots & 0 & 0 \\ 0 & 0 & 1 & \ldots & 0 & 0 \\ \cdot & \cdot & \cdot & \cdot\cdot\cdot & \cdot & \cdot \\ \cdot & \cdot & \cdot & \cdot\cdot\cdot & \cdot & \cdot \\ \cdot & \cdot & \cdot & \cdot\cdot\cdot & \cdot & \cdot \\ 0 & 0 & 0 & \ldots & 1 & 0 \end{vmatrix}, \text{ and}$$

$$\mathbf{w}_k = [\, r_k \, 0 \, 0 \, 0 \, \cdots \, 0 \,]^T.$$

It is seen that the size of the state vector is given by $n = M+p+q-1$, and that the states at time instant t_k are simply the speech samples at the previous $M+p+q-1$ instants. The state transition matrix Φ_k and the process noise vector $\mathbf{w}_k$ are seen to have very simple structures which facilitate the implementation of the Kalman algorithm. The speech signal itself can now be expressed as an output of the above process, i.e., as a linear combination of states. For example, the speech signal with a single time delay is given by

$$s_{k-1} = [\, 1 \, 0 \, 0 \, 0 \ldots 0 \,] \, \mathbf{x}_k.$$

Having obtained a state-space description of the random speech process, the use of the Kalman filter at the decoder to synthesize speech through state estimation is fairly straightforward. As pointed out earlier, the following information is required: the measurements $\mathbf{z}_k$, the system matrices Φ_k and $\mathbf{H}_k$, the noise covariance matrices $\mathbf{Q}_k$ and $\mathbf{R}_k$, and suitable values for $\hat{\mathbf{x}}_0^-$ and $\mathbf{P}_0^-$.

In order to reduce the bit rate requirement of the coder, the measurements $\mathbf{z}_k$ are not provided at every sampling instant, but only at every L^{th} sampling instant, where $L > 1$. The effect of these *sparse* measurements on the Kalman algorithm is that (9) and (10) are executed L times before looping back to (6).

The state transition matrix Φ_k is completely specified by the combined predictor coefficients γ_l, $l = 1,2, \cdots ,M+p+q-1$ which, in turn, are completely specified by the short-delay predictor coefficients α_i, $i = 1,2, \cdots ,p$, the long-delay predictor coefficients β_j, $j = 1$, $2, \cdots ,q$, and the approximate pitch period M. These $p+q+1$ parameters are obtained from the original speech through LP analysis, coded, and transmitted to the decoder.

Since the process noise vector $\mathbf{w}_k$ has only one nonzero element, the process noise covariance matrix $\mathbf{Q}_k = [q_{ij}] = E[\mathbf{w}_k \mathbf{w}_k^T]$ also has only one nonzero element, viz., $q_{11} = E[r_k^2]$. An estimate of q_{11} is obtained by inverse filtering the original speech and computing the time average of the squared values of the residual samples, which is then coded and transmitted to the decoder.

The measurement matrix $\mathbf{H}_k$ can be arbitrarily chosen since all the states of the process (which are actually speech samples) are accessible at the encoder. This degree of freedom can be used to improve the performance of the Kalman estimator. In [9], we have shown that the choice of $\mathbf{H}_k$ as the eigenvector corresponding to the largest eigenvalue of the *a priori* error covariance matrix $\mathbf{P}_k^-$ minimizes the trace of the *a posteriori* error covariance matrix $\mathbf{P}_k$ given that the measurement noise covariance matrix $\mathbf{R}_k$ is either zero or equal to $\lambda_k \mathbf{H}_k \mathbf{P}_k^- \mathbf{H}_k^T$ where $\lambda_k > 0$ is a scalar constant. If $\mathbf{H}_k$ is to be chosen as above, then the Kalman estimator must be executed at the encoder as well, and information about $\mathbf{H}_k$ need not be transmitted explicitly. While executing the Kalman algorithm at the encoder, some additional bit rate savings can be achieved if we transmit the "innovations" $\mathbf{H}_k(\mathbf{x}_k - \hat{\mathbf{x}}_k^-) = \mathbf{H}_k \mathbf{e}_k^-$ instead of the (noiseless) measurements $\mathbf{H}_k \mathbf{x}_k$ since the variance of the innovation can be expected to be smaller than the variance of the noiseless measurement if the *a priori* estimate $\hat{\mathbf{x}}_k^-$ is reasonably good. The variance of the innovation can be expressed as $E[\mathbf{H}_k \mathbf{e}_k^- \mathbf{e}_k^{-T} \mathbf{H}_k^T] = \mathbf{H}_k \mathbf{P}_k^- \mathbf{H}_k^T$.

The measurement noise term $\mathbf{v}_k$ arises in the proposed coder because of the need to quantize the innovation and thereby keep the bit rate requirement low. If no quantization of the innovation is performed, then $\mathbf{v}_k = 0$ and $\mathbf{R}_k = 0$. Otherwise, the quantization error variance provides an estimate of $\mathbf{R}_k$. In quantizing the innovation, we treat it as a zero mean gaussian random variable with variance $H_k P_k^- H_k^T$, and choose the quantizer to be a pdf-optimized, nonuniform quantizer. In this case, the measurement noise variance can be expressed as $R_k = \lambda H_k P_k^- H_k^T$, where λ is a constant that depends on the bit allocation B of the quantizer. Since R_k can be computed as above, there is no need to transmit this information explicitly.

Both $\hat{x}_0^-$ and P_0^- can be taken to be zero if we assume that the execution of the Kalman algorithm can be started during a silent period of the given speech signal.

A schematic block diagram of the proposed coder is shown in Figure 2. For simplicity, all the details discussed above are not shown. For example, the Kalman estimator at the encoder is not shown. Similarly, the measurements are shown directly as linear combinations of speech samples and not as derived quantities from innovations.

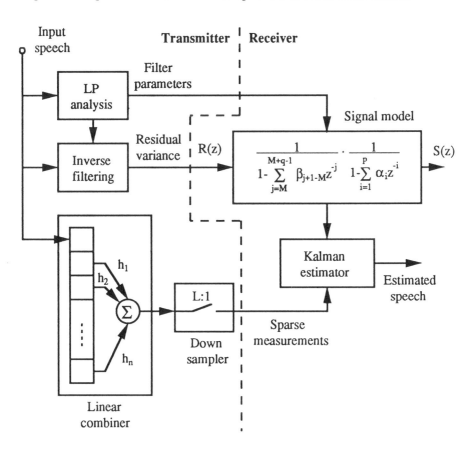

Figure 2. Schematic block diagram of the proposed coder

EXPERIMENTAL RESULTS

The proposed coder was implemented in software and several experiments were performed to study the effect of various parameters and their quantization on the coder performance. Speech sentences sampled at 8 KHz were used in the experiments and the coder performance was evaluated primarily using the objective SEGSNR measure. In analyzing the original speech to extract the signal model parameters, the short-delay predictor coefficients were computed first using the weighted, stabilized covariance method with high-frequency correction [7][8]. The order p of the short-delay predictor was set between 10 and 16, and the coefficients were updated every 10 or 20 ms. The long-delay predictor coefficients, the approximate pitch period, and the residual variance were computed next. The order q of the long-delay predictor was set at 3, and the approximate pitch period M was chosen from the range of 5 to 13.75 ms (40 to 110 samples at 8 KHz). These parameters were updated every 5 or 10 ms. The important results obtained from the experiments are summarized below.

By choosing different values for the measurement interval L, the effect of the frequency of measurements f_m (8000/L) on the coder performance was first studied. As one would expect, the performance was found to improve with increasing values of f_m, and the dependence was found to be quite strong. The effect of changing the order p of the short-delay predictor was also studied. The performance improved slightly as p was changed from 10 to 16.

The effect of quantizing the measurements was next studied using different numbers of bits B for the quantizer of the innovations. It was found that the performance of the coder did not deteriorate too much until B was reduced to about 2. The number of bits per second allocated to the measurements depends both on f_m and B. In order to study the relative effect of quantity (f_m) versus quality (B) of measurements, an experiment was conducted in which f_m was changed to different values, while keeping $f_m \times B$ constant. It was found that it is better to have a greater number of approximate measurements than a fewer number of accurate measurements, provided B is at least 2.

The advantage of using the eigenvector of P_k^- corresponding to its largest eigenvalue as the measurement matrix H_k (rather than a fixed measurement matrix) was also established through an experiment. In addition to improving the coder performance significantly, the optimal selection yielded a nearly white reconstruction error spectrum.

While studying the effect of quantizing the signal model parameters, it was found that the performance deteriorated much more quickly with the quantization of the model parameters than with that of the measurements.

The relative influence of the model parameters and the measurements on the coder performance was then studied by designing and testing several coders with bit rates in the range of 9 - 10 Kbps. Among those tested, the following coder with a bit rate of 9934 bps was found to be the best. The number of bps assigned to the model parameters of this coder was 4600 and the number of bps assigned to the measurements was 5334. The order p of the short-delay predictor was chosen as 12, and the corresponding coefficients were updated every 20 ms. Other model parameters were updated every 10 ms. The measurement interval L was selected as 3 ($f_m \approx 2667$) and B as 2.

The performance of this coder for different speech sentences are shown in Table 1. In general, the coder performance was found to be quite good for voiced sounds and poor to average for unvoiced sounds. In a subjective evaluation using informal listening tests, the coder performance was found to be slightly better than that of a 5.5 bit μ-law PCM coder. It was also noted that the type of distortion introduced by the proposed coder is quite different from that of a μ-law coder.

Table 1. Performance of 9934 bps coder for a few speech sentences

Sentence (M/F)	SEGSNR (dB)	SNR (dB)
Cats and dogs each hate the other (M)	14.0	17.2
Add the sum to the product of these three (M)	12.6	15.2
Add the sum to the product of these three (F)	13.0	15.0
Mabel stood on the rock (M)	14.9	15.5
The ripe taste of cheese improves with age (M)	13.7	14.8

CONCLUSIONS

A new approach to speech coding using the Kalman state estimation technique is described. Simulation results indicate that a coder based on this approach can provide high-quality speech at low bit rates. Like the other low bit rate coders, the proposed coder employs a two-stage, time-varying, all-pole filter in the speech synthesis model. Instead of excitation information, however, the difference between actual and predicted values of a linear combination of speech samples is provided at sparse intervals.

The proposed coder is computationally quite expensive; on a HDS AS/9180 computer, it took approximately 180 sec to process 1 sec of

358

speech. Most of the time was spent on updating the error covariance matrix in the Kalman estimation algorithm. No major effort was taken however to optimize the program to run faster. The total delay (encoding and decoding) involved with the proposed coder is of the order of 50 ms.

Several issues remain open for investigation in the new approach. An important issue is the shaping of the reconstruction error spectrum to take advantage of the masking phenomenon in speech perception. Some preliminary studies have been made in this regard [10]. The effect of channel errors on the coder performance is another important issue to be studied. Other issues are the use of irregular measurements, use of an improved signal model, and reduction of implementation complexity.

ACKNOWLEDGEMENTS

We would like to thank Dr. Jim Mills of Tellabs, Inc. for his help in performing the informal listening tests.

REFERENCES

1. B.S. Atal and J.R. Remde, "A New Model of LPC Excitation for Producing Natural-Sounding Speech at Low Bit Rates", *Proc. Int. Conf. on ASSP*, pp. 614-617, April 1982.
2. M.R. Schroeder and B.S. Atal, "Code-Excited Linear Prediction: High Quality Speech at Very Low Bit Rates", *Proc. Int. Conf. on ASSP*, pp. 937-940, April 1985.
3. R.C. Rose and T.P.Barnwell III, "The Self-Excited Vocoder - An Alternate Approach to Toll Quality at 4800 bps", *Proc. Int. Conf. on ASSP*, pp. 453-456, April 1986.
4. T.W. Parsons, "Voice and Speech Processing", *McGraw Hill*, 1987.
5. R.E. Kalman, "A New Approach to Linear Filtering and Prediction Problems", *Trans. ASME J. Basic Eng.*, pp. 35-45, March 1960.
6. R.G. Brown, "Introduction to Random Signal Analysis and Kalman Filtering", *John Wiley & Sons*, 1983.
7. B.S. Atal, "Predictive Coding of Speech Signals and Subjective Error Criteria", *IEEE Trans. on ASSP*, pp. 247-254, June 1979.
8. S. Singhal and B.S. Atal, "Improving Performance of Multi-Pulse LPC Coders at Low Bit Rates", *Proc. Int. Conf. on ASSP*, pp. 1.3.1-1.3.4, March 1984.
9. T.V. Ramabadran and D. Sinha, "On the Selection of Measurements in Least-Squares Estimation", *Proc. IEEE Int. Conf. on Systems Eng.*, Dayton, Ohio, pp. 221-226, August 1989.
10. D. Sinha, "A New Approach to Speech Coding Using Sparse Measurements and Least-Squares Estimation", *M.S. Thesis*, Dept. of EE & CprE, Iowa State University, 1989.

34

QUALITATIVE ANALYSIS AND ENHANCEMENT OF SINE TRANSFORM CODING

Hyokang Chang and Yi-sheng Wang

Hughes Network Systems
11717 Exploration lane
Germantown, MD 20876

INTRODUCTION

In the past several years, significant progress has been made in the STC (sine transform coder) approach in terms of compression gain and implementation complexity [1]-[3]. As a result, it now appears to be a viable alternative to the CELP in the range of 2.4 kbps to 9.6 kbps. In contrast to the CELP, which is a time-domain, search-oriented technique, the STC compresses speech directly in the frequency domain using a harmonic model without decomposing it into excitation and time-varying filter parts. Because of its simplicity in speech modelling, at least in principle, it provides a tractable means of gaining insight into the process of speech compression.

Harmonic modelling can be viewed as an approximation problem of short- to medium-term speech spectrum using only a set of harmonic components. In the time domain, it is equivalent to replacing quasi-periodic speech segments with periodic ones. Model parameters are estimated based on approximately two and half times the pitch period and they are updated at a 50 Hz frame rate. During speech synthesis, these parameters are interpolated by a factor of two to make the transition from one frame to next gradual, effectively yielding a model parameter update rate of 100 Hz. To further reduce the boundary effect, an overlap-add technique is incorporated.

In this paper, we first attempt to analyze various types of distortions in the STC qualitatively and then propose potential enhancement to the algorithm. The reference algorithm employed in our analysis is the one presented in [1] and [2] with a frame size of 20 ms and a frame rate of 100 Hz. Major sources of distortions in the STC are: the imposition of a harmonic model that allows only harmonically related spectral components, pitch estimation errors, the boundary effect due to block processing, and the marginal effectiveness of a model for unvoiced or transient

segments. The focus of our analysis will be on the subjective as well as objective
effects for each type of distortion discussed above.

QUALITATIVE ANALYSIS

Although it sounds restrictive, a harmonic model is very effective in
representing voiced sound with a small set of parameters. The effect of harmonic
modelling is shown in figure 1 for the speech waveform with a significant amount of
short-term variations. At an update rate of 100 Hz without interpolation, the model
tracks even the fast-varying fine structure very well. At a 50 Hz rate, however, the
fine structure is noticeably distorted. The perceptual effect of this type of distortion
is a slight loss in high frequencies and small increase in random noise. Since the
model tends to maintain only the steady component, it is not surprising that fast-
varying fine structure undergoes more distortions than the steady one. For low-
pitched speech, this type of distortion is almost negligible for voiced sound. This is
because with lower pitched speech there are less number of pitch periods which need
to be approximated by a single representative period than with higher pitched speech,
given a frame size. Ideally, the dependency of speech quality on the pitch can be
resolved by making the frame size variable. The added complexity, however, may
not be justified, considering the relatively minor improvement to be realized.

Pitch estimation is a critical element in the STC since all the harmonic
frequencies are derived from it. In our approach, pitch estimation is performed in the
frequency domain using 512-point FFT spectra in two steps: coarse estimation with a
fixed window (64 ms) and fine estimation with a pitch-dependent window (two and
half times the pitch period). Pitch errors occur in the form of pitch frequency offset
or pitch frequency scaling by an integer factor (submultiple of pitch frequency).
Pitch frequency doubling or tripling seems to occur rarely in the frequency domain
approach. Our subjective analysis indicates that pitch errors can be detrimental if
they cause discontinuities in the overall shape of the waveform. Pitch offset errors
thus are less serious than pitch scaling errors. In principle, however, the STC should
be tolerant of pitch scaling errors, i.e., an estimated pitch frequency being a
submultiple of the true one, because redundant harmonic components can still be
suppressed during coding. On the other hand, a pitch scaling error in the LPC
approach may result in serious degradation since it changes the excitation
characteristics significantly. The effect of pitch scaling in the STC is depicted in
figure 2 using an ideal harmonic model. It is clear from the figure that a pitch
scaling error by itself does not cause any major degradation in the synthesized
waveform. It, however, increases the number of model parameters to be coded. The
problem of pitch scaling should therefore be resolved jointly with the parameter
quantization process.

Frame interpolation by the use of a midframe is a relatively straightforward
process. A midframe maintaining a harmonic structure is generated in the frequency
domain by interpolating harmonic components. Special attention, however, is

361

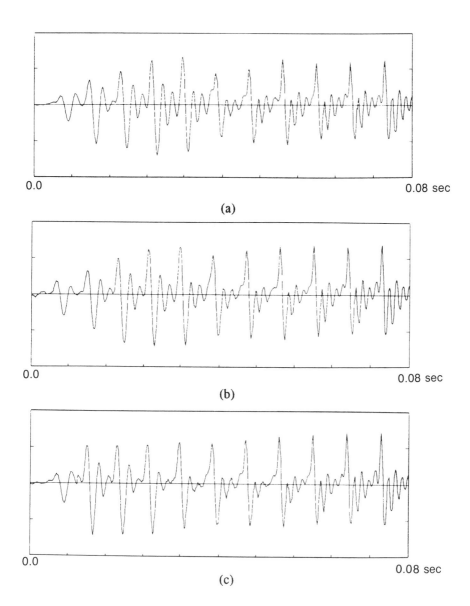

Figure 1. (a) Original speech waveform, (b) synthesized speech at a 100 Hz frame rate, (c) synthesized speech at a 50 Hz frame rate with frame interpolation.

362

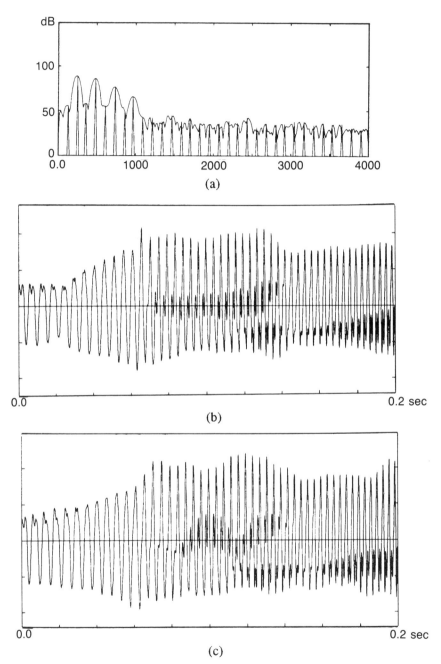

(a)

(b)

(c)

Figure 2. Effect of pitch frequency halving, (a) short-term spectrum with harmonic sampling, (b) original speech waveform, (c) synthesized speech with pitch halving.

required for the midframe generation if the pitch frequency fluctuates with an integer scaling factor. To analyze the effect of frame interpolation, we compared the synthesized speech obtained using a midframe with that obtained using a true frame. Comparisons between interpolated and true parameters are shown in figure 3 over 0 to 1.5 kHz for a typical voiced segment. While amplitude interpolation is quite accurate over the entire band, phase interpolation is effective only at low frequencies. The perceptual difference between those two schemes were already discussed earlier. To find out which interpolation error, amplitude or phase, contributes more distortion subjectively, we ran an experiment by generating a midframe with true amplitude and interpolated phase and vice versa. The result of the experiment, however, was inconclusive.

One of the important issues in harmonic modelling is how to handle unvoiced speech segments. Since unvoiced segments do not have a harmonic structure, harmonic modelling requires a separate procedure for representing them. The current STC synthesizes the unvoiced segment using only the short-term amplitude spectrum over a 20 ms window. Figure 4 shows the effect of the STC over unvoiced speech segments. The most notable effect is the spreading of unvoiced speech energy lasting only 5 to 10 ms over the entire 20 ms frame. To get the subjective feel for the distortion the unvoiced frame introduces, we compared the synthesized speech by the STC with the one generated by a reference system which employs a harmonic model only for voiced segments without disturbing unvoiced segments, i.e., no speech coding for unvoiced segments. The comparison result showed that the reference system generated noticeably clearer speech than the STC. In particular, the loss of speech onset, a short burst just before voiced sound, resulted in slight degradation in intelligibility in the STC. Our observation based on this comparative test is that the dynamic characteristic of unvoiced sound in the time domain needs to be maintained for clear and intelligible speech reproduction.

ENHANCEMENT OF STC

It was already pointed out that the current STC is not quite effective in dealing with unvoiced sound due to its inability to track fast-varying speech dynamics. An alternative for this problem is to adopt a smaller frame size, i.e., 10 ms instead of 20 ms, and not to apply frame interpolation for unvoiced segments to limit the spreading of speech energy. For voiced segments, a frame size of 20 ms is still maintained. Without frame interpolation, there will be discontinuities in the waveform to some extent. Our subjective test indicated that discontinuities within unvoiced sound is not so much noticeable, however. The use of a smaller frame size for unvoiced sound also helps the pitch estimation and coding of voiced sound because it provides more precise boundaries between voiced and unvoiced sound.

The spectrum of an unvoiced frame can be obtained by applying an 80-point DFT since there is no coherency to maintain. The DFT components can then be treated like harmonic components of a voiced frame. Unlike the current STC, this approach attempts to maintain the waveform even for unvoiced sound. The drawback of this approach is that there are only half as many bits for unvoiced frames as for voiced frames. It may not be a serious constraint, however, since an accurate representation of the spectrum is not necessary for unvoiced sound. In order to maintain continuity of the waveform, it is recommended that the first few phases be coded. The synthesized speech waveform obtained with the proposed approach employing phases of first four harmonics of unvoiced frames is shown in figure 4(c). It is clear from the figure that with the proposed approach the fast-varying speech dynamics including the speech onset is well maintained. For coding of the spectral envelope of an unvoiced frame, a codebook approach may be an attractive alternative.

REFERENCES

1. R. J. McAulay and T. F. Quatieri, "Computationally efficient sine wave synthesis and its application to sinusoidal transform coding," pp. 370-373, ICASSP 1988.

2. R. J. McAulay and T. F. Quatieri, "Multirate sinusoidal transform coding at rates from 2.4 kbps to 8 kbps," pp. 38.7.1-38.7.4, ICASSP 1987.

3. D. W. Griffin and J. S. Lim, "Multiband excitation vocoder," pp. 1223-1235, IEEE Trans. ASSP, August 1988.

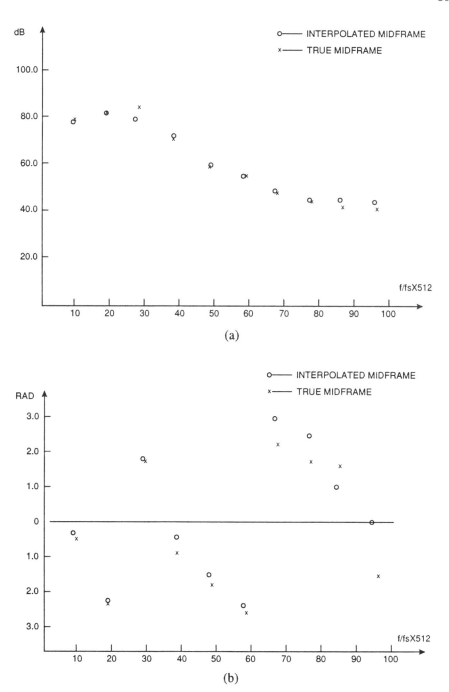

Figure 3. Comparison between true and interpolated midframe parameters (a) harmonic amplitudes, (b) harmonic phases.

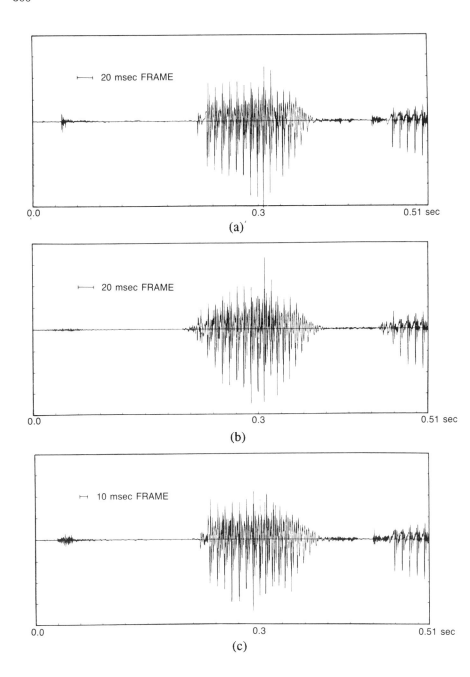

Figure 4. (a) Original speech waveform, (b) synthesized speech by current STC, (c) synthesized speech by proposed approach.

35

AN ADM SPEECH CODING WITH TIME DOMAIN HARMONIC SCALING

Jing Yuan, C.S. Chen, Hongmei Zhou

Electrical Engineering Department
University of Akron, Akron, OH 44325

ABSTRACT

A semi-waveform speech coder is presented that combines adaptive delta modulation (ADM) with time domain harmonic scaling (TDHS) to achieve good quality speech coding at 4.4 kb/s. The use of TDHS reduces the transmission bit rate by 50%, while still maintaining good speech quality. Compared to other coding schemes of the same data-rate range, the proposed method is simple and fast. Another feature of this method is the variable output data-rate which is suitable for combined source and channel coding.

INTRODUCTION

A speech coder is proposed here that combines adaptive delta modulation (ADM) with time domain harmonic scaling (TDHS) to achieve good quality speech coding at 4.4 kb/s. The TDHS algorithm was initially derived by Malah [1]. We improved this algorithm by using a new window with higher performance and slightly more computation than the triangular window used by Malah. The basic function of TDHS is to reduce the bit rate by a factor of 2. It compresses two consecutive pitch periods into one at the transmitter, and expands the transmitted signal to recover the missing pitch periods at the receiver. Fig. 1 gives a basic block diagram of our approach. The sampled speech signal is first compressed to form an intermediate signal. This processing is performed in a pitch synchronous manner by the TDHS algorithm. Pitch detection is done by a robust pitch boundary detector presented in [2]. The ADM encoder then encodes the compressed signal to form the encoded data. At the receiver, the digital signal is decoded by the ADM decoder to form the intermediate signal which is then expanded back to its original sampling rate by a TDHS expansion algorithm. The output speech signals still maintain good quality.

A high-order adaptive linear predictor is used in the ADM and the quantization step-size is chosen to be the average predicted error of each segment of signal. The detailed coding scheme is described in the following sections. Compared to other coding schemes of the same data-rate range, the proposed method is simple and fast without degrading speech quality. Another feature

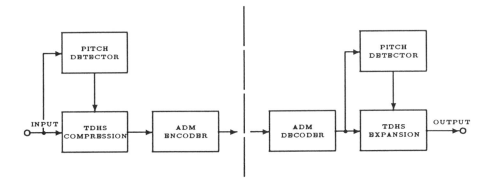

Figure 1: Basic Block Diagram of ADM Speech Coder with TDHS

of this method is the variable output data-rate which is suitable for combined source and channel coding.

THE TDHS ALGORITHM

The time-domain harmonic scaling (TDHS) algorithm consists of properly weighting several adjacent input signal segments (with pitch dependent duration) by a suitable window function, to produce an output segment [1]. In 2:1 TDHS compression, two pitch periods of speech data are weighted and added to produce one segment. Using 2:1 TDHS compression, we can reduce the data-rate by 50%. The compressed signal is then encoded and transmitted. In the receiver the signal is decoded and then expanded back to its original sampling rate by a TDHS expansion algorithm. The TDHS algorithm was derived by Malah [1]. An important factor in the implementation of the TDHS algorithm is the proper choice of the window function. We improved this algorithm by using a new window function to weight the speech signal.

To determine the proper window function $w(t)$ to be used, Malah has derived a family of window functions. For 2:1 TDHS compression, the well known Hanning window is the proper window function to be used. If pitch tracking is performed, the pitch period N_p varies from one segment to the other. The values of window function corresponding to different sampling instants vary according to different values of N_p. For the Hanning window, different values of the window function have to be stored or interpolated — this requires a large amount of computation. To simplify the computation of the TDHS algorithm, Malah used the triangular window instead of the Hanning window. However, the performance with the triangular window was lower than with the Hanning

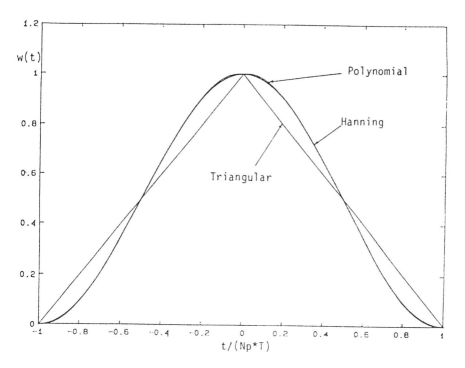

Figure 2: The window functions (Hanning, Polynomial and Triangular). Note: Hanning and polynomial functions look almost identical since the differences between them are very small compared to the height of the window functions.

window.

In our implementation, we use a third-order polynomial

$$w(t) = 1 + 0.125t - 3.375t^2 + 2.25t^3 \qquad\qquad 0 \le t \le 1 \qquad (1)$$

to obtain a good approximation of the Hanning window. The triangular window is a first-order polynomial. For the polynomial window of Eq.(1), we need just two more additions and two more multiplications than the triangular window for each sampling instant. However, the performance with the polynomial window has been found in simulations to be higher than with the triangular window. Fig. 2 shows the window functions $w(t)$, and Fig. 3 the corresponding frequency responses.

From Fig. 3 we can see that our polynomial window has the same main lobe and the first side lobe as the Hanning window; the other side lobes are slightly higher than the Hanning window. The triangular window has much higher side lobes than the polynomial window. The polynomial window provides a

370

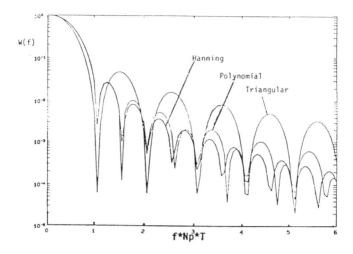

Figure 3: Frequency response of the window functions shown in Fig. 2.

better approximation to the desired ideal low-pass filter and therefore better performance than the triangular window.

Pitch detection was done by the algorithm presented in [2]. The pitch detector is particularly good for this application since it is robust under additive white noise and provides accurate pitch boundary information.

A typical waveform obtained in the TDHS algorithm is shown in Fig. 4. The speech segment shown in this figure is part of the word "mesh". The upper figure is the original signal. The lower figure is the reconstructed signal which is obtained by compressing the original and then expanding the compressed with our polynomial window. The two waveforms look similar. Very little degradation is present in the reconstructed speech signal in listening.

ADM WITH A HIGH–ORDER PREDICTOR

For a traditional delta modulator, the quantization error is proportional to the quantization step-size q. A smaller q means less quantization error, but it also results in smaller dynamic range which is not suitable for a speech signal. Many algorithms have been proposed to resolve this contradicting requirement. In this study, we propose an alternative approach to the problem. We use a high-order adaptive linear predictor to replace the fixed first order predictor. The quantization step-size is chosen to be the average predicted error of each segment of signal. The block diagram of the coder is shown in Fig. 5. The adaptive linear predictor is of 8th-order, and its coefficients are updated every 20ms. The input s_k is a speech signal sequence after time-domain harmonic scaling. The output d_k is a binary data string. The coder computes the linear

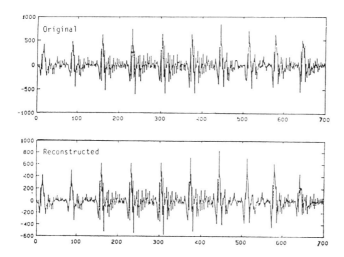

Figure 4: Original and reconstructed signals obtained with the polynomial window.

prediction error and works as follows:

$$e_k = s_k - \sum_{i=1}^{p} a_i \hat{s}_{k-i} \tag{2}$$

$$d_k = sgn(e_k) \tag{3}$$

$$\hat{s}_k = q_n d_k + \sum_{i=1}^{p} a_i \hat{s}_{k-i} \tag{4}$$

where $\{a_i\}_{i=1}^{p}$ are the linear prediction coefficients, q_n is the quantization step-size and p is the order of the predictor — $p = 8$.

The coefficients of the predictor are chosen in such a way that the following error function is minimized over the previous speech data block:

$$E_{n-1} = \sum_{s_k \in B_{n-1}} \left(\hat{s}_k - \sum_{i=1}^{p} a_i \hat{s}_{k-i} \right)^2 \tag{5}$$

where B_{n-1} denotes the previous speech block. Since speech signals are highly correlated, we do not expect the statistics of the current speech block to differ dramatically from that of the previous block. Using the set of predictive coefficients obtained from the previous speech block will keep E_n reasonably small for the current block.

The predictive coefficients are not transmitted, they can be computed at the receiver because all $\{\hat{s}_k\}$s are available. The coefficients a_i are less sensitive

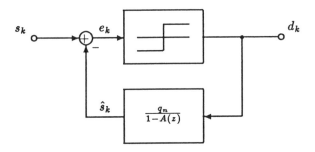

Figure 5: The ADM speech coder with an 8th-order adaptive linear predictor

to channel noise because of the least square nature of the algorithm used to compute them.

The quantization step-size, q_n, is computed as follows,

$$q_n = \frac{1}{||B_n||} \sum_{s_k \in B_n} |s_k - \sum_{i=1}^{p} a_i s_{k-i}| \qquad (6)$$

where $||B_n||$ denotes the total number of samples in block B_n. Once q_n is available, the speech coder uses it in Eq.(3) to obtain the digit string $D_n = \{d_k, d_{k+1}, \ldots, d_{k+N}\}$. The transmitted data block consists of two parts: the step-size q_n and the digit string D_n. Eq.(4) is used at the receiver to reconstruct speech signals.

IMPLEMENTATION

In mobile communication systems, the channel condition changes as the stations move. When the channel condition is good, it is a better strategy to devote more channel resources to transmitting the speech information instead of inserting low rate error control codes; when the channel condition becomes poor, one would like to sacrifice some speech information and use more channel resources for error control purpose. A varying rate speech coder is necessary in these applications [3,4].

The proposed coding scheme can be easily implemented as a varying rate coder. When more speech information is to be transmitted, one can do it by using one more bit for each speech sample:

$$d_k^* = sgn(s_k - \hat{s}_k) \qquad (7)$$

$$s_k^* = \frac{q_n}{2} d_k^* + \hat{s}_k \qquad (8)$$

The overall algorithm is implemented as follows. Speech signals are sampled at 8kHz with 12-bit precision. The sampled speech signal is compressed by the 2:1 TDHS compression algorithm with the polynomial window. The compressed signal is encoded by the ADM encoder. The encoded data are transmitted. Pitch information is not transmitted. In the receiver the digital signal is decoded by the ADM decoder to form the intermediate signal. Pitch information is extracted from this intermediate signal. Finally, the TDHS expansion algorithm with the polynomial window is performed to expand this intermediate signal back to its original sampling rate.

The block size used in the ADM coder is 80 speech samples which represent 20ms in the original signal; the data to be coded have been reduced to half of the original. For each block, 8 bits are assigned to q_n, and 80 bits to $\{d_k\}$s. This corresponds to a transmission rate of 4.4 kb/s. When a higher rate is needed, $\{d_k^*\}$s are added, and there are 80 more bits in each block for $\{d_k^*\}$s. The data rate becomes 8.4 kb/s. Generally, the algorithm reconstructs very good quality speech at 4.4 kb/s. When working at 8.4 kb/s, the output speech sounds comparable to the original speech sampled with 12-bit precision; only some minor difference is perceived.

References

[1] D. Malah, *Time-Domain Algorithms for Harmonic Bandwidth Reduction and Time Scaling of Speech Signals*, IEEE Trans. ASSP, Vol.ASSP–27, pp. 121–133, April 1979.

[2] C.S.Chen and Jing Yuan, *A Robust Pitch Boundary Detector*, Proc. ICASSP, Vol.1, pp. 366–369, April 1988.

[3] R.V. Cox, et al, *A Sub-Band Coder Designed for Combined Source and Channel Coding*, Proc. ICASSP, Vol 1, pp. 235–238, April 1988.

[4] D.J. Goodman, C.E. Surdberg, *Combined Source and Channel Coding for Variable-bit-rate Speech Transmission*, B.S.T.J. vol. 62, no. 7, pp. 2017–2036, Sept. 1983.

INDEX